Physiological and Molecular Biology of Ornamental Plants—2nd Edition

Physiological and Molecular Biology of Ornamental Plants—2nd Edition

Guest Editors

Caiyun Wang
Tuo Zeng

MDPI

Basel • Beijing • Wuhan • Barcelona • Belgrade • Novi Sad • Cluj • Manchester

Guest Editors
Caiyun Wang
Key Laboratory of
Horticultural Plant Biology
Huazhong Agricultural
University
Wuhan
China

Tuo Zeng
School of Life Sciences
Guizhou Normal University
Guiyang
China

Editorial Office
MDPI AG
Grosspeteranlage 5
4052 Basel, Switzerland

This is a reprint of the Special Issue, published open access by the journal *Horticulturae* (ISSN 2311-7524), freely accessible at: https://www.mdpi.com/journal/horticulturae/special_issues/7WE42J8RHZ.

For citation purposes, cite each article independently as indicated on the article page online and as indicated below:

Lastname, A.A.; Lastname, B.B. Article Title. *Journal Name* **Year**, *Volume Number*, Page Range.

ISBN 978-3-7258-5789-0 (Hbk)
ISBN 978-3-7258-5790-6 (PDF)
https://doi.org/10.3390/books978-3-7258-5790-6

Contents

horticulturae

Editorial

Physiological and Molecular Biology of Ornamental Plants—2nd Edition

Tuo Zeng [1,*] and Caiyun Wang [2,*]

1 Guizhou Key Laboratory of Forest Cultivation in Plateau Mountain, School of Life Sciences, Guizhou Normal University, Guiyang 550025, China
2 National Key Laboratory for Germplasm Innovation & Utilization of Horticultural Crops, College of Horticulture & Forestry Sciences, Huazhong Agricultural University, Wuhan 430070, China
* Correspondence: zengtuo@gnzu.edu.cn (T.Z.); wangcy@mail.hzau.edu.cn (C.W.)

1. Introduction

Ornamental plants, celebrated for their rich floral variation, vivid colors, and pleasant fragrances, hold enormous cultural, ecological, and economic value. Their traits are determined by intricate physiological processes and molecular mechanisms that govern pigmentation, secondary metabolism, reproductive success, and stress adaptation. However, many of these mechanisms remain only partially understood. This Special Issue brings together 14 contributions that collectively advance our understanding of ornamental plant biology. These papers cover three thematic areas, floral traits, floral scent and secondary metabolism, and adaptive responses to environmental stresses, providing a comprehensive overview of the latest progress in ornamental plant physiology and molecular biology. Collectively, these studies clarify the regulatory networks involved in floral pigmentation, secondary metabolite pathways, and stress-responsive processes, while showing how integrative approaches are advancing knowledge of ornamental trait formation and resilience.

2. Floral Traits: Color and Development

Flower color is one of the most visible and economically important traits in ornamentals [1]. *Osmanthus fragrans* is a heritage evergreen ornamental widely used in gardens and urban landscapes, valued for its compact crown and intensely fragrant blossoms across diverse cultivar groups [2]. In *O. fragrans*, two complementary studies investigated carotenoid metabolism in detail. One identified a mechanism in which delayed starch degradation leads to chromoplast structural abnormalities, preventing carotenoid cleavage and ultimately deepening floral pigmentation (contribution 1). The other focused on the red-flowered cultivar 'Yanzhi Hong' and showed that lycopene accumulation, together with reduced α-carotene synthesis due to downregulation of *LYCE* and *LYCB* genes, accounts for its distinctive rouge-red petals (contribution 2).

Anthocyanin-based pigmentation represents another dimension of floral coloration. Transcriptomic profiling of four flower colors in *Impatiens uliginosa* revealed differential expression of structural genes such as *C4H*, *FLS*, *PAL*, and *ANS*, in addition to transcription factors including *MYB* and *bHLH*, indicating that both biosynthetic genes and regulators shape anthocyanin accumulation patterns (contribution 3). Another study in *I. uliginosa* demonstrated that copper stress disrupts pigment biosynthesis pathways, leading to visible petal color changes (contribution 4).

Floral development, including organ growth and pollen fertility, is fundamental to reproductive success in ornamentals. Resource allocation plays a central role in floral growth and function. Citrus flowers combine ornamental and edible value: their highly fragrant blossoms support both landscaping use and traditional culinary applications [3]. In *Citrus maxima*, an analysis of C:N:P stoichiometry and non-structural carbohydrates across floral organs and phenophases revealed stage- and organ-specific partitioning, showing how coordination between carbon reserves and nutrient demand underpins organ development during anthesis (contribution 5). A genome-wide survey of the *DIR* gene family in *Rosa chinensis* identified 33 *RcDIRs*, including a tandem cluster, with *RcDIR12* exhibiting pollen-specific promoter activity; functional assays confirmed expression in mature pollen grains, providing evidence for pollen-specific *DIR* function in roses (contribution 6).

Together, these findings illustrate how pigment biosynthetic pathways, nutrient metabolism, and reproductive gene networks jointly shape floral coloration, organ development, and pollen function in ornamentals. Integrative analyses across these processes are providing a more coherent view of the molecular and physiological basis of floral traits.

3. Floral Scent and Secondary Metabolites

Floral fragrance is a defining trait of many ornamentals, contributing both to ecological interactions and commercial appeal. Orchids, with nearly 30,000 species worldwide, are especially noted for their diverse floral morphologies and complex fragrance profiles [4]. They also provide important model systems for understanding the molecular basis of floral scent, particularly volatile terpene biosynthesis. In *Dendrobium chrysotoxum*, genome-wide identification of *terpene synthase* (*TPS*) genes revealed 37 family members, among which three TPS-b genes encode enzymes that catalyze the production of linalool, a dominant monoterpene shaping the floral scent profile (contribution 7). Similarly, in *Cymbidium ensifolium*, 30 *TPS* genes were characterized and classified into three subfamilies, with *CeTPS1* and *CeTPS18* highly expressed in flowers and localized to chloroplasts, indicating specialized roles in terpenoid metabolism during floral development (contribution 8). Together, these studies demonstrate that orchids have evolved a diverse TPS repertoire, with distinct subfamilies differentially contributing to the unique fragrance profiles that underpin their ornamental and ecological value.

Beyond orchids, pyrethrum (*Tanacetum cinerariifolium*) exemplifies the dual roles of secondary metabolites in ornamentals. It produces two classes of bioactive compounds with distinct ecological functions: the volatile (E)-β-farnesene, which mediates plant–insect interactions, and the non-volatile pyrethrins, which act as natural insecticides [5]. Transcriptomic analysis combined with GC–MS-based metabolite profiling revealed asynchronous regulation for pyrethrins, where gene expression did not correspond with metabolite accumulation. In contrast, both the expression of (E)-β-farnesene biosynthetic genes and the accumulation of the metabolite were concentrated in peduncles. Jasmonate signaling, particularly through *MYC2*, emerged as a central hub coordinating these processes (contribution 9). This finding highlights a complex regulatory divergence between volatiles and non-volatiles within the same plant, providing valuable insights into how ornamental plants balance ecological interactions with defense-oriented chemistry.

Collectively, recent work indicates that floral fragrance and defensive chemistry are coordinated through context-dependent partitioning of biosynthetic flux between volatile and non-volatile branches. Within this architecture, the diversification of terpene synthases structures volatile terpene profiles in orchids, while non-volatile defenses are regulated in parallel, linking bouquet composition to ecological function and applied aims in ornamentals.

4. Adaptive Strategies Under Environmental Stresses

Ornamental plants are frequently exposed to a wide range of abiotic stresses, and understanding their adaptive strategies is essential for both horticultural practice and basic research. Among them, *Rhododendron*, one of the largest woody plant genera, is highly valued for its remarkable diversity of flower colors and ornamental appeal [6]. It also serves as a representative system for studying how root–microbe interactions contribute to stress responses. In *Rhododendron delavayi*, metabolomic and microbiome analyses revealed that waterlogging stress triggered the secretion of root metabolites capable of recruiting beneficial microbes, which persisted into the recovery phase and contributed to stress alleviation (contribution 10). A complementary study in *R. decorum* showed that cadmium exposure not only reshaped rhizosphere microbial communities but also suppressed the accumulation of key secondary metabolites such as flavonoids, thereby weakening both chemical defenses and symbiotic stability (contribution 11).

Beyond *Rhododendrons*, additional contributions broaden the perspective on stress adaptation in ornamentals. *Hydrangea macrophylla* represents a unique case where abiotic stress tolerance directly underpins ornamental quality. Sepal bluing depends on aluminum accumulation, making this species an important model for linking stress adaptation with visible floral traits. Transcriptome analyses of contrasting cultivars under Al treatment identified 43 differentially expressed genes, including 7 associated with Al accumulation. Among them, *HmALS3.1* emerged as a key candidate involved in Al transport (contribution 12). *Cinnamomum camphora* is a widely used evergreen ornamental and street tree throughout subtropical East Asia, prized for its dense, shade-casting crown and fragrant blossoms [7]. In *C. camphora*, transcriptomic analyses under cold stress uncovered calcium- and ethylene-mediated signaling pathways, with *ERF* transcription factors acting as pivotal regulators, thereby providing a molecular framework for understanding how woody ornamentals perceive and respond to low temperature (contribution 13). At a broader scale, a field survey in northern Italy documented the emergence of new ornamental plant diseases under climate change. Beyond reporting new pathogens, this study highlighted the role of "living-lab" gardens as early warning systems and emphasized integrated management approaches, ranging from resistant cultivars and climate-aware cultural practices to biological control and molecular diagnostics, as essential tools to address shifting pathogen pressures (contribution 14).

Collectively, these studies support a systems-level view of stress adaptation in ornamentals, where molecular regulation, secondary metabolism, and plant–microbe interactions operate together to confer resilience. The contributions further delineate practical measures, including resistant cultivars, climate-adaptive cultivation, biological control, and molecular diagnostics, for sustaining ornamental plant performance under environmental change.

5. Conclusions

This Special Issue highlights major advances in ornamental plant physiology and molecular biology. Studies on floral traits have clarified the metabolic and transcriptional networks underlying pigmentation and development, while research on floral scent and secondary metabolism has identified key gene families shaping fragrance and defense. Work on stress adaptation has revealed the complex physiological adjustments and ecological interactions that sustain ornamental performance under adverse conditions.

Across these themes, the contributions delineate coordinated regulatory networks that connect pigment and floral development with the biosynthesis and allocation of volatile and non-volatile metabolites. They also show how stress-responsive processes, together with plant–microbe interactions, add an important layer of resilience to ornamental species.

In doing so, the papers exemplify established integrative approaches that combine gene family analyses, expression profiling, metabolite measurements, and targeted functional assays to resolve mechanisms of trait formation and ecological performance.

Looking ahead, priorities include expanding research beyond a narrow set of model ornamentals to under-represented taxa and traits, such as fragrance complexity, floral longevity, and architectural features. Strengthening field and multi-environment validation of molecular findings and advancing comparative and translational studies that connect mechanism with breeding and cultivation, will be equally important. Continued attention to robust phenotyping and data standardization will further enhance reproducibility and cross-study synthesis. Taken together, the fourteen contributions deepen the scientific foundation of ornamental biology and outline practical directions for innovation in research, breeding, and sustainable cultivation.

Author Contributions: Writing—Original Draft Preparation, Writing—Review and Editing, T.Z. and C.W. All authors have read and agreed to the published version of the manuscript.

Funding: This research received no external funding.

Conflicts of Interest: The authors declare no conflicts of interest.

List of Contributions:

1. Zeng, X.; Tan, Y.; Wen, X.; He, Q.; Wu, H.; Zou, J.; Yang, J.; Cai, X.; Chen, H., Delayed starch degradation triggers chromoplast structural aberration to inhibit carotenoid cleavage: a novel mechanism for flower color deepening in *Osmanthus fragrans*. *Horticulturae* **2025**, *11*, 864.
2. Wei, S.; Wu, J.; Yu, P.; Tan, Y.; He, Q.; Yang, J.; Cai, X.; Zou, J.; Chen, H.; Zeng, X., Metabolomic and transcriptomic analysis of unique floral coloration in *Osmanthus fragrans* cultivars. *Horticulturae* **2024**, *10*, 801.
3. Zhang, X.; Tan, Y.; Li, X.; Liu, Z.; Li, F.; Huang, H.; Huang, M., Analysis of transcriptome and expression of C4H and FLS genes on four flower colors of *Impatiens uliginosa*. *Horticulturae* **2024**, *10*, 415.
4. Zhu, J.; Li, X.; Huang, H.; Huang, M., Resistances and physiological responses of *Impatiens uliginosa* to copper stress. *Horticulturae* **2024**, *10*, 751.
5. Liao, J.; Hu, S.; Kong, Y.; Pan, H.; Zhu, M.; Yu, T.; Hu, H.; Zhuang, G.; Gao, S., Variations in C:N:P stoichiometry and non-structural carbohydrates in different parts of pomelo (*Citrus maxima*) flowers at three phenophases. *Horticulturae* **2025**, *11*, 1053.
6. Dong, Q.; Yang, Q.; Wang, Z.; Zhao, Y.; Guo, S.; Peng, Y.; Li, Q.; Han, Y., Genome-wide identification and pollen-specific promoter analysis of the *DIR* gene family in *Rosa chinensis*. *Horticulturae* **2025**, *11*, 717.
7. Yang, Y.; Gong, J.; Nong, R.; Liu, Q.; Xia, K.; Qiu, S.; Wang, Z., Terpene synthase (*TPS*) family member identification and expression pattern analysis in flowers of *Dendrobium chrysotoxum*. *Horticulturae* **2025**, *11*, 566.
8. Wang, M.; Liu, B.; Li, J.; Huang, N.; Tian, Y.; Guo, L.; Feng, C.; Ai, Y.; Fu, C., Bioinformatics analysis and expression features of terpene synthase family in *Cymbidium ensifolium*. *Horticulturae* **2024**, *10*, 1015.
9. Zeng, T.; Li, J.; Wang, C.; Li, J., Comparative transcriptomic analysis of pyrethrin and EβF biosynthesis in *Tanacetum cinerariifolium* stems and flowers. *Horticulturae* **2025**, *11*, 201.
10. Tang, J.; Huang, Q.; Wang, Q.; Shan, F.; Wu, S.; Zhang, X.; Tang, M.; Yi, Y., The comprehensive root metabolite–rhizomicrobiota response patterns of *Rhododendron delavayi* to waterlogging stress and post–waterlogging recovery. *Horticulturae* **2025**, *11*, 770.
11. Tang, M.; Chen, L.; Wang, L.; Yi, Y.; Wang, J.; Wang, C.; Chen, X.; Liu, J.; Yang, Y.; Malik, K.; Gong, J., Comprehensive analysis of microbiomes and metabolomics reveals the mechanism of adaptation to cadmium stress in rhizosphere soil of *Rhododendron decorum* subsp. diaprepes. *Horticulturae* **2024**, *10*, 884.

12. Luo, S.; Li, Y.; Wan, Y.; Fan, Y.; Liu, C.; Yuan, S., Identification of key candidate genes involved in aluminum accumulation in the sepals of *Hydrangea macrophylla*. *Horticulturae* **2024**, *10*, 1180.
13. Bi, B.; Shao, L.; Xu, T.; Du, H.; Li, D., Transcriptomic analysis reveals calcium and ethylene signaling pathway genes in response to cold stress in *Cinnamomum camphora*. *Horticulturae* **2024**, *10*, 995.
14. Gullino, M. L.; Bertetti, D.; Pugliese, M.; Garibaldi, A., Emerging ornamental plant diseases and their management trends in northern Italy. *Horticulturae* **2025**, *11*, 955.

References

1. Shen, Y.; Rao, Y.; Ma, M.; Li, Y.; He, Y.; Wang, Z.; Liang, M.; Ning, G. Coordination among flower pigments, scents and pollinators in ornamental plants. *Hortic. Adv.* **2024**, *2*, 6. [CrossRef]
2. Chen, H.; Zeng, X.; Yang, J.; Cai, X.; Shi, Y.; Zheng, R.; Wang, Z.; Liu, J.; Yi, X.; Xiao, S. Whole-genome resequencing of *Osmanthus fragrans* provides insights into flower color evolution. *Hortic. Res.* **2021**, *8*, 98. [CrossRef] [PubMed]
3. Agustí, M.; Reig, C.; Martínez-Fuentes, A.; Mesejo, C. Advances in *Citrus* flowering: A review. *Front. Plant Sci.* **2022**, *13*, 868831. [CrossRef] [PubMed]
4. Yang, F.; Gao, J.; Li, J.; Wei, Y.; Xie, Q.; Jin, J.; Lu, C.; Zhu, W.; Wong, S.; Zhu, G. The China orchid industry: Past and future perspectives. *Ornam. Plant Res.* **2023**, *4*, e002. [CrossRef]
5. Zeng, T.; Li, J.; Li, J.; Hu, H.; Zhu, L.; Liu, K.; Bai, J.; Jiang, Q.; Wang, C. Pyrethrins in *Tanacetum cinerariifolium*: Biosynthesis, regulation, and agricultural application. *Ornam. Plant Res.* **2024**, *4*, e015. [CrossRef]
6. Nie, S.; Ma, H.; Shi, T.; Tian, X.; El-Kassaby, Y.A.; Porth, I.; Yang, F.; Mao, J. Progress in phylogenetics, multi-omics and flower coloration studies in *Rhododendron*. *Ornam. Plant Res.* **2024**, *4*, e003. [CrossRef]
7. Wang, X.; Xu, C.; Zheng, Y.; Wu, Y.; Zhang, Y.; Zhang, T.; Xiong, Z.; Yang, H.; Li, J.; Fu, C. Chromosome-level genome assembly and resequencing of camphor tree (*Cinnamomum camphora*) provides insight into phylogeny and diversification of terpenoid and triglyceride biosynthesis of *Cinnamomum*. *Hortic. Res.* **2022**, *9*, uhac216. [CrossRef] [PubMed]

horticulturae

MDPI

Article

Delayed Starch Degradation Triggers Chromoplast Structural Aberration to Inhibit Carotenoid Cleavage: A Novel Mechanism for Flower Color Deepening in *Osmanthus fragrans*

Xiangling Zeng [1,2,†], Yunfei Tan [1,†], Xin Wen [1], Qiang He [3], Hui Wu [3], Jingjing Zou [2], Jie Yang [2], Xuan Cai [2] and Hongguo Chen [2,*]

1 School of Pharmacy, Xianning Medical College, Hubei University of Science and Technology, Xianning 437100, China; cengxiangling@hbust.edu.cn (X.Z.); tanyunfei0926@163.com (Y.T.); 19863349824@163.com (X.W.)
2 National Forestry and Grassland Administration Engineering Research Center for *Osmanthus fragrans*, Hubei University of Science and Technology, Xianning 437100, China; silence@hbust.edu.cn (J.Z.); yj_anna@hbust.edu.cn (J.Y.); caixuan@hbust.edu.cn (X.C.)
3 Public Inspection and Testing Center, Xianning 437100, China; qianghe2015@126.com (Q.H.); 13251011983@163.com (H.W.)
* Correspondence: chen_hongguo1969@163.com
† These authors contributed equally to this work.

Abstract: The color of flowers in *Osmanthus fragrans* is regulated by carotenoid metabolism. The orange-red variety, Dangui, is believed to have evolved from the yellow variety, Jingui, through a natural bud mutation. This study uses the Jingui cultivar 'Jinqiu Gui' (JQG) and its bud mutation cultivar 'Huolian Jindan' (HLJD) as materials, combining genome resequencing, ultrastructural observation, targeted metabolomics, and transcriptomic analysis to elucidate the molecular and cellular mechanisms underlying flower color variation. Phylogenetic analysis confirms that HLJD is a natural bud mutation of JQG. Ultrastructural observations reveal that during petal development, chromoplasts are transformed from proplastids. In HLJD petals, starch granules degrade more slowly and exhibit abnormal morphology, resulting in chromoplasts displaying crystalline, tubular, and fibrous composite structures, in contrast to the typical spherical plastoglobuli found in JQG. Targeted metabolomics identified 34 carotenoids, showing significant increases in the levels of ε-carotene, γ-carotene, α-carotene, and β-carotene in HLJD petals compared to JQG, with these levels continuing to accumulate throughout the flowering process, while the levels of the cleavage products α-ionone and β-ionone decrease. Transcriptomic analysis indicates that carotenoid metabolic pathway genes do not correlate directly with the phenotype; however, 49 candidate genes significantly associated with pigment accumulation were identified. Among these, the expression of genes such as glycoside hydrolases (LYG036752, etc.), sucrose synthase (LYG010191), and glucose-1-phosphate adenylyltransferase (LYG003610) are downregulated in HLJD. This study proposes for the first time the pathway of "starch degradation delay → chromoplast structural abnormalities → carotenoid cleavage inhibition" for deepening flower color, providing a new theoretical model for the metabolic regulation of carotenoids in non-photosynthetic tissues of plants. This research not only identifies key target genes (such as glycoside hydrolases) for the color breeding of *O. fragrans* but also establishes a theoretical foundation for the color enhancement of other ornamental plants.

Keywords: *Osmanthus fragrans*; floral color; bud mutation; pastid; carotenoid; starch

1. Introduction

Osmanthus fragrans, a member of the Oleaceae family and *Osmanthus* genus, is an evergreen woody plant renowned as a traditional fragrant flower in China. This species is not only widely utilized for ornamental purposes in gardens but also possesses significant fragrance, is edible, and has medicinal values, offering immense potential for the development of deep-processing products that contribute importantly to the local economies in China. Flower color serves as a critical ornamental quality trait of *O. fragrans*, which is categorized into four groups based on flowering period and flower color traits: Sijigui, Yingui, Jingui, and Dangui, with nearly 200 cultivars currently available [1]. Carotenoids are key compounds influencing the variation in flower color among *O. fragrans* cultivars and serve as direct precursors in the synthesis of key aromatic compounds, such as α-ionone and β-ionone [2–5]. This is essential for enhancing the ornamental and economic value of *O. fragrans*. Investigating the carotenoid metabolic mechanisms in *O. fragrans* holds significant practical implications for the targeted cultivation of new cultivars with enhanced ornamental appeal (more vibrant flower colors) and higher economic value (stronger fragrance). Furthermore, this research enriches the theoretical foundation for the metabolic regulation of carotenoids in plant flowers, providing a universal model for flower color improvement and fragrance synthesis in other species.

The production of carotenoids relies on the methylerythritol phosphate (MEP) pathway, which provides the precursor geranylgeranyl pyrophosphate (GGPP) [6,7]. Phytoene synthase (PSY) catalyzes the conversion of GGPP into phytoene, which subsequently undergoes dehydrogenation and cyclization, leading to the formation of various carotenoids, including β-carotene and lutein [7,8]. These carotenoids can be cleaved by carotenoid cleavage dioxygenases into colored products, such as β-citraurin and crocin, or transformed into aromatic compounds like β-ionone [9–11]. Additionally, they may be degraded by 9-cis-epoxycarotenoid dioxygenase (NCED), resulting in the production of abscisic acid (ABA), which plays an important role in plant stress responses [12,13]. In plants, plastids serve as the sites for carotenoid synthesis and storage, categorized into proplastids, amyloplasts, chloroplasts, and chromoplasts, with chromoplasts being the primary locations for carotenoid production and accumulation [7,14]. The structural characteristics of chromoplasts vary across different species and plant organs, and they can be classified into types such as plastoglobuli, crystalline, membranous, and tubular [14]. Furthermore, chromoplasts can arise from the transformation of other plastids, with numerous studies indicating that they can develop from chloroplasts [7,15]. However, instances of amyloplasts transforming into chromoplasts are significantly less documented in petals.

Despite the conservativeness of carotenoid metabolic pathways in plants, the composition and content of carotenoids in non-green tissues across different species exhibit notable diversity [16,17]. Previous studies have indicated that α- and β-carotene predominantly accumulate in the petals of most cultivars of *O. fragrans*, with variations in their content leading to differences in flower color phenotypes, resulting in the formation of the cultivars Yingui, Jingui, and Dangui [3,18]. The evolutionary relationships among these three groups were clarified only after our research group completed whole-genome resequencing of over 100 *O. fragrans* cultivars, revealing that the flower colors of *O. fragrans* evolved from light to dark, with Dangui being the last to form, possibly originating from bud mutation [1,19,20]. The discovery of natural mutants of Dangui supports this conclusion [21]. However, to date, there have been no reports on the mechanisms underlying the formation of Dangui mutants. This study presents a typical case of a natural mutation from Jingui to Dangui, combining phenotypic comparisons with phylogenetic tree construction through genome resequencing. It demonstrates that 'Huolian Jindan' (HLJD) is a bud mutation cultivar of 'Jinqiu Gui' (JQG), thereby providing important natural mutant resources for studying the mechanisms of flower color variation in *O. fragrans*. Subsequently, ultramicroscopic observations, carotenoid metabolomics, and transcriptomic analyses of petals from JQG and HLJD at different developmental stages were conducted to explore the mechanisms of flower color mutation in *O. fragrans* and to identify key genes regulating floral color variation, thereby providing a theoretical basis and important genetic resources for the improvement of flower color quality in *O. fragrans* and other plants.

2. Materials and Methods

2.1. Plant Materials

The original mother plant (S308) and the bud mutant branch (S309) (Figure 1A) were collected from the Suzhou Lingering Garden, Jiangsu Province, China. The petals and leaves of the cultivars 'Jinqiu Gui' (JQG) (S310) and 'Huolian JinDan' (HLJD) (S314) were collected during the bud stage (S1), initial flowering stage (S2), and full flowering stage (S3) (Figure 1B) from the National *Osmanthus* Germplasm Resource Bank of the Landscaping Co., Ltd. in Yuhang District, Hangzhou City, Zhejiang Province. After sample collection, they were immediately frozen in liquid nitrogen and then transferred to a −80 °C freezer for storage.

(A)
(B)

Figure 1. Flower and leaf phenotypes of wild-type and bud mutant of *O. fragrans* (**A**), as well as JQG and HLJD cultivars of *O. fragrans* at different flowering stages (**B**).

2.2. Genome Resequencing

High-quality DNA was extracted from healthy leaves using the CTAB method, following the manufacturer's protocol. The quality of the extracted DNA was evaluated using the Qsep400 system (AutoQ Biosciences, San Diego, CA, USA). Subsequently, libraries for four samples were prepared with the TruSeq Library Construction Kit and sequenced on the MGI-SEQ 2000 platform at Frasergen Bioinformatics Co., Ltd. in Wuhan, China. To investigate genetic variations, clean reads from the resequencing data of these four samples were processed. The adapter sequences and low-quality reads were filtered out using Trimmomatic (version 0.36) [22]. BWA (version 0.7.15-r1140) [23] was then employed to map the clean reads to the "Liuyejingui" reference genome [1]. GATK software (version 4.0.9.0) [24] was subsequently used to detect SNPs, with the following filtering parameters applied: QD < 2.0, MQ < 40.0, FS > 60.0, SOR > 3.0, MQRankSum < −12.5, and ReadPosRankSum < −8.0. Furthermore, a neighbor-joining (NJ) phylogenetic tree was constructed using Treebest software (v1.9.2), with a bootstrap value of 1000. The resulting tree was visualized using iTOL (version 4) [25].

2.3. Observation of Ultrastructure

Flower petals collected at various developmental stages were placed in a 4% glutaraldehyde fixative and subsequently washed with 0.1 M phosphate-buffered saline (PBS, pH 7.4). After this, the samples were treated with a prepared 1% osmium tetroxide solution in a dark environment at room temperature for 7 h. They then underwent a stepwise dehydration procedure involving ethanol, followed by infiltration with the embedding medium. The polymerization of the embedding blocks took place in an oven at 60 °C for 48 h. Ultra-thin sections were created using an ultramicrotome and stained with a saturated alcoholic solution of 2% uranyl acetate in the dark. Finally, the samples were examined and photographed utilizing a transmission electron microscope (FEI Tecnai G^2 F20 S-TWIN).

2.4. Carotenoid Extraction and Determination

A total of 50 mg of dried petal powder was weighed and subsequently extracted twice using 0.5 mL of a mixed solution comprising hexane, acetone, and ethanol in equal volumes (1:1:1, *v*/*v*/*v*), containing 0.01% BHT (g/mL). The extracts obtained were combined and concentrated before being re-dissolved in 100 μL of a methanol and t-butyl methyl ether (MTBE) mixture (1:1, *v*/*v*). This solution was then filtered through a 0.22 μm membrane and stored in a brown injection vial for LC-MS/MS analysis utilizing a QTRAP6500+ (SCIEX, Framingham, MA, USA). Chromatographic separation was achieved using a YMC C30 column (3 μm, 100 mm × 2.0 mm i.d.). Mass spectrometry data were collected through Ultra Performance Liquid Chromatography (UPLC) utilizing an ExionLC™AD in combination with Tandem Mass Spectrometry (MS/MS) on a QTRAP®6500+. The mobile phase and analytical detection parameters were modified following the protocols established by Wei et al. (2024) [3]. Qualitative analysis of the mass spectrometry data was conducted based on the creation of a standard substance MWDB (MetwareDatabase). For quantitative analysis, the Multiple Reaction Monitoring (MRM) mode of triple quadrupole mass spectrometry was employed alongside standard curves.

2.5. Determination of Volatile Components

Frozen petals (1 g) were subjected to grinding into a fine powder with the aid of liquid nitrogen, which was then transferred into a 40 mL centrifuge tube. Following this, 5 mL of n-hexane containing the internal standard methyl laurate at a final concentration of 20 ng/μL was introduced to the extraction solution, and the mixture was gently agitated for 30 min. After this agitation, the extraction solution was decanted into a fresh 50 mL centrifuge tube. An extra 5 mL of the extraction solution was added to the residual solid. Both extraction solutions were subsequently combined and allowed to rest overnight at −20 °C. The upper layer of the solvent was cautiously removed and concentrated to 2 mL using a nitrogen blow-down concentrator. The obtained solution was dried over an anhydrous Na_2SO_4 column and then subjected to analysis via gas chromatography-mass spectrometry (GC-MS). The chromatographic separation utilized an HP-5MS capillary column (30 m × 0.25 mm × 0.25 μm, J&W Scientific), with high-purity helium (99.999%) serving as the carrier gas at a flow rate of 1.2 mL/min. Refer to Zeng et al. (2016) [5] for the conditions of chromatographic and mass spectrometry analysis. The analysis of target analytes was performed using Agilent Mass Hunter Qualitative Analysis Navigator B.08.00 software coupled with NIST MS Search 2.3. Quantitative assessments were executed in accordance with the internal standard method utilizing methyl laurate.

2.6. Transcriptome Sequencing

RNA-seq were performed by Wuhan IGENEBOOK Biotechnology Co., Ltd. (Wuhan, China) (http://www.igenebook.com accessed on 29 October 2024). Total RNA was isolated from the petals of *O. fragrans* utilizing the TRNzol Universal (Tiangen Biotech Co., Ltd., Beijing, China), followed by evaluating RNA quality with the Qsep400 system (AutoQ Biosciences, San Diego, CA, USA). For the preparation of libraries, the Novogene VAHTS mRNA-seq V8 Library Prep Kit for Illumina was used, and sequencing was performed on the Illumina Novaseq 6000 platform (Illumina, San Diego, CA, USA). Clean reads were aligned to the reference genome 'LiuyeJingui' using Hisat2 (v2.1.0), allowing for up to two mismatches. RNA-seq raw data have been deposited into the public database of the National Center of Biotechnology Information (NCBI) BioProject PRJNA1279839 (accession number SAMN49507044- SAMN49507061). Genes identified during this procedure were compared against public protein databases, such as RefSeq non-redundant proteins (NR). Transcript levels were quantified, and gene expression was normalized to fragments per kilobase of transcript per million mapped fragments (FPKM) using Feature-count (v1.6.0). Differential expression analysis was carried out with edgeR, applying a significance threshold of FDR < 0.05 and $|\log2FoldChange| > 1$. Furthermore, enrichment analyses for Gene Ontology (GO) and Kyoto Encyclopedia of Genes and Genomes (KEGG) were performed utilizing the Cluster Profiler in the R package (version 4.1.2.), with the groundwork for enrichment analysis based on hypergeometric distribution and a q value threshold of 0.05.

2.7. Data Analysis

The data analysis was carried out using the Oebiotech Cloud Platform (https://cloud.oebiotech.com/task/, accessed on 2 March 2025), which facilitated the creation of clustering heatmaps for metabolites and gene expression. Additionally, it performed PCA analysis, conducted Venn analysis, investigated co-expression correlations, and carried out variance analysis. The findings are reported as mean ± SD, derived from three biological replicates for every sample.

3. Results

3.1. Identification of a Bud Mutation Cultivar of O. fragrans

In the Liuyuan Garden of Suzhou, Jiangsu Province, a chimeric *O. fragrans* tree has been documented for many years, where bud mutation branches of the orange-red cultivar (Dan Gui) have emerged from the golden cultivar (Jin Gui). The parent tree has been preliminarily identified as 'Jinqiu Gui' (JQG) based on its phenotype. The bud mutation branches are primarily characterized by a change in flower color phenotype, exhibiting orange-red petals on the inside and golden edges on the outside (as shown in Figure 1A).

This phenotype closely resembles that of 'Huolian Jindan' (HLJD), which is preserved in the National *Osmanthus* Germplasm Resource Bank of the Landscaping Co., Ltd. in Yuhang District, Hangzhou City, Zhejiang Province, China. To further confirm the phylogenetic relationships among JQG (S310), HLJD (S314), the parent tree from Suzhou Liuyuan Garden (S308), and the bud mutation branch (S309), DNA was extracted from the leaves of the four samples for genome resequencing. This data was subsequently utilized to construct a phylogenetic tree alongside resequencing data from 119 *O. fragrans* cultivars previously analyzed by our research group. As illustrated in Figure 2, *O. fragrans* var. Dangui is categorized into four distinct branches. The Suzhou Liu Garden bud mutation branch (S309) and HLJD (S314) cluster closely with the Suzhou Liu Garden parent (S308) and JQG (S310), as well as with the JQG cultivar sourced from germplasm banks in Wuhan, Hubei (S432) and Xianning, Hubei (S122). In contrast, these do not share the same branch as other *O. fragrans* cultivars of Dangui. This observation supports the conclusion that HLJD is indeed a bud mutation cultivar of JQG, and that the formation mechanism of HLJD is distinct from that of other *O. fragrans* cultivars of Dangui.

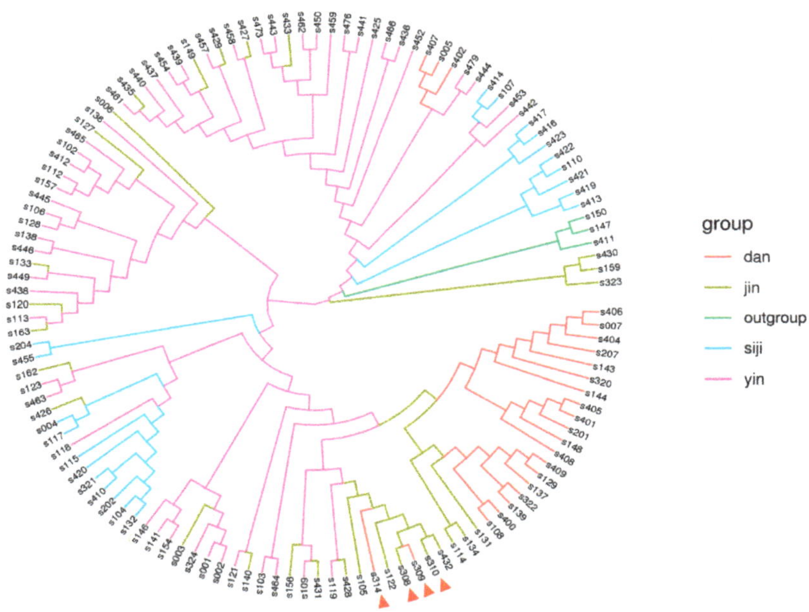

Figure 2. Phylogenetic tree of four samples and 119 *O. fragrans* cultivars. The red triangles represent the four samples that underwent genome resequencing this study.

3.2. Ultrastructural Analysis of Plastids in O. fragrans Petals

Further observations of the ultrastructure of plastids in the petals of JGQ and HLJD at various flowering stages were conducted using transmission electron microscopy (Figure 3). During the bud stage, both JGQ and HLJD exhibited plastids containing a significant number of starch grains. At the initial flowering and full-flowering stages, the large starch grains within the plastids were observed to disappear. In JGQ, a considerable number of chromoplasts with plastoglobuli were identified, featuring round white starch grains at the center of these plastoglobuli. In contrast, HLJD displayed a highly complex internal structure of chromoplasts, characterized by a small number of irregularly shaped plastoglobuli, along with crystalline, tubular, and fibrous structures within the same chromoplast. The remaining starch grains also exhibited irregular shapes. Notably, the rate of starch grain disappearance in the petals of HLJD during the initial- and full-flowering stages was significantly slower than that in JGQ. It is speculated that the chromoplasts in *O. fragrans* petals have evolved from amyloplasts, and the normal hydrolysis of starch grains facilitates the formation of plastoglobuli in chromoplasts.

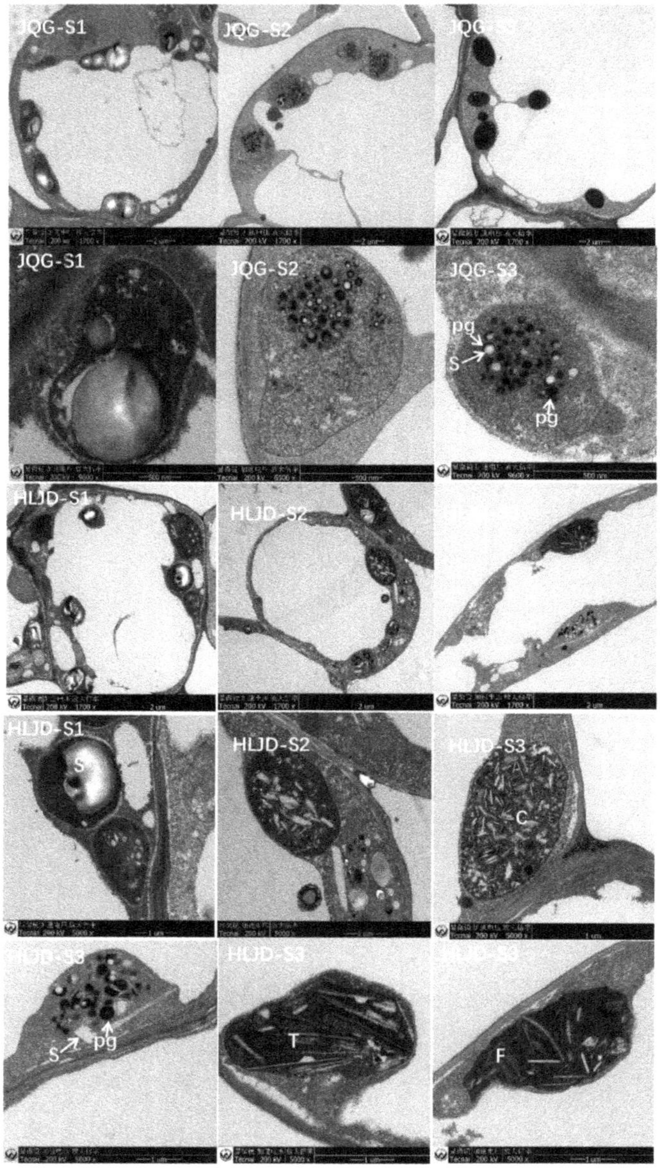

Figure 3. Observation of the ultrastructure of petal chromoplasts in JQG and HLJD cultivars of *O. fragrans* at different flowering stages. S—Starch granules; pg—plastoglobuli structure; T—tubular structure; C—crystalline structure; F—fibrous structure.

3.3. Quantitative Evaluation of Carotenoids in *O. fragrans* Petals

The carotenoid composition and content of petals from JQG and HLJD at different flowering stages were analyzed using UPLC-MS/MS, leading to the identification of 34 carotenoids: 6 types of carotenes, 10 types of xanthophylls, and 18 types of xanthophyll esters (see Supplementary Materials Table S1). Principal component analysis (PCA) indicated (Figure 4A) that the S1, S2, and S3 stages of HLJD are situated near the first quadrant and its boundary, whereas JQG's S1 is located in the third quadrant, S2 in the second quadrant, and S3 in the fourth quadrant. This suggests significant differences between HLJD and JQG, as well as notable variations in the petals of JQG across different stages. From Figure 4B, it is evident that the relative content of carotene in HLJD during the three stages accounts for 60–80%, exhibiting an increasing trend as the flowers open. In JQG, the relative content of xanthophyll is highest during the S1 stage but decreases as the flowers open, while the carotene content increases. Cluster heatmap analysis reveals (Figure 4C) that the 34 carotenoids are classified into three branches: Clade I exhibits the highest content in S3; Clade II shows higher content in S2; and Clade III has the highest content in S1. Based on the floral color phenotypes of JQG and HLJD at various stages, the color deepens progressively with flower opening, with HLJD displaying a deeper flower color than JQG. The contents of β-carotene, α-carotene, and ε-carotene correspond to the observed phenotypic changes.

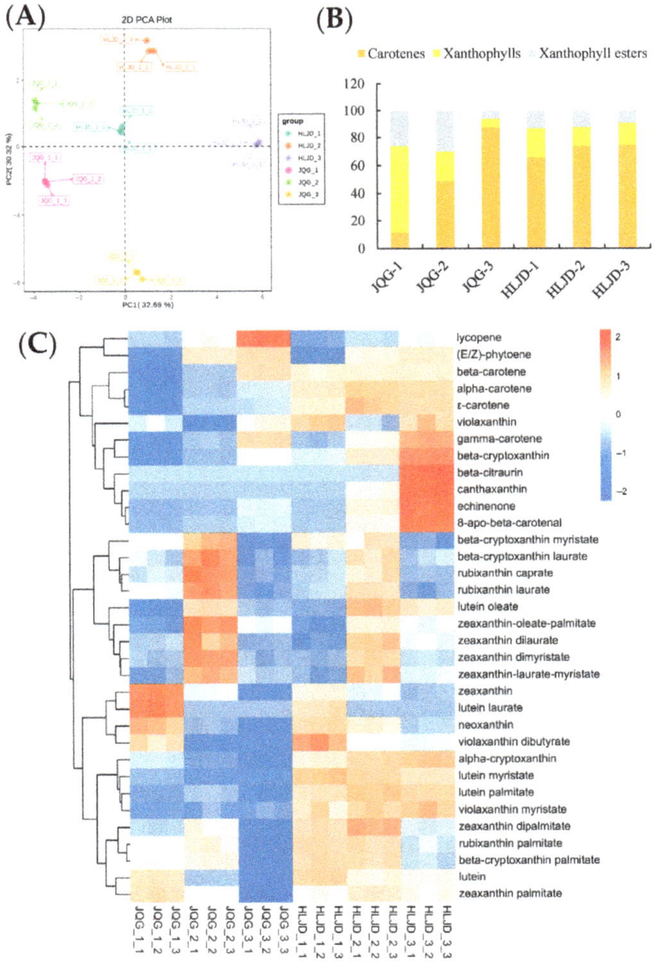

Figure 4. Analysis of carotenoid metabolites in petals of *O. fragrans* JQG and HLJD cultivars at different flowering stages. (**A**) Principal component analysis; (**B**) Proportion of carotenoid components in petals; (**C**) Carotenoid metabolite clustering heat map.

3.4. Transcriptome Sequencing and DEGs Analysis Between JQG and HLJD Petals of O. fragrans

To investigate the molecular mechanism underlying changes in flower color in *O. fragrans*, petals from the JQG and HLJD cultivars at three flowering stages were employed to create libraries for high-throughput sequencing. The quality metrics, Q20 and Q30, surpassed 95% and 90%, respectively, affirming the sequencing data's reliability for subsequent analysis. Results from PCA analysis (Figure 5A) reveal distinct differences among varying cultivars and flowering stages. A differential gene expression analysis was performed, using FDR < 0.05 and | log2FoldChange | > 1 as criteria for selecting genes expressed at different flowering stages. The findings show (Figure 5B) that in the JQG cultivar, there were 3865 upregulated differentially expressed genes (DEGs) and 4301 downregulated DEGs in S2 compared to S1; S3 exhibited 5324 upregulated DEGs and 7133 downregulated DEGs; and in the comparison of S3 to S2, there were 2222 upregulated DEGs and 2936 downregulated DEGs. For the HLJD cultivar, relative to S1, S2 revealed 3036 upregulated DEGs and 3770 downregulated DEGs; in S3, there were 3952 upregulated DEGs and 5100 downregulated DEGs; and when comparing S3 with S2, 2337 upregulated DEGs and 2643 downregulated DEGs were identified. When comparing the DEGs between JQG and HLJD across the same stages, the results indicated 3660 upregulated DEGs and 1799 downregulated DEGs in JQG for the S1 stage; 2850 upregulated DEGs and 1875 downregulated DEGs in S2; and 1117 upregulated DEGs and 750 downregulated DEGs in S3. A Venn analysis of both upregulated and downregulated DEGs across the three stages between JQG and HLJD was performed (Figure 5C), yielding 98 and 51 DEGs, respectively.

3.5. Association Analysis Between Metabolome and Transcriptome of the Important Genes in O. fragrans

The content of various components in the carotenoid metabolic pathway across different varieties and periods was visually illustrated (Figure 6A). Throughout all three stages, the levels of (E/Z)-phytoene and lycopene were consistently greater in JQG compared to HLJD, displaying an upward trend in content as the flowers opened. Conversely, the concentrations of ε-carotene, γ-carotene, α-carotene, α-cryptoxanthin, and β-cryptoxanthin increased with flower unfolding, with HLJD consistently exhibiting higher levels than JQG during the entire flowering period. Notably, β-carotene concentrations were significantly elevated in HLJD compared to JQG during the S1 and S2 stages, whereas JQG surpassed HLJD in the S3 stage. Additionally, lutein levels were higher in HLJD than in JQG, although they gradually declined as the flowers progressed in opening. Zeaxanthin levels also diminished with the opening of flowers; however, during the S1 phase, JQG presented notably higher levels than HLJD. Considering the content and color characteristics of the various carotenoid components, it can be inferred that ε-carotene, γ-carotene, α-carotene, and β-carotene may play crucial roles in flower color mutations. Furthermore, GC-MS was utilized to analyze the contents of carotenoid cleavage products, specifically α-ionone and β-ionone, in the petals (Figure 6A). The findings demonstrated that the levels of these two critical components related to *O. fragrans* fragrance were more significant in JQG petals than in HLJD. A heatmap analysis of the expression levels of carotenoid metabolism pathway genes derived from transcriptome data (Figure 6B) showed no genes correlated with changes in carotenoid component levels, suggesting that genes associated with the carotenoid metabolic pathway are not the primary contributors to variations in flower color.

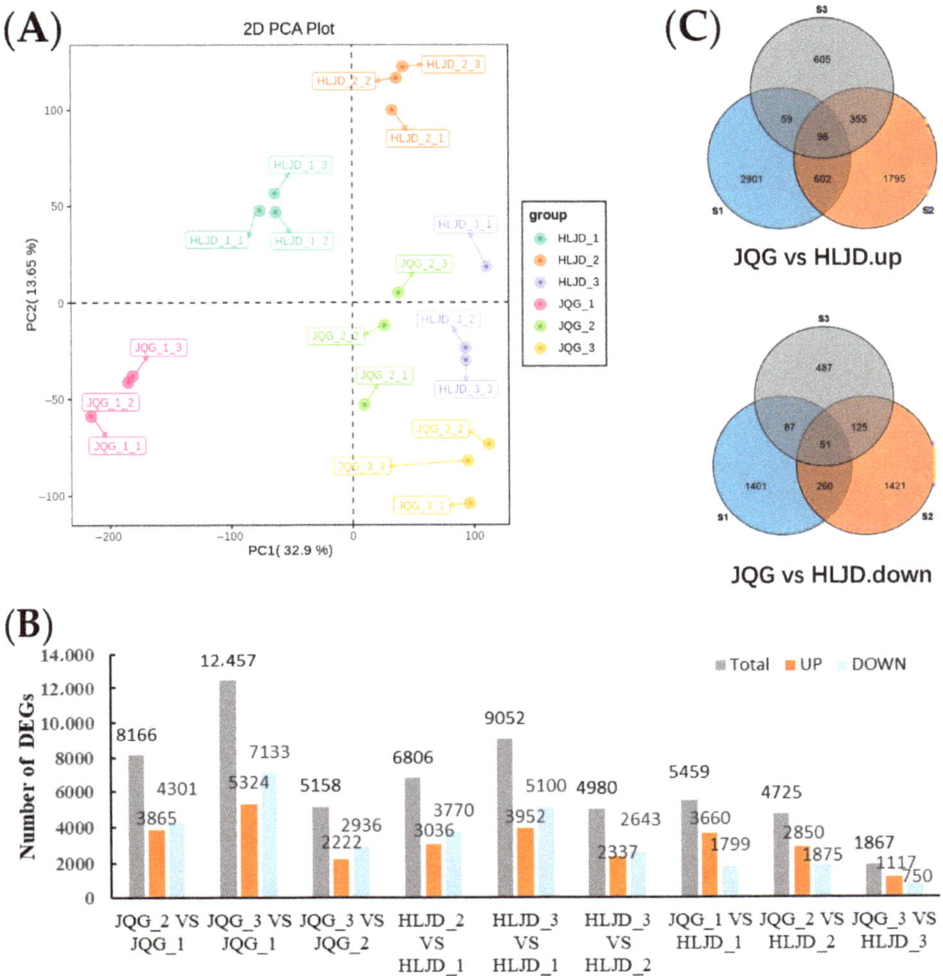

Figure 5. Transcriptomic features for petals of *O. fragrans* JQG and HLJD cultivars at different flowering stages. (**A**) Principal component analysis; (**B**) All DEGs statistics; (**C**) Venn diagram for up or down DEGs of JQG vs. HLJD at stage S1, S2, and S3.

A correlation analysis was performed on 98 upregulated and 51 downregulated differentially expressed genes (DEGs) in relation to ε-carotene, γ-carotene, α-carotene, and β-carotene, using a significance threshold of $p < 0.05$. The results revealed that 68 of these genes were associated with ε-carotene, 82 with γ-carotene, 99 with α-carotene, and 60 with β-carotene. Additionally, it was found that 49 genes were related to all four carotenoid types (Figure 6C). To evaluate the relationships between these carotenoid components and the identified 49 genes, Pearson's correlation method was utilized, as illustrated in Figure 6D. Functional annotations for these genes are detailed in Supplementary Table S2, which contains 14 genes with unknown functions. Within the gene ontology (GO) classifications, 'hydrolase activity' in the 'molecular_function' category was predominant, including four genes: *LYG036752* (beta-glucosidase), *LYG035009* (beta-glucosidase), *LYG007921* (alpha-glucosidase), and *LYG016487* (beta-galactosidase).

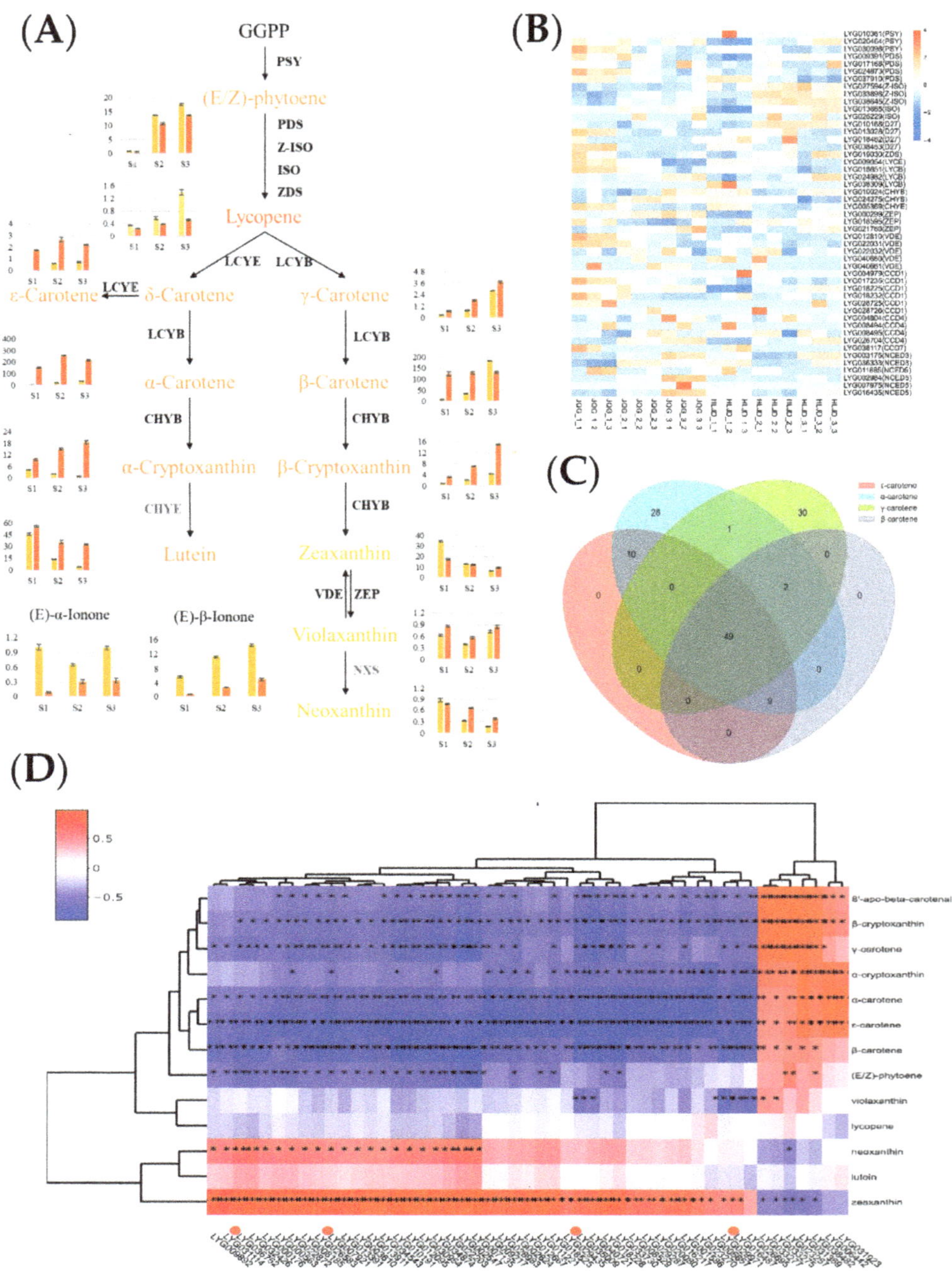

Figure 6. Conjoint analysis of carotenoid metabolome and transcriptome in *O. fragrans* JQG and HLJD petals. (**A**) Content of various components in the carotenoid metabolic pathway; (**B**) heat map of carotenoid metabolism pathway genes; (**C**) Venn map of ε-carotene, γ-carotene, α-carotene, and β-carotene related genes; (**D**) cluster heatmap of the co-expression correlation between carotenoid components and the 49 related genes. The correlation analysis was conducted using Pearson's correlation method (* $p \leq 0.05$, ** $p \leq 0.01$, *** $p \leq 0.001$).

4. Discussion

This research initially establishes that HLJD represents a natural bud mutation variety of JQG. This finding provides direct support for the evolutionary model of floral color in *O. fragrans*, suggesting a transition from light to dark and reinforcing the proposition that the Dangui group evolved through bud mutations [1,21]. The distinctive phenotype of HLJD, characterized by orange-red inner petals and golden outer petals, serves as a valuable natural mutant for investigating the genetic basis of floral color development in plants. Examination of plastid structures revealed a significant presence of starch-containing plastids within the petals of both cultivars during the bud phase, with chromoplast characteristics being faint. As the plants progressed to the initial- and full-flowering stages, starch granules gradually diminished, while chromoplast features became more pronounced. However, notable structural differences in the chromoplasts were observed between the two cultivars: in JQG, only chromoplasts containing plastoglobuli were present, featuring round starch granules encased in plastoglobuli, whereas HLJD exhibited a diverse array of chromoplast structures, including crystalline, tubular, fibrillar, and membrane forms, along with a limited presence of plastoglobuli. Concurrently, the rate of starch granule degradation in HLJD was slower compared to JQG, resulting in irregular forms. This suggests that chromoplasts in the petals of *O. fragrans* likely originate from the transformation of amyloplasts, and that the typical hydrolysis of starch is crucial for the formation of spherical plastoglobuli within chromoplasts. Previous studies have documented the transformation of amyloplasts into chromoplasts in the roots, tubers, or fruits of plant species such as cassava, potatoes, citrus, and kiwifruit, indicating that starch degradation promotes the accumulation of the sugar precursors necessary for carotenoid synthesis [26–30]. In plant petals, previous studies have reported that colored plastids predominantly arise from the transformation of chloroplasts or from the direct differentiation from proplastids [31,32]. This developmental mechanism encompasses the remodeling of cellular structures, regulation of pigment metabolism, and precise coordination of gene expression [33]. This study is the first to demonstrate that the development of colored plastids in plant petals originates from amyloplasts. However, the underlying mechanisms facilitating this process require further elucidation.

Targeting metabolomics of carotenoids reveals that the concentrations of ε-carotene, γ-carotene, α-carotene, and β-carotene in HLJD petals are significantly higher compared to those in JQG, and these concentrations further increase as the flowers open. Conversely, the levels of carotenoid cleavage products, such as α-ionone and β-ionone, are markedly lower in HLJD than in JQG. This observation implies that the mutant may enhance the development of dark phenotypes by directing metabolic flow towards pigment production while simultaneously suppressing the cleavage pathway. Previous research has indicated that among the three groups of *O. fragrans*, the Dangui cultivar exhibits elevated levels of α-carotene and β-carotene [1,18]; however, its fragrance is less pronounced than that of Jingui and Yingui due to reduced concentrations of α-ionone and β-ionone [4,34]. It has been demonstrated that low expression or functional impairment of the *CCD4* gene inhibits carotenoid cleavage [1,35,36]. In our previous investigation of the red-flowered cultivar 'Yanzhi Hong', decreasing expression of LYCE obstructed the synthesis of downstream α-carotene, leading to an accumulation of red lycopene [3]. This study, which integrates metabolomics and transcriptomics analyses, indicates that the genes associated with carotenoid synthesis and metabolism do not have a direct connection with the floral color mutation phenotype. This suggests that pigment accumulation may be influenced by indirect factors, such as substrate availability or conditions within storage microenvironments. Furthermore, it is noted that starch granule degradation is slower in HLJD petals compared to JQG, characterized by irregular forms, with significant crystalline, tubular, and fibrous structures present within the chromoplasts instead of plastoglobuli.

It is hypothesized that the irregular degradation of starch hinders the development of spherical chromoplasts, which facilitates the cleavage of carotenoids into apocarotenoids, while crystalline, tubular, and fibrous chromoplasts are more advantageous for carotenoid storage without degradation.

A total of 49 candidate genes were identified through the correlation analysis between metabolites and transcriptomes. The genes that are highly correlated with carotenoid accumulation are associated with starch degradation and soluble sugar synthesis, including four glycoside hydrolase genes (e.g., LYG036752, LYG035009, LYG007921, LYG016487), one sucrose synthase (LYG010191), and one glucose-1-phosphate adenylyltransferase (LYG003610). These genes exhibited significantly lower transcription levels in HLJD compared to JQG. Notably, the allelic variation of the Ven1 gene in maize endosperm regulates the contents of polar and non-polar carotenoids. The increase in non-polar carotenoids delays the degradation of the amyloplast membrane, thereby hindering the interaction between protein bodies and starch granules, which affects the formation of hard endosperm [37]. This study indirectly supports the association between carotenoids and starch metabolism; however, direct evidence linking the two remains elusive. Based on our combined analysis of metabolomics and transcriptomics, we focused on the glycoside hydrolase genes involved in starch degradation. Subsequent functional analyses of these genes will provide evidence for the direct association between carotenoid and starch metabolism. Integrating these findings, we arrive at the following inference (Figure 7): the decreased expression of candidate genes, such as glycoside hydrolases, leads to a delay in the hydrolysis of starch granules. This delay subsequently causes abnormal development of chromoplast structures (e.g., atypical crystallization, tubular, and fibrous structures). Such structural abnormalities compel chromoplasts to prioritize storage over the cleavage of carotenoids. Consequently, compared to chromoplasts including spherical plastoglobuli, the likelihood of carotenoid cleavage is significantly diminished. Additionally, the cleavage pathways, including the synthesis of α-ionone and β-ionone synthesis, are inhibited, ultimately resulting in a substantial accumulation of pigments and a deepening of flower colors.

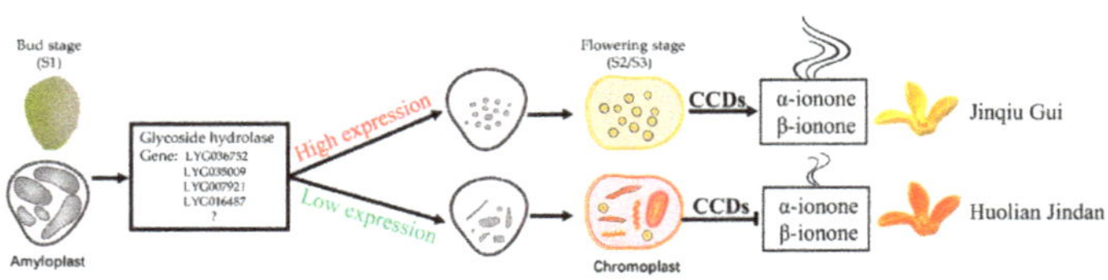

Figure 7. A proposed model on floral color formation of *O. fragrans*.

5. Conclusions

This study, based on the microstructure and multi-omics analysis of natural flower color mutants of *O. fragrans*, reveals for the first time a key pathway driven by delayed starch degradation that deepens flower color: "Delayed starch degradation → Abnormal chromoplast structure → Inhibition of carotenoid cleavage." The elucidation of this pathway provides new insights into the metabolic regulation of carotenoids in the non-photosynthetic tissues of plants. Given the current relative scarcity of research on the interrelationship between starch metabolism and carotenoid metabolism, along with the limited available cases for reference, exploring the deeper molecular mechanisms of their interaction will be an important direction for future research. This study clarifies the potential core role of glycoside hydrolase genes in this pathway. Subsequent research will focus on in-depth functional validation of these genes to elucidate their specific roles in regulating starch degradation and carotenoid homeostasis. The relevant research findings are expected to identify key targets for the molecular breeding of flower color in *O. fragrans* and other ornamental plants, thereby promoting the breeding of high ornamental varieties.

Supplementary Materials: The following supporting information can be downloaded at: https://www.mdpi.com/article/10.3390/horticulturae11070864/s1, Table S1: Components and contents of carotenoids in petals of JQG and HLJD at three stages; Figure S1: GC chart of volatile components detection in JQG and HLJD petals; Table S2: Functional annotation of genes related to ε-carotene, γ-carotene, α-carotene, and β-carotene in *O. fragrans*.

Author Contributions: Conceptualization, H.C.; methodology, X.Z., Y.T., and X.W.; software, J.Y.; validation, X.C.; formal analysis, X.W.; investigation, J.Z.; resources, X.Z.; data curation, H.W.; writing—original draft, Y.T.; writing—review and editing, X.Z.; visualization, H.W.; supervision, Q.H.; project administration, X.Z. and H.C.; funding acquisition, X.Z., J.Z., J.Y., X.C. and H.C. All authors have read and agreed to the published version of the manuscript.

Funding: This research was funded by the National Natural Science Foundation of China (32101581, 32271951), Hubei Province Natural Science Foundation (2025AFB857, 2024AFB1057, 2023AFB1063) and Hubei University of Science and Technology Innovation Team (2022T02).

Data Availability Statement: Whole Genome Resequencing and RNA-seq raw data have been deposited into the public database of the National Center of Biotechnology Information (NCBI) BioProject PRJNA1279839. RNA-seq raw data were also deposited under these NCBI accession number SAMN49507044- SAMN49507061).

Acknowledgments: The authors express their sincere gratitude to Baichun Shen from the Department of Forestry of Zhejiang Province, China, for his invaluable assistance in sampling the Osmanthus fragrans 'JQG' and 'HLJD'.

Conflicts of Interest: The authors declare no conflicts of interest.

References

1. Chen, H.; Zeng, X.; Yang, J.; Cai, X.; Shi, Y.; Zheng, R.; Wang, Z.; Liu, J.; Yi, X.; Xiao, S.; et al. Whole-genome resequencing of *Osmanthus fragrans* provides insights into flower color evolution. *Hortic. Res.* **2021**, *8*, 98. [CrossRef] [PubMed]
2. Zou, J.; Cai, X.; Yang, J.; Zeng, X.; Liu, D.; Huang, S.; Chen, X.; Yang, Q.; Wang, C.; Chen, H. DNA hypomethylation mediates flower opening and senescence in sweet osmanthus through auxin and ethylene responsive pathways. *Postharvest Biol. Technol.* **2023**, *198*, 112250. [CrossRef]
3. Wei, S.; Wu, J.; Yu, P.; Tan, Y.; He, Q.; Yang, J.; Cai, X.; Zou, J.; Chen, H.; Zeng, X. Metabolomic and Transcriptomic Analysis of Unique Floral Coloration in *Osmanthus fragrans* Cultivars. *Horticulturae* **2024**, *10*, 801. [CrossRef]
4. Cai, X.; Mai, R.; Zou, J.; Zhang, H.; Zeng, X.; Zheng, R.; Wang, C. Analysis of aroma-active compounds in three sweet osmanthus (*Osmanthus fragrans*) cultivars by GC-olfactometry and GC-MS. *J. Zhejiang Univ.-Sci. B Biomed. Biotechnol.* **2014**, *15*, 650–660. [CrossRef] [PubMed]
5. Zeng, X.; Liu, C.; Zheng, R.; Cai, X.; Luo, J.; Zou, J.; Wang, C. Emission and accumulation of monoterpene and the key terpene synthase (TPS) associated with monoterpene biosynthesis in *Osmanthus fragrans* Lour. *Front. Plant Sci.* **2016**, *6*, 1232. [CrossRef] [PubMed]
6. Iorizzo, M.; Ellison, S.; Senalik, D.; Zeng, P.; Satapoomin, P.; Huang, J.; Bowman, M.; Iovene, M.; Sanseverino, W.; Cavagnaro, P.; et al. A high-quality carrot genome assembly provides new insights into carotenoid accumulation and asterid genome evolution. *Nat. Genet.* **2016**, *48*, 657–666. [CrossRef] [PubMed]
7. Villwock, S.S.; Li, L.; Jannink, J.L. Carotenoid-carbohydrate crosstalk: Evidence for genetic and physiological interactions in storage tissues across crop species. *New Phytol.* **2024**, *244*, 1709–1722. [CrossRef] [PubMed]
8. Zhou, X.; Rao, S.; Wrightstone, E.; Sun, T.; Lui, A.C.W.; Welsch, R.; Li, L. Phytoene synthase: The key rate-limiting enzyme of carotenoid biosynthesis in plants. *Front. Plant Sci.* **2022**, *13*, 884720. [CrossRef] [PubMed]
9. Sun, Q.; He, Z.; Wei, R.; Ye, J.; Chai, L.; Cheng, Y.; Xu, Q.; Deng, X. Red peel regulator 1 links ethylene response factor 25 and β-citraurin biosynthetic genes to regulate ethylene-induced peel reddening in citrus. *Plant Cell* **2024**, *37*, 10. [CrossRef] [PubMed]
10. Zheng, X.; Mi, J.; Balakrishna, A.; Liew, K.X.; Ablazov, A.; Sougrat, R.; Al-Babili, S. Gardenia carotenoid cleavage dioxygenase 4a is an efficient tool for biotechnological production of crocins in green and non-green plant tissues. *Plant Biotechnol. J.* **2022**, *20*, 2202–2216. [CrossRef] [PubMed]

11. Baldermann, S.; Kato, M.; Kurosawa, M.; Kurobayashi, Y.; Fujita, A.; Fleischmann, P.; Watanabe, N. Functional characterization of a carotenoid cleavage dioxygenase 1 and its relation to the carotenoid accumulation and volatile emission during the floral development of *Osmanthus fragrans* Lour. *J. Exp. Bot.* **2010**, *61*, 2967–2977. [CrossRef] [PubMed]
12. Frey, A.; Effroy, D.; Lefebvre, V.; Seo, M.; Perreau, F.; Berger, A.; Sechet, J.; To, A.; North, H.M.; Marion-Poll, A. Epoxycarotenoid cleavage by NCED5 fine-tunes ABA accumulation and affects seed dormancy and drought tolerance with other NCED family members. *Plant J.* **2012**, *70*, 501–512. [CrossRef] [PubMed]
13. Gavassi, M.A.; Silva, G.S.; Silva, C.M.S.; Thompson, A.J.; Macleod, K.; Oliveira, P.M.R.; Cavalheiro, M.F.; Domingues, D.S. Habermann, G. NCED expression is related to increased ABA biosynthesis and stomatal closure under aluminum stress. *Environ Exp. Bot.* **2021**, *185*, 104404. [CrossRef]
14. Sun, T.; Yuan, H.; Cao, H.; Yazdani, M.; Tadmor, Y.; Li, L. Carotenoid metabolism in plants: The role of plastids. *Mol. Plant* **2018** *11*, 58–74. [CrossRef] [PubMed]
15. Ling, Q.; Sadali, N.M.; Soufi, Z.; Zhou, Y.; Huang, B.; Zeng, Y.; Rodriguez-Concepcion, M.; Jarvis, R.P. The chloroplast-associated protein degradation pathway controls chromoplast development and fruit ripening in tomato. *Nat. Plants* **2021**, *7*, 655–666 [CrossRef] [PubMed]
16. Luan, Y.; Fu, X.; Lu, P.; Grierson, D.; Xu, W. Molecular mechanisms determining the differential accumulation of carotenoids in plant species and varieties. *Crit. Rev. Plant Sci.* **2020**, *39*, 125–139. [CrossRef]
17. Hermanns, A.; Zhou, X.; Xu, Q.; Tadmor, Y.; Li, L. Carotenoid pigment accumulation in horticultural plants. *Hortic. Plant J.* **2020** *6*, 343–360. [CrossRef]
18. Wang, Y.; Zhang, C.; Dong, B.; Fu, J.; Hu, S.; Zhao, H. Carotenoid Accumulation and Its Contribution to Flower Coloration of *Osmanthus fragrans*. *Front. Plant Sci.* **2018**, *9*, 1499. [CrossRef] [PubMed]
19. Xiang, Q.; Liu, Y. *An Illustrated Monograph of the Sweet Osmanthus Cultivars in China*; Zhejiang Science and Technology Press Hangzhou, China, 2008.
20. Yang, K. *Chinese Osmanthus*; China Forestry Publishing House: Beijing, China, 2020.
21. Wang, Y.; Luo, Y.; Zhang, C.; Fu, J.; Hu, S.; Zhao, H. Flower Color and Pigment Composition in the Petals of Bud Mutation and its Stock Plant of *Osmanthus fragrans* 'Jingui'. *Acta Hortic. Sin.* **2017**, *44*, 528–536.
22. Simon, A. *FastQC: A Quality Control Tool for High Throughput Sequence Data*; Version 0.10.1; Babraham Bioinformatics: Cambridge UK, 2010.
23. Li, H.; Durbin, R. Fast and accurate short read alignment with Burrows-Wheeler transform. *Bioinformatics* **2009**, *25*, 1754–1760 [CrossRef] [PubMed]
24. McKenna, A.; Hanna, M.; Banks, E.; Sivachenko, A.; Cibulskis, K.; Kernytsky, A.; Garimella, K.; Altshuler, D.; Gabriel, S.; Daly, M. et al. The Genome Analysis Toolkit: A MapReduce framework for analyzing next-generation DNA sequencing data. *Genome Res* **2010**, *20*, 1297–1303. [CrossRef] [PubMed]
25. Letunic, I.; Bork, P. Interactive Tree of Life (iTOL) v4: Recent updates and new developments. *Nucleic Acids Res.* **2019**, *47* W256–W259. [CrossRef] [PubMed]
26. Drapal, M.; Gerrish, C.; Fraser, P.D. Changes in carbon allocation and subplastidal amyloplast structures of specialised Ipomoea batatas (sweet potato) storage root phenotypes. *Phytochemistry* **2022**, *203*, 113409. [CrossRef] [PubMed]
27. Gutschker, S.; Ruescher, D.; Rabbi, I.Y.; Rosado-Souza, L.; Pommerrenig, B.; Pauly, M.; Robertz, S.; van Doorn, A.M.; Schlereth, A. Neuhaus, H.E.; et al. Carbon usage in yellow-fleshed *Manihot esculenta* storage roots shifts from starch biosynthesis to cell wall and raffinose biosynthesis via the myo-inositol pathway. *Plant J.* **2024**, *119*, 2045–2062. [CrossRef] [PubMed]
28. Zhou, X.; Mcquinn, R.; Fei, Z.; Wolters, A.A.; van Eck, J.; Brown, C.; Giovannoni, J.J.; Li, L. Regulatory control of high levels of carotenoid accumulation in potato tubers. *Plant Cell Environ.* **2011**, *34*, 1020–1030. [CrossRef] [PubMed]
29. Zeng, Y.; Du, J.; Wang, L.; Pan, Z.; Xu, Q.; Xiao, S.; Deng, X. A comprehensive analysis ofchromoplastdifferentiationrevealscomplexproteinchangesassociated with plastoglobule biogenesis and remodeling of protein systems in sweet orange flesh. *Plant Physiol.* **2015**, *168*, 1648–1665. [CrossRef] [PubMed]
30. Li, A.; Lin, J.; Zeng, Z.; Deng, Z.; Tan, J.; Chen, X.; Ding, G.; Zhu, M.; Xu, B.; Atkinson, R.G.; et al. The kiwifruit amyloplast proteome (kfALP): A resource to better understand the mechanisms underlying amyloplast biogenesis and differentiation. *Plant J.* **2024**, *118*, 565–583. [CrossRef] [PubMed]
31. Liu, Z.; Ye, K.; Du, X.; Peng, Y.; Qin, J.; Zeng, L. Correlation analysis of petal morphology and pigment content of Rosa 'Angela' *J. Beijing Univ. Agric.* **2022**, *37*, 1–10.
32. Yuan, Y.; Li, X.; Yao, X.; Fu, X.; Cheng, J.; Shan, H.; Yin, X.; Kong, H. Mechanisms underlying the formation of complex color patterns on *Nigella orientalis* (Ranunculaceae) petals. *New Phytol.* **2023**, *237*, 2450–2466. [CrossRef] [PubMed]
33. May, T.S.Y.; Edwige, M. Under the rainbow: Novel insights on the mechanisms driving the development and evolution of petal pigmentation. *Curr. Opin. Plant Biol.* **2025**, *86*, 102743. [CrossRef] [PubMed]
34. Xin, H.; Wu, B.; Zhang, H.; Wang, C.; Li, J.; Yang, B.; Li, S. Characterization of volatile compounds in flowers from four groups of sweet osmanthus (*Osmanthus fragrans*) cultivars. *Can. J. Plant Sci.* **2013**, *93*, 923–931. [CrossRef]

5. Han, Y.; Wang, X.; Chen, W.; Dong, M.; Yuan, W.; Liu, X.; Shang, F. Differential expression of carotenoid-related genes determines diversified carotenoid coloration in flower petal of *Osmanthus fragrans*. *Tree Genet. Genomes* **2014**, *10*, 329–338. [CrossRef]
6. Han, Y.; Lu, M.; Yue, S.; Li, K.; Dong, M.; Liu, L.; Wang, H.; Shang, F. Comparative methylomics and chromatin accessibility analysis in *Osmanthus fragrans* uncovers regulation of genic transcription and mechanisms of key floral scent production. *Hortic. Res.* **2022**, *9*, 96. [CrossRef] [PubMed]
7. Wang, H.; Huang, Y.; Xiao, Q.; Huang, X.; Li, C.; Gao, X.; Wang, Q.; Xiang, X.; Zhu, Y.; Wang, J.; et al. Carotenoids modulate kernel texture in maize by influencing amyloplast envelope integrity. *Nat. Commun.* **2020**, *11*, 5346. [CrossRef] [PubMed]

horticulturae

MDPI

Article

Metabolomic and Transcriptomic Analysis of Unique Floral Coloration in *Osmanthus fragrans* Cultivars

Shenni Wei [1], Jiang Wu [2], Ping Yu [3], Yunfei Tan [3], Qiang He [2], Jie Yang [1], Xuan Cai [1], Jingjing Zou [4], Hongguo Chen [4] and Xiangling Zeng [1,*]

1 National Forestry and Grassland Administration Engineering Research Center for Osmanthus Fragrans, Hubei University of Science and Technology, Xianning 437100, China; weishenni2024@163.com (S.W.); yj_anna@hbust.edu.cn (J.Y.); caixuan@hbust.edu.cn (X.C.)

2 Public Inspection and Testing Center, Xianning 437100, China; wujiang2718@163.com (J.W.); qianghe2015@126.com (Q.H.)

3 School of Pharmacy, Xianning Medical College, Hubei University of Science and Technology, Xianning 437100, China; yuping202407@163.com (P.Y.); tanyunfei0926@163.com (Y.T.)

4 Research Center for Osmanthus Fragrans, Xianning Research Academy of Industrial Technology of Osmanthus Fragrans, Xianning 437100, China; silence@hbust.edu.cn (J.Z.); chen_hongguo1969@163.com (H.C.)

* Correspondence: cengxiangling@hbust.edu.cn

Abstract: The floral color phenotypes of *Osmanthus fragrans* cultivars range from light yellow to orange yellow, with 'Yanzhi Hong' being the only reported cultivar with a red color. However, the underlying reason for this unique floral coloration remains unclear. The study conducted targeted metabolomics and transcriptomics analyses on the petals of 'Yanzhi Hong' at both initial and peak flowering stages. Candidate gene expression was validated, and expression levels of the petals of three cultivars were compared using RT-qPCR. The results revealed the presence of 27 components in the petals of 'Yanzhi Hong', including 5 carotenoids, 8 xanthophylls, and 14 xanthophyll esters. Notably, lycopene was detected in abundance for the first time in *O. fragrans* cultivars. Carotenes accounted for $78.82 \pm 3.17\%$ and $91.19 \pm 1.69\%$ of the total carotenoid content in petals during the initial and peak flowering stages, respectively, with all carotene contents increasing during the peak flowering period. β-carotene, lycopene, and γ-carotene were identified as the top three carotene components in petals during both initial and full flowering stages. The unique blush red color of 'Yanzhi Hong' petals could be attributed to the low content of α-carotene and the rich accumulation of lycopene. Furthermore, a total of 1550 differentially expressed genes (DEGs) were identified in petals at the peak flowering stage relative to the initial flowering stage, with 1003 genes being downregulated and 547 genes being upregulated during the full flowering stage. There are 926 differentially expressed genes (DEGs) annotated in the Gene Ontology (GO) database. Among these DEGs, those that were downregulated and upregulated during the peak flowering period showed significant enrichment in carbohydrate metabolism and oxidation–reduction processes, respectively. The Kyoto Encyclopedia of Genes and Genomes (KEGG) analysis identified 14 structural genes associated with phenylpropanoid biosynthesis and 7 structural genes linked to carotenoid biosynthesis. Expression levels of candidate genes involved in carotenoid biosynthesis were examined in the petals of three cultivars ('Yanzhi Hong', 'Liuye Jingui', and 'Gecheng Dangui') at both the initial and peak flowering stages. The results indicated that the decreased expression of *LYG009054* (*LYCE*) and *LYG018651* (*LYCB*) in 'Yanzhi Hong' resulted in higher lycopene accumulation and lower α-carotene content in the petals. This study offers valuable insights into the mechanisms underlying the unique flower color phenotype of *O. fragrans*, proving a basis for further research on carotenoid metabolism pathways and the breeding of new cultivars with a variety of flower colors in *O. fragrans*.

Keywords: *Osmanthus fragrans*; floral color; carotenoid

1. Introduction

Osmanthus fragrans Lour, a traditional Chinese flower renowned for its fragrance, belongs to the Oleaceae family and *Osmanthus* genus. Originating in southwest China, it boasts a cultivation history spanning over 2500 years. Widely grown in the southern region of the Yellow River Basin in China, it stands out as a superior tree species for landscaping purposes. Moreover, *O. fragrans* holds significant economic value in China, finding application in essential oils, beverages, and various food products. Key production areas for *O. fragrans* include Xianning in Hubei, Chengdu in Sichuan, Guilin in Guangxi, Suzhou in Jiangsu, and Hangzhou in Zhejiang [1]. Cultivars of *O. fragrans* are classified into four groups—Semperflorens, Albus, Luteus, and Aurantiacus—based on their flowering season and color. The Albus group is characterized by pale yellow blooms, the Luteus group flaunts yellow flowers, and the Aurantiacus group presents orange-yellow blossoms [2,3]. The limited color spectrum of *O. fragrans* cultivars is due to the absence of hues other than yellow.

The formation of plant flower color is determined by three major pigments: anthocyanins, carotenoids, and betaine [4,5]. Anthocyanins, a type of flavonoid compound, are synthesized in the cytoplasm through the phenylpropanoid metabolism pathway and stored in vacuoles, resulting in a broad spectrum of colors ranging yellow, red, or blue. Carotenoids, lipophilic C40 terpenoids, are synthesized and stored in plastids through the isoprene metabolic pathway, leading to a color range from yellow to red. Betaine, a nitrogen-containing compound, is found only in a few plants. In certain plants like chrysanthemums and roses, anthocyanins and carotenoids can coexist in the same tissue, producing a variety of flower colors [6,7]. Studies have shown that *O. fragrans*, while abundant in flavonoids, primarily contains light yellow compounds such as quercetin, naringenin, and apigenin, with no anthocyanin components detected [8–10]. The main pigments responsible for color changes in different *O. fragrans* cultivars are carotenoids, particularly α-carotene and β-carotene [10–12]. Consequently, owing to its high concentration of orange-yellow carotenoids and absence of anthocyanins, *O. fragrans* flowers display a color transition from yellow-white to orange-yellow, with limited variation in color cultivars.

'Yanzhi Hong' is a recently registered cultivar of *O. fragrans*, distinguished as the sole red cultivar known to date. Its flowers transition from light yellow to pink and ultimately to blush red during different flowering stages [13,14]. The specific mechanism behind this color transformation remains unknown. In a study by Chai et al. [14], a metabolomics analysis on 'Yanzhi Hong' petals identified 304 metabolites including iridoids and flavonoids, but not anthocyanins. This study focused on the initial and peak flowering stages to explore the carotenoid composition, gene expression patterns, and mechanisms underlying the red floral color of 'Yanzhi Hong', aiming to provide insights for breeding *O. fragrans* cultivars with diverse flower colors.

2. Materials and Methods

2.1. Plant Materials

Samples of the 'Yanzhi Hong' (YZH) cultivar were collected from Huaan Garden Nursery Base in Jinhua City, Zhejiang Province (119°38′43″ E, 28°58′16″ N), specifically focusing on two developmental stages with distinct flower colors: the initial flowering stage (S1) and the peak flowering stage (S2) (refer to Figure 1). Two other cultivars, 'Liuye Jingui' (LYJG) and 'Gecheng Dangui' (GCDG), previous studied [15], were used for RT-qPCR analysis of genes for lycopene cyclase and carotene hydroxylase and were obtained from the nursery at Huazhong Agricultural University campus (114°21′3″ E, 30°28′43″ N). Petals weighing 2 g were collected from each stage, with three biological replicates per experiment. The samples were immediately frozen in liquid nitrogen and stored at −80 °C.

Figure 1. Comparison of petal colors between 'Yanzhi Hong' and Albus, Luteus, and Aurantiacus cultivars. (**A**) The initial flowering stage of 'Yanzhi Hong' (YZH1); (**B**) The peak flowering stage of 'Yanzhi Hong' (YZH2); (**C**) The comparison of different cultivars, from left to right, is Albus cultivar, Luteus cultivar, 'Yanzhi Hong' cultivar and Aurantiacus cultivar.

2.2. Extraction of Carotenoids

Freeze-dried samples were ground into powder at 30 Hz for 1 min. Subsequently, 50 mg of the ground sample was accurately weighed and extracted with a 0.5 mL mixture of n-hexane/acetone/ethanol (1:1:1, $v/v/v$) containing 0.01% butylated hydroxytoluene (BHT) (g/mL). After swirling at room temperature for 20 min, the solutions were centrifuged at 4 °C at 13,400× g for 5 min to obtain the supernatant. The residue was re-extracted by repeating the above steps again under the same conditions and the supernatants were merged. Following the concentration of the extraction solution using evaporation under vacuum, the samples were redissolved in a 100 μL methanol/methyl tert-butyl ether (MTBE) mixed solution (1:1, v/v), filtered through a 0.22 μm membrane filter, and stored in a brown injection bottle for LC-MS/MS analysis (QTRAP6500+, SCIEX, Framingham, MA, USA, https://sciex.com.cn/ (accessed on 13 March 2024)).

2.3. Detection Conditions of Carotenoids

Sample separation and mass spectrometry data acquisition were carried out using ultra performance liquid chromatography (UPLC) system and Tandem Mass Spectrometry (MS/MS) with the QTRAP® 6500+ instrument (https://sciex.com.cn/ (accessed on 13 March 2024)). The separation utilized a C30 chromatographic column (100 mm × 2.0 mm, 3 μm; YMC, Kyoto, Japan). The mobile phase for phase A consisted of methanol/acetonitrile in a 1:3 volumetric ratio, supplemented with 0.01% BHT and 0.1% formic acid, while phase B contained MTBE with 0.01% BHT. The gradient elution method followed these time points: initial 0 min (A/B 100:0), 3 min (100:0), 5 min (30:70), 9 min (5:95), 10 min (100:0), and final 11 min (100:0). The flow rate was maintained at 0.8 mL/min, column temperature set at 28 °C, and injection volume fixed at 2 μL. Linear ion trap (LIT) and triple quadrupole (QQQ) scans were performed on a QTRAP mass spectrometer (QTRAP® 6500+ LC-MS/MS System) (https://sciex.com.cn/ (accessed on 13 March 2024)) using an APCI Heated Nebulizer in positive ion mode. Instrument control was handled with Analyst 1.6.3 software (https://sciex.com.cn/ (accessed on 13 March 2024)). The

APCI source settings included an APCI+ ion source, a source temperature of 350 °C, and a curtain gas (CUR) pressure of 25.0 psi.

2.4. Qualitative and Quantitative Analysis of Carotenoids

Carotenoids underwent analysis utilizing scheduled multiple reaction monitoring (MRM), while metabolite quantities were evaluated via data acquisition and analysis with Analyst 1.6.3 and Multi-quant 3.0.3 software. The results of hierarchical cluster analysis (HCA) were displayed as heatmaps accompanied by dendrograms, and Pearson correlation coefficients (PCC) among samples were determined using the cor function in R (v3.5.1) and visualized in heatmaps. Both HCA and PCC analyses were conducted using the pheatmap R package (2.7.1.1009). The HCA visualization showcased the normalized signal intensities of metabolites (normalized by unit variance) using a diverse color spectrum. Identification of significantly regulated metabolites between groups was based on absolute Log2 fold changes (Log2FC). Metabolites were annotated with the KEGG compound database http://www.kegg.jp/kegg/compound/ (accessed on 13 March 2024) and aligned with the KEGG Pathway database http://www.kegg.jp/kegg/pathway.html (accessed on 13 March 2024). Pathways containing significantly regulated metabolites were subjected to metabolite sets enrichment analysis (MSEA), with statistical significance assessed using *p*-values obtained from the hypergeometric test.

2.5. Transcriptome Sequencing and Differential Expression Gene Screening

RNA isolation of 'Yanzhi Hong' flower petal specimens was performed with the Trizol method, followed by Qsep400 system evaluation for quality (AutoQ Biosciences, San Diego, CA, USA). Utilizing the VAHTS mRNA seq V8 Library Prep Kit from Vazyme (Nanjing, China), the Illumina sequencing library (Illumina, San Diego, CA, USA) was prepared, starting with 1 µg of total RNA. Library preparation included polyA RNA selection, RNA fragmentation, reverse transcription using random hexamers, and sequencing on the Illumina Novaseq 6000 platform with 150 nt paired-ends. Adapters were removed and low-quality reads filtered using cutadapt (v1.11). The clean reads were aligned to the 'Liuye Jingui' reference genome [15] using Hisat2 (v2.1.0), allowing for up to two mismatches. And the raw sequencing data had been uploaded to NCBI (accession number SAMN42797811-SAMN42797816). Identified genes were compared against public protein databases like NR (RefSeq non-redundant proteins). Transcript levels were estimated and gene expression normalized as FPKM (Fragments per kilobase of transcript per million fragments mapped) with Feature-count (v1.6.0). Differential expression analysis was carried out with edgeR, applying a threshold of FDR < 0.05 and $|\log2FoldChange| > 1$. Gene Ontology (GO) and Kyoto Encyclopedia of Genes and Genomes (KEGG) enrichment analyses were performed using the Cluster Profiler in the R package, with enrichment analysis based on hypergeometric distribution and a q value cutoff of 0.05.

2.6. RT-qPCR Analysis of Genes Related to Carotenoid Synthesis

The Trizol method was utilized to extract total RNA from *O. fragrans* petals, followed by cDNA synthesis using the StarScript III All-in-one RT Mix with gDNA Remover kit (TransGen Biotech, Beijing, China). RT-qPCR was then performed with *OfRANI* as an internal reference gene (Supplementary Table S1) using 2x RealStar Fast SYBR qPCR Mix (GenStar, Beijing, China). The reaction system comprised 5 µL of 2x RealStar Fast SYBR qPCR Mix, 0.25 µL each of forward and reverse primers, 0.5 µL of cDNA, and water added to reach a final volume of 10 µL. The reaction procedure included a 4 min pre-denaturation at 94 °C, followed by denaturation at 94 °C for 10 s, annealing and extension at 60 °C for 30 s, repeated for a total of 40 cycles. PCR amplification was conducted on the Gentier 96R instrument (Tianlong, Xi'an, China), and the data were analyzed using the $2^{-\Delta\Delta Ct}$ algorithm.

2.7. Data Analysis

Metabolite clustering heatmap, co-expression correlation, linear regression analysis, and analysis of variance were conducted using the Omicshare Cloud tool https://www.omicshare.com/tools/ (accessed on 22 May 2024). The data was shown as mean ± SD, with three biological replicates for each sample.

3. Results

3.1. Difference in Carotenoids in 'Yanzhi Hong' Petals at the Initial and Peak Flowering Stages

The petal phenotypes of 'Yanzhi Hong' exhibited a noticeable color transition from the initial to peak flowering periods, as illustrated in Figure 1A,B. Different from the typical white, yellow, and orange hues of *O. fragrans* cultivars, 'Yanzhi Hong' displayed a reddish tone during the peak flowering stage, as depicted in Figure 1C. The UPLC-MS/MS analysis revealed the presence of 27 carotenoid components in 'Yanzhi Hong', including five carotenes (such as (E/Z)-phytoene, α-carotene, β-carotene, γ-carotene, and lycopene), eight xanthophylls (such as zeaxanthin, lutein, β-cryptoxanthin, etc.), and 14 xanthophyll esters (Table 1). Carotenes were the most abundant, making up 78.82% and 91.19% of total carotenoids during the initial and peak flowering stages, with higher levels of β-carotene and lycopene. Xanthophylls accounted for 13.24% and 5.61%, with zeaxanthin and lutein being the predominant. Xanthophyll esters comprised 7.94% and 3.20% of total carotenoids, with relatively high levels of β-cryptoxanthin palmate and rubixanthin palmate (Figure 2).

Table 1. Components and contents of carotenoid in petals of 'Yanzhi Hong' at initial and peak flowering stages.

No	RT	Compound Name	YZH1 Mean μg/g DW	YZH2 Mean μg/g DW
Carotenes				
1	5.00	(E/Z)-phytoene	3.72 ± 0.07 b	15.72 ± 0.42 a
2	5.92	α-carotene	1.93 ± 0.15 b	3.40 ± 0.21 a
3	6.28	β-carotene	33.73 ± 1.33 b	95.02 ± 4.37 a
4	7.38	γ-carotene	5.31 ± 0.11 b	17.35 ± 0.27 a
5	8.33	lycopene	48.84 ± 2.59 b	63.23 ± 3.31 a
			93.53 ± 4.25 b	194.72 ± 8.58 a
Xanthophylls				
6	1.58	violaxanthin	0.12 ± 0.00 a	0.09 ± 0.01 b
7	1.93	neoxanthin	0.21 ± 0.01 a	0.10 ± 0.01 b
8	4.04	lutein	3.48 ± 0.20 a	2.51 ± 0.05 b
9	4.51	8′-apo-β-carotenal	0.01 ± 0.00 b	0.08 ± 0.01 a
10	4.63	zeaxanthin	10.66 ± 0.40 a	5.77 ± 0.14 b
11	5.08	α-cryptoxanthin	0.41 ± 0.01	0.42 ± 0.02
12	5.52	β-cryptoxanthin	0.80 ± 0.06 b	2.96 ± 0.13 a
13	5.54	echinenone	0.01 ± 0.00 b	0.04 ± 0.00 a
			15.71 ± 0.69 a	11.97 ± 0.36 b
Xanthophyll esters				
14	6.46	lutein oleate	0.05 ± 0.00	0.05 ± 0.00
15	6.87	lutein palmitate	0.36 ± 0.03	0.36 ± 0.03
16	7.18	zeaxanthin palmitate	1.09 ± 0.07 a	0.73 ± 0.07 b
17	7.38	zeaxanthin dilaurate	0.04 ± 0.01	0.03 ± 0.01
18	7.48	zeaxanthin-laurate-myristate	0.03 ± 0.00	0.02 ± 0.00
19	7.63	zeaxanthin dimyristate	0.31 ± 0.01 a	0.24 ± 0.00 b
20	7.96	zeaxanthin dipalmitate	0.43 ± 0.02 a	0.26 ± 0.03 b
21	7.77	zeaxanthin-oleate-palmitate	0.20 ± 0.01 a	0.14 ± 0.01 b
22	6.95	β-cryptoxanthin laurate	0.09 ± 0.02	0.07 ± 0.01
23	7.18	β-cryptoxanthin myristate	0.16 ± 0.00 a	0.10 ± 0.01 b
24	7.32	β-cryptoxanthin palmitate	3.06 ± 0.09 a	2.25 ± 0.08 b
25	6.54	rubixanthin caprate	0.05 ± 0.00	0.04 ± 0.01
26	6.79	rubixanthin laurate	0.64 ± 0.03 a	0.50 ± 0.01 b
27	7.33	rubixanthin palmitate	2.92 ± 0.13 a	2.05 ± 0.10 b
			9.42 ± 0.43 a	6.82 ± 0.35 b

Different letters represent significant differences ($p < 0.05$) based on Tukey's multiple-range test using one-way ANOVA.

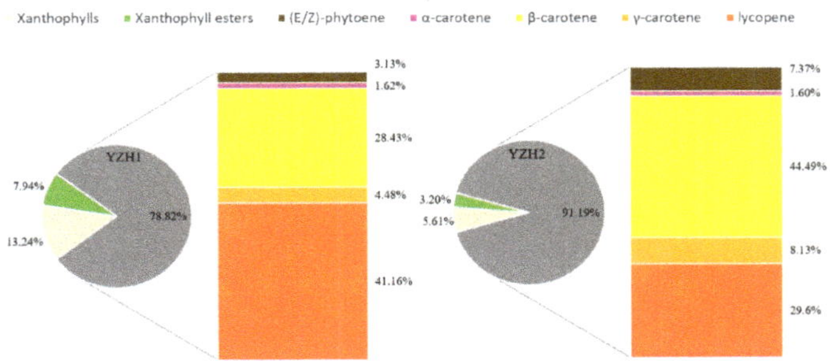

Figure 2. Proportion of carotenoid components in petals of 'Yanzhi Hong' at the initial and peak flowering stages.

Table 1 and Figure 3 illustrate the presence of identical carotenoid components in the petals of 'Yanzhi Hong' during both the initial and peak flowering stages. However, notable differences exist in the content of these components between the two stages. Specifically, the carotene content in the petals is significantly higher at the peak flowering stage compared to the initial stage. At the peak flowering stage, the levels of β-carotene, γ-carotene, and (E/Z)-phytoene are 95.02 ± 4.37 μg/g DW, 17.35 ± 0.27 μg/g DW, and 15.72 ± 0.42 μg/g DW, respectively, representing a threefold increase from the initial flowering stage. The highest lycopene content of 48.84 ± 2.59 μg/g DW is observed at the initial flowering stage, while during peak flowering, it is 63.23 ± 3.31 μg/g DW, ranking second only to β-carotene. Among xanthophylls, only the levels of β-cryptoxanthin, 8′-apo-β-carotenal, and echinenone increase during peak flowering, whereas zeaxanthin, lutein, neoxanthin, and violaxanthin decrease significantly. Additionally, most xanthophyll esters also decrease during peak flowering. These results suggest that the color transformation in 'Yanzhi Hong' during peak flowering is primarily due to an elevation in carotene content, particularly the three pigments β-carotene, lycopene, and γ-carotene.

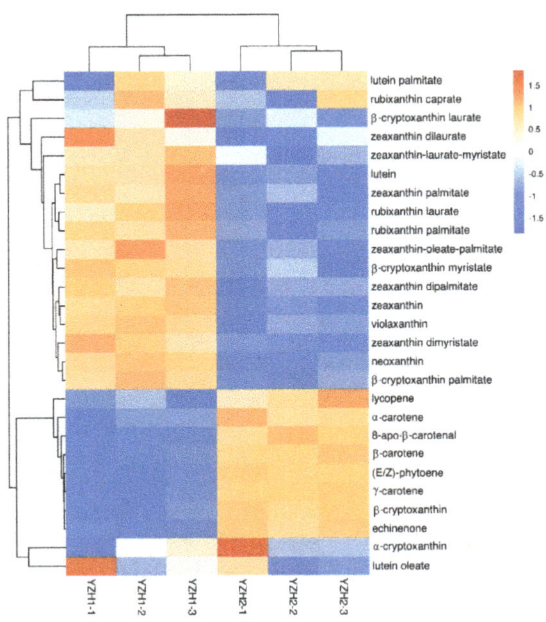

Figure 3. Carotenoid metabolite clustering heat map at the initial and peak flowering stages in 'Yanzhi Hong' petals. Color scales in heatmap represent log2(YZH2/YZH1) values.

3.2. Transcriptome Sequencing and DEGs Analysis of 'Yanzhi Hong' Petals at the Initial and Peak Flowering Stages

The transcriptome sequencing data of 'Yanzhi Hong' petals was analyzed at the initial and peak flowering stages, as detailed in Table 2. A total of 39.94 Gb of valid data was collected, with each sample yielding effective readings ranging from 35,777,926 to 52,210,754. The quality metrics, Q20 and Q30, were found to be high at 96.84% and 97.24%, and 91.53% and 92.34%, respectively, indicating the reliability of the sequencing data for further analysis. Differential gene expression analysis was conducted using FDR < 0.05 and | log2FoldChange | > 1 as selection criteria for genes expressed at different developmental stages. A total of 1550 DEGs were identified, with 1003 downregulated and 547 upregulated at the peak flowering stage (refer to Figure 4).

Table 2. Transcriptome sequencing data statistics in petals of 'Yanzhi Hong' at the initial and peak flowering stages.

Sample	RawReads	RawBases	CleanReads	CleanBases	CleanRatio	Q20	Q30	GC
YZH1_1	36,021,782	5.40	35,777,926	5.30	99.32%	96.86%	91.54%	42.72%
YZH1_2	50,681,302	7.60	50,373,046	7.48	99.39%	97.19%	92.17%	42.80%
YZH1_3	42,561,224	6.38	42,258,728	6.28	99.29%	96.95%	91.75%	42.77%
YZH2_1	43,146,876	6.47	42,876,012	6.36	99.37%	97.24%	92.34%	42.53%
YZH2_2	46,010,634	6.90	45,692,536	6.77	99.31%	96.99%	91.81%	42.53%
YZH2_3	52,623,910	7.89	52,210,754	7.75	99.21%	96.84%	91.53%	42.48%

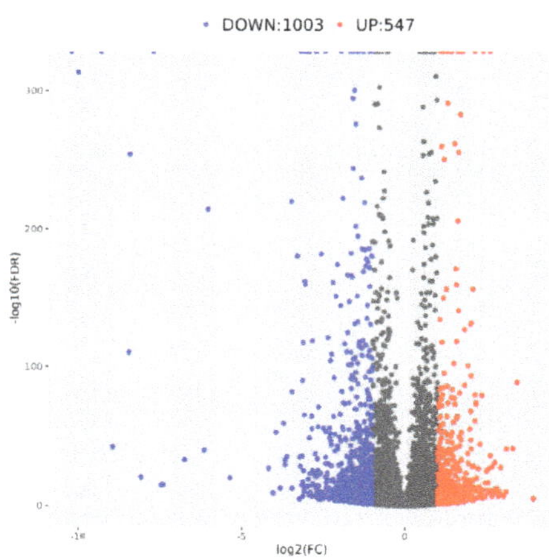

Figure 4. Volcano map of differential expression genes at the peak vs. initial flowering stage of *O. fragrans* 'Yanzhi Hong'.

GO enrichment analysis was conducted on 1550 identified DEGs, which were categorized into three groups by GO: biological processes, cellular components, and molecular functions (see Figure 5). The most notable enrichments in biological processes were observed in oxidation–reduction and carbohydrate metabolism processes, with 111 and 51 DEGs, respectively. Additionally, 85 DEGs related to membrane components showed significant enrichment in cellular composition. In terms of molecular function, genes associated with hydrolase activity and iron ion binding exhibited the highest abundances. Among the DEGs downregulated during the flowering period, significant enrichments were observed in carbohydrate metabolism, cell membrane composition, and hydrolytic enzyme activity. Conversely, upregulated DEGs during the flowering period showed significant enrichment in oxidation–reduction processes.

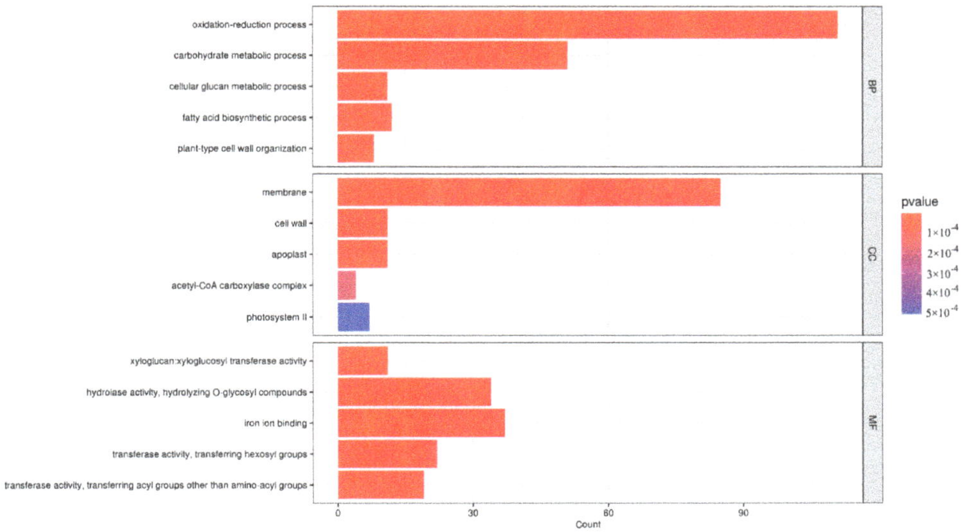

Figure 5. Bar chart of DEGs GO enrichment in *O. fragrans* 'Yanzhi Hong' at the peak vs. initial flowering stage.

KEGG enrichment analysis revealed that a total of 259 DEGs were annotated in the KEGG database, with significant enrichment in pathways related to plant pathogen interactions, phenylpropanoid biosynthesis, photosynthesis, and fatty acid biosynthesis (Figure 6). Downregulated DEGs during the peak flowering period were primarily associated with photosynthesis and fatty acid synthesis, while upregulated DEGs were linked to MAPK signaling, indole alkaloids, and terpenoid metabolism. Within the enriched DEGs, 14 structural genes were identified in phenylpropanoid biosynthesis (Table S2), including key enzymes such as peroxidase, HCT, CAD, COMT, F5H, 4CL, and PAL. Additionally, seven DEGs were involved in carotenoid biosynthesis, encompassing NCED, LCYE, VDE, PSY, CYP707A, and AAO3 (Table S2).

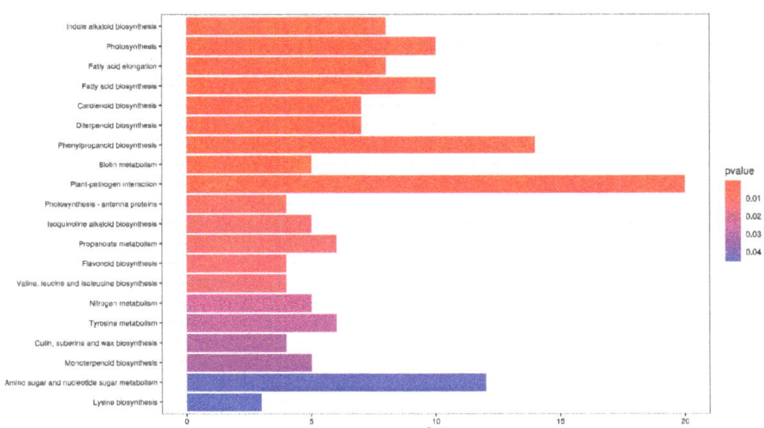

Figure 6. KEGG enrichment bar chart of DEGs in *O. fragrans* 'Yanzhi Hong' at the peak vs. initial flowering stage.

3.3. Association Analysis of Metabolic and Transcriptomic Date of Carotenoid Metabolism Pathway Genes

We identified 62 genes involved in carotenoid metabolism based on the functional annotation results from 'Liuye Jingui' (Table S3). These genes include four phytoene synthase (*PSY*), four 15-cis-phytoene desaturase (*PDS*), three 15-cis-zeta-carotene isomerase (*Z-ISO*), two prolycopene isomerase (*ISO*), four beta-carotene isomerase D27 (ISO D27), one zeta-carotene desaturase (*ZDS*), and one lycopene epsilon cyclase (*LYCE*), among others. The significant accumulation of carotenoid components like β-carotene, lycopene, γ-carotene, (E/Z)-phytoene, α-carotene, and β-cryptoxanthin is the main factor driving the floral color change during peak flowering (Table 1 and Figure 3). Pearson's method was employed to assess the correlation between these carotenoid components and metabolic pathway genes, illustrated in Figure 7. Notably, these components exhibit significant positive correlations with certain genes such as *LYG020464* (*PSY*), *LYG013665* (*ISO*), *LYG024982* (*LYCB*), *LYG026704* (*CCD4*), *LYG003175* (*NCED3*), and *LYG016435* (*NCED5*), while showing significant negative correlations with genes like *LYG030098* (*PSY*), *LYG024873* (*PDS*), *LYG033898* (*Z-ISO*), and several others.

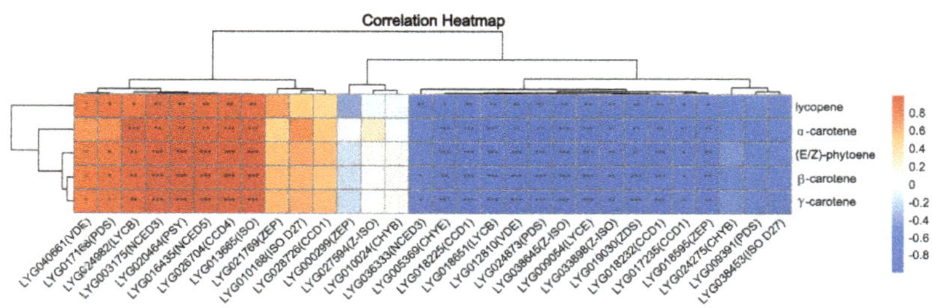

Figure 7. Cluster heatmap of co-expression correlation between carotenoid metabolic and metabolism pathway genes at the initial and peak flowering stages in 'Yanzhi Hong' petals. The correlation analysis was conducted using Pearson's correlation method (* $p \leq 0.05$, ** $p \leq 0.01$, *** $p \leq 0.001$).

3.4. RT-qPCR Validation of Carotenoid Metabolism Pathway Genes

To ensure the accuracy of transcriptome data, RT-qPCR was conducted on five differentially expressed genes (DEGs) involved in carotenoid metabolism pathways. These genes included *LYG030098* (*PSY*), *LYG016435* (*NCED5*), *LYG003175* (*NCED3*), *LYG012810* (*VDE*), and *LYG009054* (*LYCE*) (Figure 8A). The expression levels of *LYG030098*, *LYG09054*, and *LYG012810* were notably higher during the initial flowering stage compared to the peak flowering stage. No significant difference was observed in the expression level of *LYG003175* and *LYG016435* between the two stages. 'Yanzhi Hong' and other common *O. fragrans* cultivars displayed variations in carotenoid composition, leading to increased lycopene accumulation and decreased α-carotene accumulation in petals. Consequently, two *LYCB* genes (*LYG018651* and *LYG024982*), two *CHYB* genes (*LYG024275* and *LYG010024*), and one *CHYE* gene (*LYG005369*) were also analyzed using RT-qPCR. A strong correlation with an R^2 value of 0.89 was observed between the RT-qPCR and FPKM data (Figure 8B), indicating a high level of reliability in the expression profiles of the 10 genes.

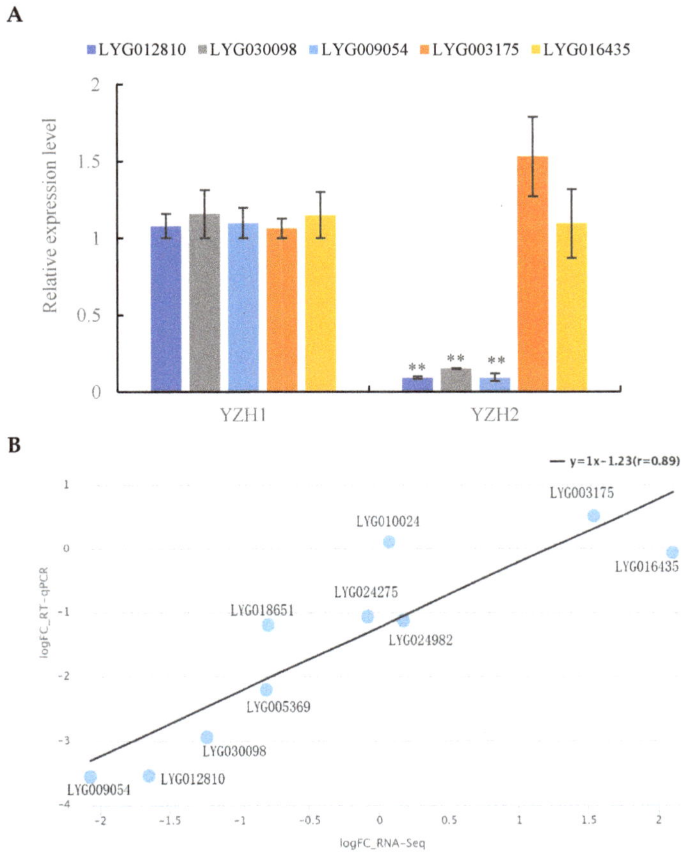

Figure 8. RT-qPCR analysis of carotenoid metabolism pathway genes in the initial and peak flowering stages of 'Yanzhi Hong'. (**A**) The expression levels of DEGs. Asterisks indicate the significant difference between the initial and peak flowering stages revealed by *t*-test: ** $p < 0.01$. (**B**) A linear fitting analysis was performed on the FPKM and RT-qPCR data.

Previous studies emphasized the abundance of α-carotene and β-carotene in various cultivars, particularly in the Luteus and Aurantiacus groups of *O. fragrans* [11,16]. So, this research focused on comparing the gene expression levels of *LYG009054* (*LYCE*), *LYG005369* (*CHYE*), *LYG018651* (*LYCB*), *LYG024982* (*LYCB*), *LYG024275* (*CHYB*), and *LYG010024* (*CHYB*) in the petals of 'Yanzhi Hong', 'Liuye Jingui', and 'Gecheng Dangui' at the initial and peak flowering stages (Figure 9). The results revealed that *LYG009054* (*LYCE*) expression was significantly lower in 'Yanzhi Hong' at both flowering stages compared to the other cultivars, showing higher expression during the initial flowering stage across all three. *LYG018651* (*LYCB*) expression was highest in 'Gecheng Dangui', followed by 'Liuye Jingui', and lowest in 'Yanzhi Hong', with higher levels at the peak flowering stage in all three. *LYG005369* (*CHYE*) expression was higher in 'Liuye Jingui' compared to 'Yanzhi Hong' and 'Gecheng Dangui', although there was no significant difference between 'Yanzhi Hong' and 'Gecheng Dangui' at the initial flowering stage, but higher in 'Gecheng Dangui' at the peak flowering stage. *LYG024982* (*LYCB*) expression showed minimal differences among the three cultivars at the peak flowering stage. *LYG010024* (*CHYB*) expression in 'Liuye Jingui' and 'Gecheng Dangui' was higher than in 'Yanzhi Hong', with no substantial difference among the three during this period. *LYG024257* (*CHYB*) gene expression was significantly higher in 'Yanzhi Hong' and 'Liuye Jingui' compared to 'Gecheng Dangui', although the difference between 'Yanzhi Hong' and 'Liuye Jingui' at the peak flowering stage was not significant. The decline in expression levels of *LYG009054* and *LYG018651* could potentially impact the increased lycopene accumulation and decreased α-carotene content in the petals of 'Yanzhi Hong'.

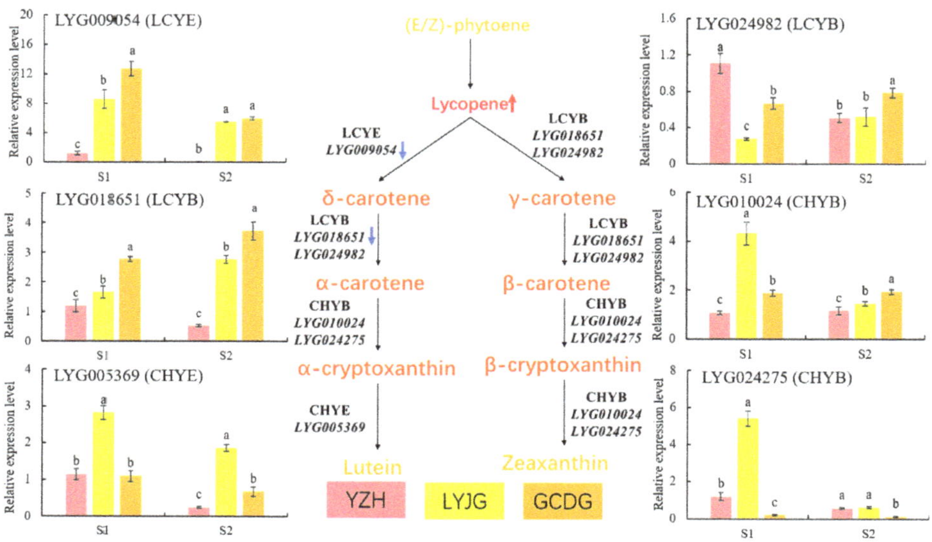

Figure 9. RT-qPCR analysis of lycopene cyclase and carotene hydroxylase genes in the petals of three *C. fragrans* cultivars at the initial (S1) and peak flowering (S2) stages. Different letters represent significant differences ($p < 0.05$) based on Tukey's test using one-way ANOVA. The blue arrow represents the gene expression change in the 'YZH' cultivar. The red arrow represents the metabolite content change in the 'YZH' cultivar.

4. Discussion

This study conducted a comprehensive analysis of carotenoid components in the petals of 'Yanzhi Hong' during initial and peak flowering stages using targeted metabolomics. A total of 27 carotenoid components were identified, including 5 carotenes, 8 xanthophylls, and 14 xanthophyll esters. Particularly noteworthy was the detection of xanthophyll esters, resulting from the fusion of xanthophylls and fatty acids, in *O. fragrans* cultivars for the first time, with zeaxanthin esters being the most prevalent among them. While the carotenoid components remained consistent between the initial and peak flowering stages, there was a significant difference in their content. Notably, the content of five carotene components exhibited a substantial increase at the peak flowering stage compared to the initial flowering stage. Carotenes constituted the majority of the carotenoids, representing 78.82% in the initial flowering stage and 91.19% in the peak flowering stage, highlighting their crucial role in influencing the color of *O. fragrans* flowers as previously reported [10,11,16,17]. Conversely, most xanthophylls and xanthophyll esters either decreased or remained stable during the peak flowering period. Given that the petals of 'Yanzhi Hong' transition to rouge-red at the peak flowering stage, the results suggest that the increase in carotene content is the primary factor contributing to the floral color change at this stage.

Lycopene is the most abundant component in the petals of 'Yanzhi Hong' at the initial flowering stage, followed by β-carotene and γ-carotene. However, at the peak flowering stage, β-carotene becomes the most abundant, followed by lycopene and γ-carotene. This indicates that β-carotene, lycopene, and γ-carotene are the top three components in the petals at both stages. Previous studies [10,11,16,17] have not highlighted the presence of lycopene in *O. fragrans* cultivars, making 'Yanzhi Hong' the first cultivar identified to be rich in lycopene, a red pigment commonly found in mature red tomato fruits. Previous research indicates that β-carotene and α-carotene, orange-yellow pigments, play a significant role in influencing flower color in *O. fragrans* cultivars [11]. However, 'Yanzhi Hong' exhibits relatively low levels of α-carotene in its petals, at 1.62% and 1.60% during the initial and full flowering stages, respectively. This suggests that the red hue of 'Yanzhi Hong' is likely due to its lower α-carotene content and higher lycopene accumulation in the petals. Despite the presence of abundant orange-yellow β-carotene in the petals, the combination of β-carotene and lycopene results in the appearance of a rouge-red color.

The cyclization of lycopene, catalyzed by lycopene cyclase, is a crucial step in the carotenoid biosynthesis pathway [18,19]. This process results in the formation of β-carotene from two β rings, and α-carotene from one α ring and one β ring. Further conversion of β-carotene and α-carotene leads to the production of zeaxanthin and lutein through the action of carotenoid hydroxylase. In the petals of 'Yanzhi Hong', there is a significant accumulation of lycopene and β-carotene, while the levels of α-carotene, which contains the α ring, are relatively low. This suggests a possible blockage in the synthesis of the α-carotene branch. Transcriptome analysis revealed the presence of one *LYCE*, one *CHYE*, two *LYCB*, and two *CHYB* genes. A comparison of gene expression levels in the petals of 'Yanzhi Hong', 'Liuye Jingu', and 'Gecheng Dangui' at different flowering stages indicated that the reduced expression of the *LYCE* gene (*LYG009054*) and *LYCB* gene (*LYG018651*) in 'Yanzhi Hong' might be responsible for the accumulation of lycopene and the low content of α-carotene. Functional validation studies on the *LYCE* and *LYCB* genes of *O. fragrans* showed that LYCE can convert lycopene into ε-carotene with two α rings, while LYCB can convert lycopene into β-carotene with two β rings [20]. The lower transcription levels of *LYCE* and *LYCB* in the petals of 'Yanzhi Hong' compared to other *O. fragrans* cultivars contribute to its unique flower color phenotype. However, further investigation is needed to understand the mechanism behind the decreased transcription levels of these genes.

The diverse pigment components in plants are crucial for creating a variety of flower colors. Carotenoids play a significant role in both color formation and the development of the characteristic aroma in *O. fragrans* [11,21]. In common cultivars ranging from light yellow to orange, the levels of α-carotene and β-carotene determine the intensity of the flower color. However, due to minimal variation in the main pigment components, the color appears uniform. Since *O. fragrans* germplasm resources have limited anthocyanins, diversifying carotenoid types is suggested as a breeding method to enhance flower color. Nevertheless, as *O. fragrans* is well-known for its distinct aroma, α-carotene and β-carotene act as direct precursors for key aroma components such as α-ionone and β-ionone [22,23]. Modifying the carotenoid composition could potentially decrease the production of these important aroma compounds, resulting in a loss of fragrance. In a study conducted by Chai et al., an analysis of aroma components in 'Yanzhi Hong' at different developmental stages revealed a deficiency in essential aroma compounds like linalool, linalool oxides, α-ionone, β-ionone, and dihydro-β-ionone [14,24–26]. Consequently, the petals of 'Yanzhi Hong' lack the typical aroma associated with *O. fragrans*. To create new *O. fragrans* cultivars with vivid flower colors and characteristic scents, investigating the key factors that hinder anthocyanin synthesis is a crucial area for future research.

Supplementary Materials: The following supporting information can be downloaded at: https://www.mdpi.com/article/10.3390/horticulturae10080801/s1, Figure S1: RT-qPCR primers; Table S1: RT-qPCR primers; Table S2: Annotated results of DEGs related to carotenoid and flavonoid metabolism; Table S3: Gene information of carotenoid metabolism pathway.

Author Contributions: Conceptualization, X.Z.; methodology, S.W., J.W. and P.Y.; software, J.Y.; validation, X.C.; formal analysis, Q.H.; investigation, J.Z.; resources, H.C.; data curation, Y.T.; writing—original draft, S.W.; writing—review and editing, X.Z.; visualization, P.Y.; supervision, J.W.; project administration, J.Z. and H.C.; funding acquisition, X.Z., X.C. and J.Y. All authors have read and agreed to the published version of the manuscript.

Funding: This research was funded by the National Natural Science Foundation of China (32101581) and Hubei Province Natural Science Foundation (2023AFB1063 and 2024AFB1057).

Data Availability Statement: Data are contained within the article and Supplementary Materials.

Acknowledgments: The authors express their sincere gratitude to Baichun Shen from Department of Forestry of Zhejiang Province, China, for his invaluable assistance in sampling the *Osmanthus fragrans* 'Yanzhi Hong'.

Conflicts of Interest: The authors declare no conflicts of interest.

References

1. Liu, Y. Cultivar classification of *Osmanthus fragrans* Lour. and the development of germplasm resources of *Osmanthus* Lour. *J. Plant Resour. Environ.* **1993**, *2*, 44–48.
2. Xiang, Q.; Liu, Y. An Illustrated Monograph of the Sweet Osmanthus Cultivars in China. *Chin. Landsc. Archit.* **2008**, *9*, 96.
3. Yang, K. *Chinese Osmanthus*; China Forestry Publishing House: Beijing, China, 2020.
4. Tanaka, Y.; Sasaki, N.; Ohmiya, A. Biosynthesis of plant pigments: Anthocyanins, betalains and carotenoids. *Plant J.* **2008**, *54*, 733–749. [CrossRef] [PubMed]
5. Zhang, H.; Yuan, X.; Gao, J.; Wang, X.; Wang, H.; Li, Y.; Wang, L.; Fu, Z.; Li, B. Mechanism of Flower Petal Coloration and Molecular Breeding. *Biotechnol. Bull.* **2023**, *39*, 23–31.
6. Song, X.; Tian, Y.; Gao, K.; Li, J.; Li, Y.; Wang, J.; Deng, C.; Zhang, F.; Kong, D.; Fan, G.; et al. Genetic and QTL analysis of flower color and pigments in small-flowered chrysanthemum based on high-density genetic map. *Ornam. Plant Res.* **2023**, *3*, 17. [CrossRef]
7. Wen, J.; Wang, C.; Feng, H.; Li, S.; Wang, L.; Wu, R.; Zhao, S. Research Progress on Flower Color of Rose. *Acta Hortic. Sinica* **2021**, *48*, 2044–2056.
8. Cai, X.; Su, F.; Jin, H.; Yao, C.; Wang, C. Components and extraction methods for petal pigments of *Osmanthus fragrans* 'Siji Gui'. *J. Zhejiang For. Coll.* **2010**, *27*, 559–564.
9. Zou, J.; Zeng, X.; Chen, H.; Cai, X.; Wang, C. Analysis on characteristic color compounds in different varieties of *Osmanthus fragrans* Lour. during flowering and senescence. *J. South. Agric.* **2017**, *48*, 1683–1690.
10. Wang, Y.; Luo, Y.; Zhang, C.; Fu, J.; Hu, S.; Zhao, H. Flower Color and Pigment Composition in the Petals of Bud Mutation and its Stock Plant of *Osmanthus fragrans*'Jingui'. *Acta Hortic. Sinica* **2017**, *44*, 528–536.
11. Wang, Y.; Zhang, C.; Dong, B.; Fu, J.; Hu, S.; Zhao, H. Carotenoid Accumulation and Its Contribution to Flower Coloration of *Osmanthus fragrans*. *Front. Plant Sci.* **2018**, *9*, 1499. [CrossRef]
12. Gu, H.; Yang, X.; Wang, L. Progress in Molecular Biological Studies of *Osmanthus fragrans*. *J. Trop. Subtrop. Bot.* **2024**, *in press*.
13. Xiang, Q.; Wang, X.; Liu, Y. Annual Report ICRCO 2014(2) Three new cultivars of *Osmanthus fragrans*. *J. Nanjing For. Univ.* **2014**, *38*, 2+181.
14. Chai, Z.; Zhang, M.; Li, L.; Zhang, C.; Duan, Y. Non-Targeted Metabolomics Analysis of the Petals of *Osmanthus fragrans* 'Yanzhi Hong' in Different Developmental Phases. *J. Northwest For. Univ.* **2022**, *37*, 107–113.
15. Chen, H.; Zeng, X.; Yang, J.; Cai, X.; Shi, Y.; Zheng, R.; Wang, Z.; Liu, J.; Yi, X.; Xiao, S.; et al. Whole-genome resequencing of *Osmanthus fragrans* provides insights into flower color evolution. *Hortic. Res.* **2021**, *8*, 98. [CrossRef]
16. Zeng, X. Research of TPS and CCD Function Analysis and Their Influence on Petal Color and Scent in *Osmanthus fragrans* Lour. Ph.D. Dissertation, Huazhong Agricultural University, Wuhan, China, 2020.
17. Wang, Y.; Zhang, C.; Xu, B.; Fu, J.; Du, Y.; Fang, Q.; Dong, B.; Zhao, H. Temperature regulation of carotenoid accumulation in the petals of sweet osmanthus via modulating expression of carotenoid biosynthesis and degradation genes. *BMC Genom.* **2022**, *23*, 418. [CrossRef] [PubMed]
18. Lu, C.; Gao, Y.; Huang, H.; Dai, S. Carotenoid Metabolism and Regulation in Plants. *Acta Hortic. Sinica* **2022**, *49*, 2559–2578.
19. He, J.; Fan, Y. Progress in Composition and Metabolic Regulation of Carotenoids Related to Floral Color. *Acta Hortic. Sinica* **2022**, *49*, 1162–1172.
20. Qing, H.; Chen, J.; Jiang, L.; Qian, J.; Fu, J.; Zhang, C. Functional Characterization of Two Lycopene Cyclases from Sweet Osmanthus (*Osmanthus fragrans*). *Sci. Hortic.* **2022**, *299*, 111062. [CrossRef]
21. Cai, X.; Mai, R.; Zou, J.; Zhang, H.; Zeng, X.; Zheng, R.; Wang, C. Analysis of aroma-active compounds in three sweet osmanthus (*Osmanthus fragrans*) cultivars by GC-olfactometry and GC-MS. *J. Zhejiang Univ.-Sci. B* **2014**, *15*, 638–648. [CrossRef]
22. Huang, F.C.; Molnár, P.; Schwab, W. Cloning and functional characterization of carotenoid cleavage dioxygenase 4 genes. *J. Exp. Bot.* **2009**, *60*, 3011–3022. [CrossRef]
23. Baldermann, S.; Kato, M.; Kurosawa, M.; Kurobayashi, Y.; Fujita, A.; Fleischmann, P.; Watanabe, N. Functional characterization of a carotenoid cleavage dioxygenase 1 and its relation to the carotenoid accumulation and volatile emission during the floral development of *Osmanthus fragrans* Lour. *J. Exp. Bot.* **2010**, *61*, 2967–2977. [CrossRef] [PubMed]
24. Xin, H.; Wu, B.; Zhang, H.; Wang, C.; Li, J.; Yang, B.; Li, S. Characterization of volatile compounds in flowers from four groups of sweet osmanthus (*Osmanthus fragrans*) cultivars. *Can. J. Plant Sci.* **2013**, *93*, 923–931. [CrossRef]

25. Zou, J.; Cai, X.; Zeng, X.; Zheng, R.; Wang, C. Changes of Aroma-active Compounds in Different Cultivars of *Osmanthus fragrans* during Flowering. *Acta Hortic. Sin.* **2017**, *44*, 1517–1534.
26. Xia, K.; Jiang, B.; Zhao, Z.; Fan, J.; Wen, G.; Li, F.; Gao, L.; Jiang, Q.; Qiu, S. Comparative analysis of aromatic components from different cultivars of *Osmanthus fragrans* in Guilin. *Guihaia* **2018**, *38*, 1493–1504.

horticulturae

MDPI

Article

Analysis of Transcriptome and Expression of *C4H* and *FLS* Genes on Four Flower Colors of *Impatiens uliginosa*

Xiaoli Zhang †, Yi Tan †, Xinyi Li, Zengdong Liu, Fan Li, Haiquan Huang * and Meijuan Huang *

College of Landscape Architecture and Horticulture Sciences, Southwest Research Center for Engineering Technology of Landscape Architecture (State Forestry and Grassland Administration), Yunnan Engineering Research Center for Functional Flower Resources and Industrialization, Research and Development Center of Landscape Plants and Horticulture Flowers, Southwest Forestry University, Kunming 650224, China; m18468178186@163.com (X.Z.); tanyi@swfu.edu.com (Y.T.); xinyili@swfu.edu.cn (X.L.); 18287762119@163.com (Z.L.); lifan@swfu.edu.cn (F.L.)

* Correspondence: haiquan_huang@swfu.edu.cn (H.H.); meijuanhuang@swfu.edu.cn (M.H.)

† These authors contributed equally to this work.

Abstract: Flower color is a major feature of ornamental plants, and the rich flower color of plants is an important factor in determining their ornamental and economic values, so flower color is an important research target for gardening and horticulture breeders at home and abroad. Our research group collected four colors of *Impatiens uliginosa* (white, pink, red, and deep red) during the collection of germplasm resources in the field. In this study, we analyzed the transcriptomes of the four flower colors of *I. uliginosa* by using RNA-Seq technology. The transcriptomes were screened to identify candidate genes related to flower color, and the coloring mechanisms of four flower colors were revealed at the molecular level. The main findings were as follows: (1) The number of the four different transcripts ranged from 64,723 to 93,522 and contained a total of 100,705 unigenes. (2) The analysis of differentially expressed genes revealed structural genes including *C4H*, *FLS*, *PAL*, and *ANS* and transcription factors including *MYB*, *MYB*-related, *AP2-EREBP*, and *bHLH*. (3) Among the four flower colors of *I. uliginosa*, the *C4H1* gene had the highest expression in pink flowers, and the *C4H2* gene had the highest expression in red flowers. This indicated that *C4H* genes positively regulated the red flower color of *I. uliginosa*. However, *FLS* expression was the highest in white flowers, and with deepening flower color, *FLS* gene expression gradually weakened, acting as a negative regulator. The results of this study could lay the theoretical foundation for investigating the mechanism of coloration and flower color variation in *I. uliginosa*.

Keywords: *Impatiens uliginosa*; flower color; transcriptome analysis; *C4H* gene; *FLS* gene

1. Introduction

Flower color is an important ornamental trait of garden plants, which determines the economic and ornamental value of plants and has important biological significance such as attracting insect pollinators, reducing stress, and perpetuating the race [1]. The main coloring pigments of plant flowers and fruits are carotenoids, flavonoids, and betalains [2]. The formation of flower color is the result of a combination of many factors. pH, pigment type, auxin effect, metal ions, petal epidermal cell structure, and pigment number affect the coloration and variation mechanism of plant flower color [3,4].

There are large differences in the coloration mechanism of anemones originating from different regions and with different flower colors, including pigment types, mineral elements, and developmental stages. Researchers have isolated kaempferol, quercetin, geranium pigments, cornflower pigments, delphinium pigments, and their derivatives from the petals of red *Impatiens balsamina* [5,6]. Clevenger found that cornflower pigments were the most widely distributed, followed by mallow pigments, by determining the anthocyanins of 19 species of *Impatiens* [7]. Hasan identified nine flavonoids, three of which were new, by determining the leaves of the *Impatiens* bicolor [8]. Tatsuzawa et al. isolated mallow pigments and their derivatives from purple *Impatiens textori* Miqs [9]. Isolation of two new quercetin glycoside compounds from *Impatiens balsamina* was achieved by Lei et al. [10]. Mariana found dihydromyricetin, quercetin, kaempferol, and their derivatives by determining the flavonoids in the petals of *Impatiens glandulifera* [11].

Genes play an important role in the biosynthesis of anthocyanin glycosides, and the expression of these genes is closely related to anthocyanin content, thus affecting the flower color of plants [12]. Cinnamate-4-hydroxylase (*C4H*) alias trans-cinnamate-4-monooxygenase is an important enzyme for step 2 in the synthesis of anthocyanin. It can form a complex with *PAL* and *4CL* and then immobilize it through the N-terminal hydrophobic helix region, which plays an important role in the chemical transfer and subcellular localization of the electron chain [13], catalyzing the conversion of trans-cinnamic acid to 4-coumaric acid, which then efficiently modulates carbon metabolism pathways [14,15]. C4H is a member of the cytochrome P450 oxidase (Cytochrome P450, CYP450) family and belongs to the CYP73 subfamily [16]. The first P450 enzyme in plants to be clonally characterized and functionally determined is *C4H*, which maintains high activity in plant tissues compared to other P450s [17]. It has been shown that the protein activity of *C4H* directly affects the synthesis of flavonoid compounds, lignin, aromatic compounds, etc., in plants [18] and is considered to be an important regulatory point. Baek found that the changes in *C4H* gene expression during the development of blackberries (Korean black raspberry) were the same as the changes in flavonoid accumulation [19]. Studies on the expression pattern of the *C4H* gene in plants inferred that its expression level is also closely related to lignin biosynthesis [20].

In addition, the flavonol synthase gene (*FLS*) has an effect on the accumulation of anthocyanin glycosides and the synthesis of flavonols as well as the color presentation of plants [21], and the *FLS* gene determines the production of plant flavonoids. The *FLS* gene dihydroflavonols was catalyzed by *FLS* to produce flavonols. *FLS* is a member of the 2-ketoglutarate-dependent dioxygenase (2-ODD) family of enzymes, to which anthocyanidinsynthase (ANS) and flavonol synthase (FLS) in the flavonoid metabolic pathway belong [22]. Tanaka isolated and cloned the *FLS* gene from *Petunia hybrida* and also enabled its successful expression in yeast [23]. Subsequently, isolation and cloning of *FLS* genes have been reported in a large number of plants such as *Solanum tuberosum*, *Arabidopsis thaliana*, *Ginkgo biloba*, strawberry (*Fragaria* × *ananassa*), etc. [24–28]. It has also been explored that the reaction substrates of the *ANS* gene and the *FLS* gene are identical, i.e., there was competition and the amino acid sequence similarity between the two was high [29], which further confirms that the *FLS* gene has an effect on the synthesis of both phytoflavonols and anthocyanins.

The group collected *I. uliginosa*, which belongs to the genus *Impatiens* of the family Balsaminaceae of the herbaceous plants, mainly distributed in Yunnan, Guizhou, Guangxi, and other places in China in the early stage of the collection of germplasm resources in the field [30]. There are four different flower colors of *I. uliginosa*, deep red, red, pink, and white, in different habitats or in the same habitat, and the color is bright. The study of the coloring mechanism of this mutant strain is of great significance to the molecular breeding of flower color in *Impatiens*. Currently, the study of flower color in *Impatiens* is mainly focused on the determination of flavonoids in petals and the cloning and expression of structural genes in the synthesis pathway of anthocyanin, while the transcriptional regulation of the molecular mechanism of flower color presentation in the same species with multiple colors has not yet been resolved. Therefore, in this study, we mined the transcriptome data of four different flower colors of *I. uliginosa* to screen the genes related to the regulation of flower color.

2. Materials and Methods

2.1. Plant Materials

The test material used in this experiment was *I. uliginosa*, a wild species of *Impatiens*, which has four different flower colors: white, pink, red, and deep red. They were collected from Aziying, Laoyuhe Wetland Park, Kunming City, Yunnan Province, and Dadieshui, Shilin Yi Autonomous County, Yunnan Province, respectively. Five plants of each of the four flower colors of *I. uliginosa* were selected to be robust and free of pests and diseases, and the whole flower organ of one flower in the state of full bloom was extracted from each plant, which was separated from the plant and quickly put into liquid nitrogen. The four flower colors were mixed separately, labeled, and then stored at −70 °C (Figure 1).

Figure 1. Blooming period of four different color flowers of *I. uliginosa*. Bars = 1 cm.

2.2. Transcriptome Sequencing and Analysis

2.2.1. RNA Extraction and Detection

Total RNA was extracted from 1 mg of the combined samples by a grinder using the OMEGA Quick RNA Isolation Kit according to the instructions. Contaminating genomic DNA was removed using RNAase-free DNase I. RNA purity was determined using a spectrophotometer, and RNA integrity was verified using an agarose gel. Library and RNA-Seq construction were performed by UW Genetics; the constructed libraries were sequenced using the Illumina sequencing platform after passing quality control.

2.2.2. Second-Generation Sequencing and Assembly

The cDNA library was constructed by Illumina's NEB Next®Ultra™ RNA Library Prep to obtain raw data, followed by QC to obtain clean data to assemble a non-redundant unigene from scratch. In this study, we sequenced four different flower colors and assembled the *I. uliginosa* flower color transcriptomes named IUDR (deep red), IUR (red), IUP (pink), and IUW (white), respectively.

2.2.3. Sequence Analysis and Annotation

After the sequencing data were offloaded, it was necessary to assemble the nucleotide sequences of unigenes of *I. uliginosa* through quality control, filtering, and assembly steps. In order to fully obtain the functional annotation of unigene, this study used the comparison software BLASTX v2.2.23 (BasicLocal Alignment Search Tool) (E-value $\leq 1 \times 10^{-5}$) to compare the unigene sequences with the open databases, so that they can be functionally annotated and categorized. The public alignment databases include GO (Gene Ontology), KEGG (Kyoto Encyclopedia of Genes and Genomes, http://www.genome.jp/kegg, accessed on 3 December 2023), Swiss-Prot (A manually annotated and reviewed protein sequence database, http://www.genome.jp/kegg, accessed on 4 December 2023), and the Unigene Sequence Search Tool (E-value $\leq 1 \times 10^{-5}$). And the reviewed protein sequence database (http://www.ebi.ac.uk/uniprot/, accessed on 6 December 2023), COG/KOG (EuKaryotic Orthologous Group, http://ftp.ncbi.nih.gov/pub/COG/KOG, accessed on 7 December 2023), eggNOG (evolutionary genealogy of genes: Non-supervised Orthologous Groups, http://eggnogdb.embl.de/, accessed on 10 December 2023), Pfam (a database of conserved Protein families or domains), and NR (NCBI non-redundant proteins, https://blast.ncbi.nlm.nih.gov/Blast.cgi, accessed on 11 December 2023) databases were compared (E-value $\leq 1.0 \times 10^{-5}$) [30].

2.2.4. Screening and Enrichment Analysis of Differentially Expressed Genes (DEGs)

The assembled sequences were compared with the protein database for Blastx (http://www.ncbi.nlm.nih.gov/BLAST/, accessed on 15 December 2023), and based on the comparison results, the expression levels of the genes were quantified using the software RSEM v1.2.12, after which the expression levels of the genes were normalized using the FPKM method, and the differential genes were screened on the basis of the expression amount.

2.2.5. Related Gene Expression and Analysis

Primers were designed according to the sequences of *IuC4H1*, *IuC4H2*, and *IuFLS*, and *IuActin* was used as the internal reference gene (Table S1). The expression differences of *IuC4H1*, *IuC4H2*, and *IuFLS* were analyzed by qRT-PCR on the bloom samples of four flower colors of *I. uliginosa*, and three biological replicates were set for each sample. qRT-PCR was carried out on a Roche: LightCycler® 480 II Fluorescent Quantitative PCR Instrument. The qRT-PCR reaction system (20 μL) included the following: 10 μL of qPCR SYBR Green Master Mix, 7 μL of double-distilled water, 1 μL each of forward and reverse primers, and 1 μL of template cDNA. The specific program was as follows: predenaturation at 95 °C for 15 s, denaturation at 60 °C for 30 s, and annealing at 72 °C for 1 min, for 40 cycles. The relative expression of genes was analyzed by using the $2^{-\Delta\Delta Ct}$ calculation method. Data were analyzed with the help of SPSS software 27.0 and plotted by Origin software 64.

3. Results

3.1. RNA Sequencing and Transcriptome De Novo Assembly

In order to understand the molecular mechanism of flower color variation in *I. uliginosa*, the transcripts were assembled from scratch using a reference-free genome approach to construct four high-throughput sequencing libraries of four different flower colors of *I. uliginosa*, namely, IUDR (deep red), IUR (red), IUP (pink), and IUW (white). After removing the articulated subsequence, ambiguous reads, and low-quality reads, the sequencing depths from the transcriptome data of IUDR, IUR, IUP, and IUW, respectively, ranged from 6.65 to 6.77 GB with an average depth of 6.5 GB. The Q30 values ranged from 95.94 to 96.09% with an average of 96.00%, and the net number of reads ranged from 44.35 to 45.11 M, with a net reads ratio of 83.02~87.71% (Table 1).

Table 1. Sequencing quality of transcripts of four different colors.

Sample	Total Raw Reads (Mb)	Total Clean Reads (Mb)	Total Clean Bases (Gb)	Clean Reads Q20 (%)	Clean Reads Q30 (%)	GC (%)	Clean Reads Ratio (%)
IUDR	52.69	44.62	6.69	98.78	96.09	41.46	84.68
IUP	54.33	45.11	6.77	98.75	96	42.77	83.02
IUR	52.69	44.35	6.65	98.78	96.11	42.2	84.17
IUW	51.04	44.77	6.71	98.72	95.94	41.71	87.71

A count of the length and number of transcripts revealed that the number of the four different transcripts ranged from 64,723 to 93,522, with an average length of around 900 bp and an N50 of around 1500 bp (Table 2). The total sample of the four different transcriptomes of unigenes contained 100,705 full-length cDNA sequences with 115,842,379 bases, of which the average length of transcript unigene was 1150 bp (Figure 2), the N50 was 1890 bp, and the highest number of fragments distributed at 300 bp was 22,613, which was 22,613, which accounted for a total unigene percentage of 22.45%. The number of unigenes larger than 3000 bp accounted for 6.54% (6592 fragments) of the total number (Table 3).

Table 2. Quality indicators for transcripts of four different colors.

Sample	Total Number	Total Length	Mean Length	N50	N70	N90	GC (%)
IUDR	79,277	69,699,393	879	1545	946	336	41.46
IUP	91,109	76,919,883	844	1521	868	318	42.77
IUR	93,522	80,820,572	864	1530	903	329	42.2
IUW	64,723	59,663,736	921	1600	982	361	41.71

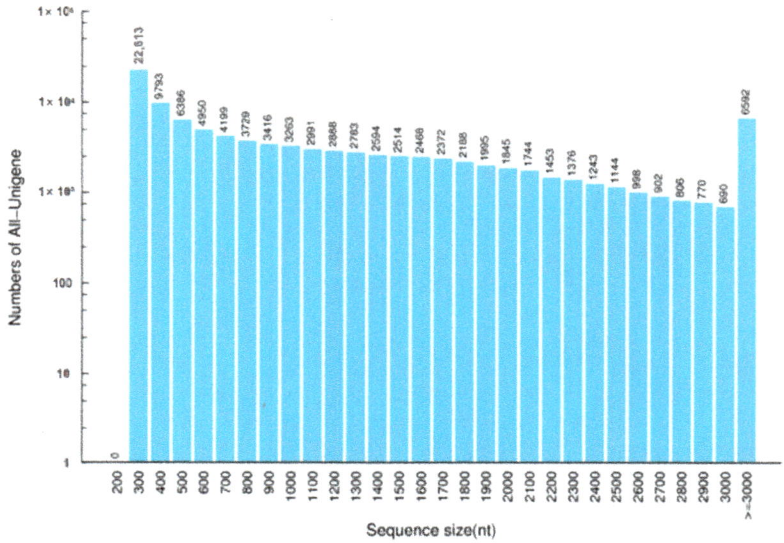

Figure 2. Length distribution of All-Unigene.

Table 3. Unigene data summary of four different colors.

Sample	Total Number	Total Length	Mean Length	N50	N70	N90	GC (%)
IUDR	55,003	58,568,930	1064	1689	1135	462	41.44
IUP	67,735	66,343,559	979	1646	1016	390	42.76
IUR	70,685	70,639,298	999	1640	1046	405	42.2
IUW	48,345	51,364,234	1062	1702	1121	453	41.68
All-Unigene	100,705	115,842,379	1150	1890	1267	495	41.98

3.2. Functional Annotation of Transcriptome Genes of *I. uliginosa*

The results of sequence match analysis with other species showed that *Vitis vinifera* had 10,213 homologous genes, accounting for 14.32% of the total, followed by *Sesamum indicum* with 4318 homologous genes, accounting for 6.06% of the total, *Coffea canephora* with 3458 homologous genes, accounting for 4.85% of the total, and *Solanum tuberosum* with 2983 homologous genes, accounting for 4.18% of the total (Figure 3). A total of 100,705 unigenes were successfully annotated by searching five public databases based on BLASTx (E-value < 1×10^{-5}), among which the NCBI-NR database contained 71,307 unigenes, Swiss-Prot contained 50,638 unigenes, the KEGG database contained 55,276 unigenes, the InterPro database contained 59,058 unigenes, and the KOG database contained 32,641 unigenes (Figure 4, Table 4).

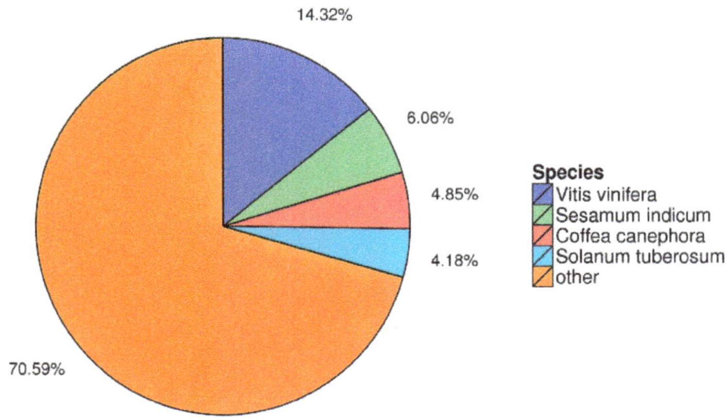

Figure 3. Percentages of unigenes that align with sequences from top five species in the NR database.

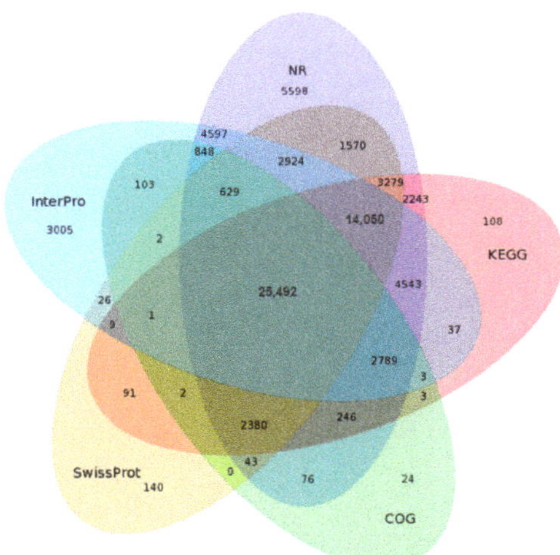

Figure 4. Annotation from five databases. Note: KEGG: Kyoto Encyclopedia of Genes and Genomes; COG: Clusters of Orthologous Groups of proteins; NR: NCBI non-redundant protein sequences.

Table 4. Annotation of unigenes searched against NR, Nt, Swiss-Prot, KEGG, KOG, InterPro, and GO databases.

Values	Number	Percentage
NR-Annotated	71,307	70.81%
Nt-Annotated	59,001	58.59%
Swissprot-Annotated	50,638	50.28%
KEGG-Annotated	55,276	54.89%
COG-Annotated	32,641	32.41%
Interpro-Annotated	59,058	58.64%
GO-Annotated	29,473	29.27%
Overall	77,877	77.33%
Total	100,705	100%

3.3. GO (Gene Ontology) and COG (Clusters of Orthologous Group of Proteins) Annotation Analysis of *I. uliginosa* Transcriptome Data

The GO gene pool contained three gene attribute sections: molecular function, biological process, and cellular component. The transcriptome of *I. uliginosa* contains 71,307 GO-annotated genes, accounting for 70.81% of the whole annotation, and consists of 58 different annotated gene functional regions. Among them, the biological process section contained 14,490 genes, the metabolic process contained 14,948 genes, and the single-organism process contained 9082 genes; the molecular function section contained 13,079 genes and the catalytic activity process contained 14,757 genes; the cellular component section contained 9905 genes, the cell cellular component contained 9905 genes, the cell part contained 11,934 genes, and the cell contained 12,094 genes. In the three categories of "translation regulator activity", "locomotion", and "cell killing", the unigenes were annotated to 1, 2, and 4 in each category (Figure 5).

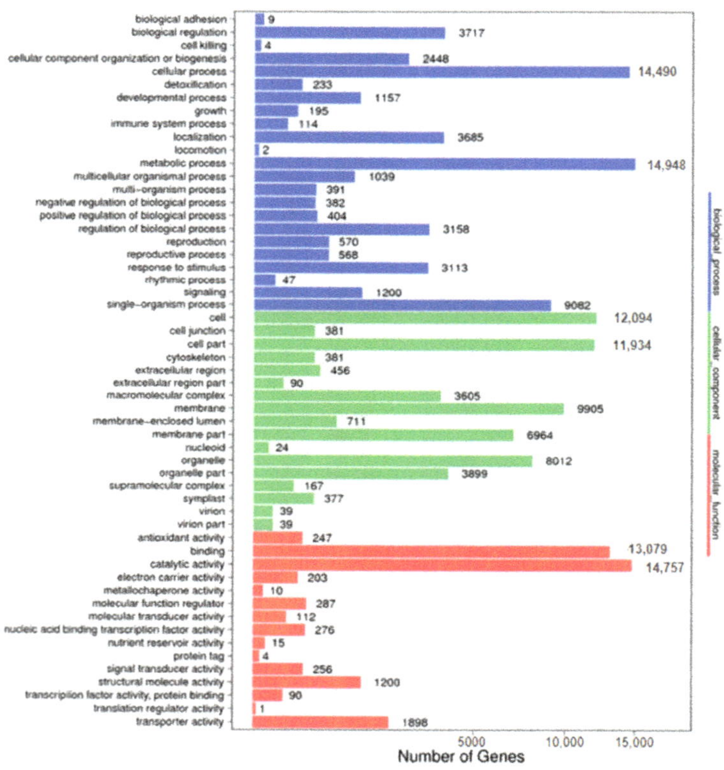

Figure 5. Annotation of unigenes searched against GO databases.

The COG database had the ability to infer unknown gene functions and compare homologous protein sequences, which accounted for 32.41% of the total annotated genes in the *I. uliginosa* transcriptome COG database, of which a total of 32,641 genes were annotated. The annotated genes were categorized into 25 classes according to their functions. General function prediction only contained a total of 9925 genes; replication, recombination, and repair contained 4816 genes; transcription contained 5406 genes. Post-translational modification, protein turnover, and chaperones contained 4029 genes (Figure 6).

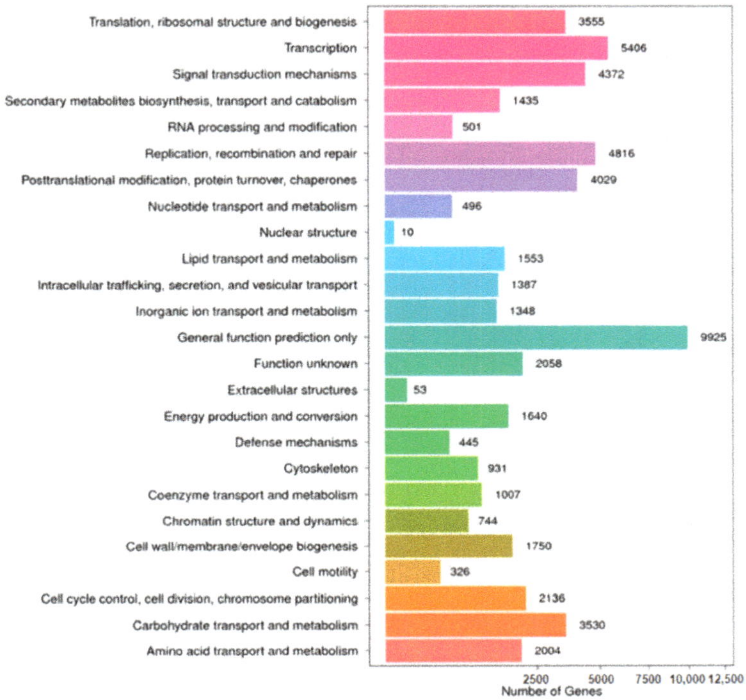

Figure 6. Annotation of unigenes searched against KOG databases.

3.4. Identification and Functional Annotation of Differential Genes (DEGs) in *I. uliginosa*

Based on the results of the gene expression levels of each sample, the number of genes obtained for each different flower color was analyzed using a Venn diagram, and the results showed that there was a large gap in the expression of genes in the four samples of different flower colors, with a total of 1402 genes between IUDR and IUP; a total of 1639 genes between IUDR and IUR; and a total of 1842 genes between IUDR and IUW; a total of 992 genes between IUP and IUR. The number of genes between IUP and IUP totaled 9929; the number of genes between IUP and IUW totaled 1157; and the number of genes between IUR and IUW totaled 1087. There were a total of 38,191 genes between the four samples, namely, IUDR, IUP, IUW, and IUR. Among them, the *C4H* and *FLS* genes were better expressed (Figure 7A).

Differentially expressed genes (DEGs) were detected using the DEseq2 algorithm, and the results of DEG detection are shown in Figure 7B. Subsequently, gene expression in the IUDR, IUP, IUW, and IUR samples was compared two-by-two, and the results showed that downregulated expression of differentially expressed genes (DEGs) predominated in the comparison of the four different flower colors (Figure 7B). A total of 6197 DEGs were expressed in IUDR vs. S IUW, of which 6098 were upregulated and 14,650 were downregulated. A total of 6197 DEGs were expressed in IUDR vs. IUP, of which 15,519 were upregulated and 13,894 were downregulated. A total of 6197 DEGs were expressed in IUW vs. IUP, of which 6234 were upregulated and 13,790 were downregulated. There were a total of 6197 DEGs in IUDR vs. IUR, with 14,353 upgrades and 15,166 downgrades; a total of 6197 DEGs in IUR vs. IUP, of which 5844 were upgraded and 7216 were downgraded; and a total of 6197 DEGs in IUR vs. IUW, with 6496 upgrades and 16,618 downgrades. In the IUDR vs. IUP, IUDR vs. IUR, and IUR vs. IUP two-by-two comparisons, there was little difference between the number of upward and downward adjustments in DEGs, while the number of downward adjustments was greater than the number of upward adjustments in the other three comparisons.

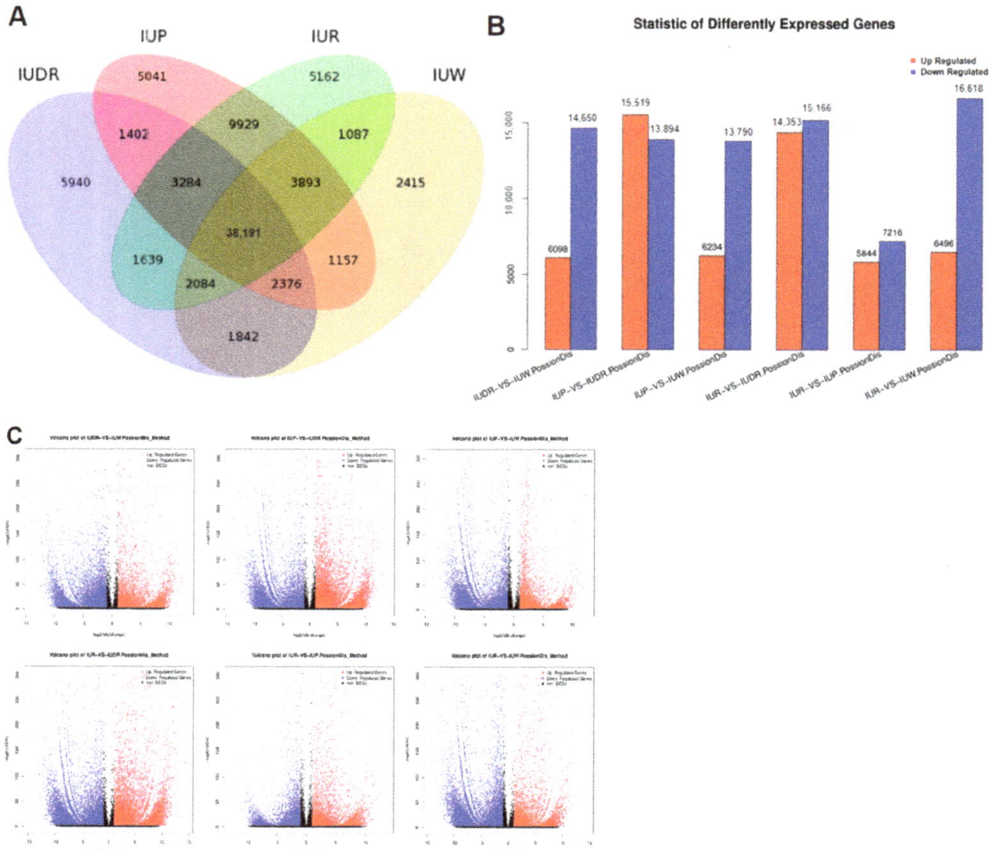

Figure 7. The expressed genes analysis between petals at four different stages: (**A**) Venn diagram shows the expression of genes in IUDR, IUP, and IUR samples. (**B**) Venn diagram showing the DEGs between IUDR vs. IUP, IUDR vs. IUR, and IUP vs. IUR. (**C**) The numbers of up- and downregulated genes between IUDR vs. IUP, IUDR vs. IUR, and IUP vs. IUR. Volcano plots of transcriptome between IUDR vs. IUP, IUDR vs. IUR, and IUP vs. IUR. Red dots represent upregulated DEGs. Blue dots represent downregulated DEGs. Black points represent non-DEGs.

The distribution of DEGs was also demonstrated using a volcano plot. In the IUDR vs. IUR comparison, the number of DEGs had a statistically significant change in the expression level in the volcano plot over the other comparisons (Figure 7C).

3.5. *Identification and Expression Analysis of Transcription Factors in I. uliginosa*

Transcription factors in plants are involved in the regulation of various metabolic pathways in plants, which enable plants to have a wider range of adaptations, and transcription factors were thought to have a wider regulation in anthocyanin biosynthesis than structural genes. Therefore, we predicted the unigenes encoding transcription factors from the transcriptome data of four different flower colors of *I. uliginosa*, and a total of 3493 unigenes were identified, which could be classified into 59 families (Figure 8), most of which were from the "*MYB*" family (429 unigenes), which accounted for 12.47% of the total number of unigenes, and the "MYB-unigenes" family (12.47% of the total number of unigenes). The "*MYB*-related" family (355 unigenes) accounted for 10.32% of the total, the "*AP2*-EREBP" family (227 unigenes) accounted for 6.60% of the total, and the "*bHLH*" family (227 unigenes) accounted for 6.60% of the total. The "*bHLH*" family (221 unigenes) accounted for 6.42% of the total. *MYB* and *MYB*-related transcription factors are essential for the plant flavonoid metabolic pathway.

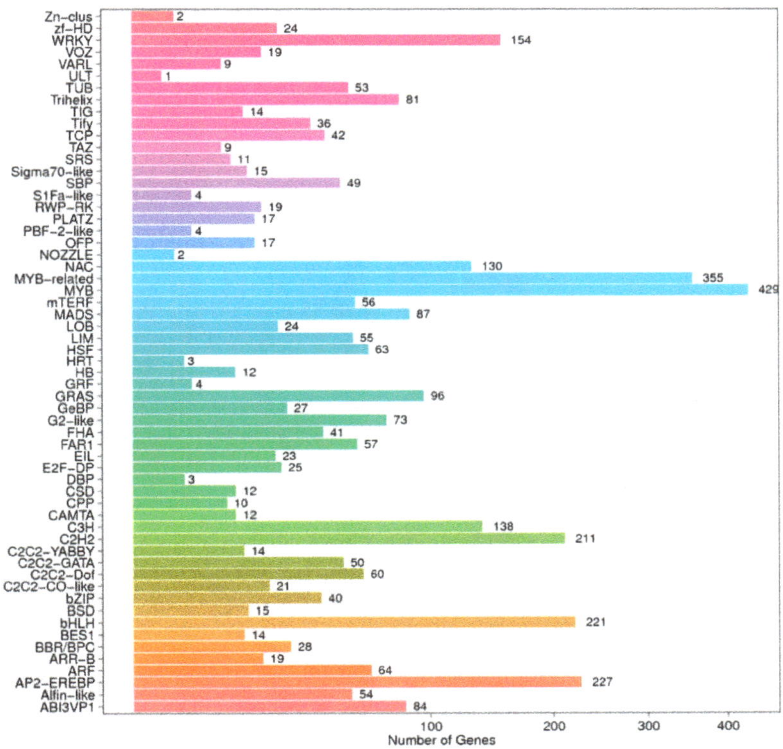

Figure 8. Transcription factor family classification of unigenes.

3.6. *qRT-PCR Validation of the Candidate Genes*

Two candidate genes were selected for qRT-PCR to verify the accuracy of the transcriptome data (Figure 9). There were significant differences in the expression of *IuC4H1*, *IuC4H2*, and *IuFLS* genes in different flower colors of *I. uliginosa*. The results showed that the *C4H1* gene was most expressed in pink flowers, slightly more in white flowers than in deep red flowers, and least in red flowers, with a significant difference in expression between pink and red. Expression of the *C4H2* gene was highest in red flowers, slightly higher in pink flowers than in deep red flowers, and weaker in white flowers, significantly different from the other three flower colors. *FLS* expression was highest in white flowers and lowest in deep red flowers and was significantly different from the other two flower colors. As the flower color deepened, the expression of *FLS* gradually weakened. Thus, among the four flower colors of *I. uliginosa*, the *C4H1* gene was expressed most frequently in pink flowers, the *C4H2* gene was expressed most frequently in red flowers, and the *FLS* was expressed most frequently in white flowers, suggesting that both the *C4H* and *FLS* genes are involved in the formation of *I. uliginosa* flower color.

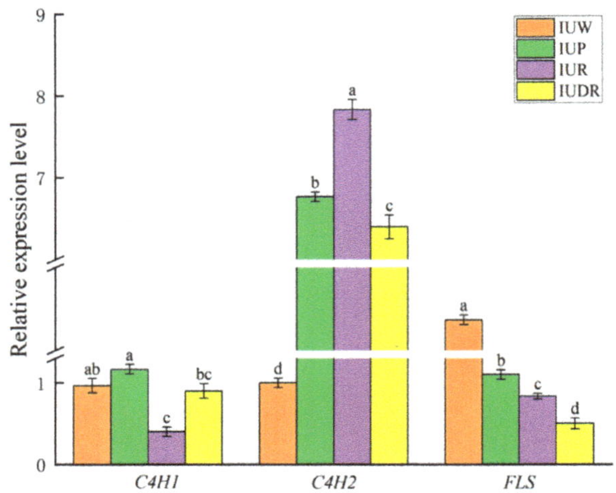

Figure 9. The relative expression of *C4H1*, *C4H2*, and *FLS* genes in four flower colors of *I. uliginosa*. Data are shown as the mean ± standard error of the mean, based on three replicates. Different letters indicate significant differences at $p < 0.05$ according to Duncan's multiple range test.

4. Discussion

The origin and evolution of flowers is a key issue in exploring the origin and evolution of angiosperm species. Meanwhile, flower color is an epigenetic and evolutionarily adapted phenotypic trait in ornamental plants, and the quality of flower color determines the ornamental and economic value of plants. Flowering plants will show a wide range of flower colors, with multiple colors of the same species and multiple species of the same color [2,31]. From previous studies, the genus *Rhododendron* (*Ericaceae*) has more than 1000 flower colors with unique ornamental value and a model system for flower color studies. Nie et al. investigated the differences in color-presenting mechanisms of two rhododendron species, yellow-flowered *R. molle* and red-flowered *R. simsii*, through a multilocus study [32]. For plants of the same genus, differences in key enzymes or gene regulatory mechanisms for pigment synthesis can also result in differences in flower color [33]. Researchers also explored genetically related genes for flower color and pigmentation by studying small-flowered chrysanthemums with different flower colors [34]. Studies on the flower color of homozygous ornamental plants are still scarce.

In this study, four flower colors of *I. uliginosa* were used as research materials, and transcriptome sequencing was performed to lay the foundation for investigating the molecular mechanism of flower color synthesis in *I. uliginosa*. The length and number of transcripts were counted, and it was found that the number of different transcripts of the four flower colors of *I. uliginosa* ranged from 64,723 to 93,522, with an average length of about 900 bp and an N50 of about 1500 bp. The total sample of the four different transcriptomes of unigenes contained 100,705 full-length cDNA sequences with 115,842,379 bases, of which the average length of transcript unigene was 1150 bp, the N50 was 1890 bp, and the highest number of fragments distributed in the 300 bp was 22,613 fragments, which accounted for 22.45% of the total number of unigenes. The number of unigenes larger than 3000 bp was 6592, accounting for 6.54% of the total number of unigenes.

The *C4H* gene is a key enzyme gene in the second step of the anthocyanin biosynthesis pathway [35], and also, the activity of this enzyme can affect the synthesis of compounds such as lignin and aromatic species, which has an important role in plant secondary metabolism [36]. In the four flower colors of *I. uliginosa* at full bloom, *C4H1* was expressed highest in pink and lowest in red, whereas *C4H2* was expressed highest in red and lowest in white, and thus it was hypothesized that the *C4H2* gene had a stronger positive effect than *C4H1* on flower color formation in *I. uliginosa*, in the same way that only one of the two copies of osmanthus is highly expressed in petals [37] and may separately fulfill their respective functions.

Flavonols are one of the major co-pigments that enhance the stability of anthocyanin glycosides and influence flower color formation. In the metabolic pathway of flavonoid compounds, flavonols share a presynthesis pathway with anthocyanin glycoside biosynthesis. Flavonol synthase (FLS) catalyzes the conversion of dihydroflavonols to flavonols [38]. The relative expression of the *FLS* gene in the four flower colors differed more significantly, and the highest expression was found in white flowers of *I. uliginosa*. This inference was consistent with the results of Ping in his functional validation of the *FLS* gene in roses, where they found that transgenic tobacco had increased flavonol content and flower differentiation to white [39]. Suppression of the *FLS* gene in *Chinese bellflower* was followed by substantial red coloration at the bud stage of the transgenic strain and pigment accumulation in the pistil, a phenotype that was evident in the progeny after crosses [40]. It can be hypothesized that the *FLS* gene may play an important role in the formation of white flowers of *I. uliginosa*.

5. Conclusions

In this study, four flower colors (white, pink, red, and deep red) of *I. uliginosa* were used as materials, and transcriptome sequencing was performed on the petals of *I. uliginosa* at the full bloom stage to investigate the intrinsic mechanism of four flower colors in *I. uliginosa* from the point of view of molecular biology. At present, there are still few studies on the specific color formation mechanism of the four flower colors of *I. uliginosa*, and the previous studies mainly focused on physiological and biochemical experiments, such as the determination of anthocyanin and pH. This study is the first to perform transcriptome sequencing of the four flower colors of *I. uliginosa*, and it can lay a theoretical foundation for further study on the molecular mechanism of different flower colors of *I. uliginosa* and other ornamental plants.

Supplementary Materials: The following supporting information can be downloaded at: https://www.mdpi.com/article/10.3390/horticulturae10040415/s1, Table S1: Sequences of primers related to qRT-PCR amplification.

Author Contributions: Software, F.L.; Investigation, Z.L.; Data curation, X.L.; Writing—original draft, X.Z.; Writing—review & editing, Y.T.; Supervision, H.H. and M.H. All authors have read and agreed to the published version of the manuscript.

Funding: This work was carried out with the support of the National Natural Science Foundation of China [32060364, 32060366, 31860230], Joint Special Key Project for Agricultural Basic Research in Yunnan Province(202101BD070001-018), Major scientific and technological projects in Yunnan Province [202102AE090052], and Program for Doctoral Supervisors Team in Genetic Improvement and High-efficient Propagation of Landscape Plants in Yunnan Province.

Data Availability Statement: Data are contained within the article and Supplementary Materials.

Acknowledgments: Thanks to Southwest Forestry University for providing experimental instruments and equipment for this experiment, and also thanks to teachers Haiquan Huang and Meijun Huang for their patient guidance during the experiment.

Conflicts of Interest: The authors declare no conflict of interest.

References

1. Kellenberger, R.T.; Glover, G.B. The evolution of flower colour. *Curr. Biol.* **2023**, *33*, R484–R488. [CrossRef] [PubMed]
2. Li, H. The of Anthocyanidin, Metal Elementsand pH on the Colour Formation of Blue Flowers from Monocots. Master's Thesis, North West Agriculture and Forestry University, Xianyang, China, 2013; pp. 52–56.
3. Han, Y.; Huang, K.; Liu, Y.; Jiao, T.; Ma, G.; Qian, Y.; Wang, P.; Dai, X.; Gao, L.; Xia, T. Functional Analysis of Two Flavanone-3-Hydroxylase Genes from *Camellia sinensis*: A Critical Role in Flavonoid Accumulation. *Genes* **2017**, *8*, 300. [CrossRef] [PubMed]
4. Lou, Q.; Liu, Y.; Qi, Y.; Jiao, S.; Tian, F.; Jiang, L.; Wang, Y. Transcriptome sequencing and metabolite analysis reveals the role of delphinidin metabolism in flower colour in grape hyacinth. *J. Exp. Bot.* **2014**, *65*, 3157–3164. [CrossRef] [PubMed]

5. Hagen, C.W. The differentiation of pigmentation in flower parts I. The flavonoid pigments of *Impatiens balsamina*, genotype *llHHPrPr*, and their distribution within the plant. *Am. J. Bot.* **1966**, *53*, 46–54. [CrossRef]
6. Miles, C.D.; Hagen, C.W. The differentiation of pigmentation in flower parts, IV. Flavonoid elaborating enzymes from petals of Impatiens balsamina. *Plant Physiol.* **1968**, *43*, 1347–1354. [PubMed]
7. Clevenger, S. Anthocyanidins of some *Impatiens* species. *Evolution* **1971**, *25*, 669–677. [PubMed]
8. Hasan, A.; Tahir, M.N. Flavonoids from the leaves of *Impatiens* bicolor. *Turk. J. Chem.* **2005**, *29*, 65–70.
9. Tatsuzawa, F.; Saito, N.; Mikanagi, Y.; Shinoda, K.; Toki, K.; Shigihara, A.; Honda, T. An unusual acylated malvidin 3-glucoside from flowers of *Impatiens textori* Miq. (*Balsaminaceae*). *Phytochemistry* **2009**, *70*, 672–674. [CrossRef]
10. Lei, J.; Qian, S.H.; Jiang, J.Q. Two new flavone glycosides from the seeds of *Impatiens balsamina* L. *J. Asian Nat. Prod. Res.* **2010**, *12*, 1033–1037. [CrossRef]
11. Mariana, N.V.; Peter, W.; Gerold, J. Flavonoids from the flowers of *Impatiens glandulifera* Royle isolated by high performance countercurrent chromatography. *Phytochem. Anal. PCA* **2016**, *27*, 116–125.
12. Zhao, L.Q.; Liu, Y.; Huang, Q.; Gao, S.; Huang, M.J.; Huang, H.Q. Effects of cell morphology, physiology, biochemistry and *CHS* genes on four flower colors of Impatiens uliginosa. *Front. Plant Sci.* **2024**, *15*, 151343830. [CrossRef] [PubMed]
13. Liu, K.; Zhao, Z.; Gao, A.; Chen, Y.; Huang, J.; Dang, Z.; Luo, R. Cloning of mango *C4H* gene and its expression analysis. *Jiangsu Agric. Sci.* **2017**, *45*, 8–12.
14. Ma, J.; Su, L.; Yuan, M.; Ji, H.; Li, A.; Wang, X. Cloning and expression studies of peanut *C4H* and *ANR* genes. *J. Nucl. Agric. Sci.* **2012**, *26*, 43–48.
15. Yao, S.; Wang, W.; Li, M.; Xu, Y.; Wang, Y.; Liu, Y.; Gao, L.; Xia, T. The Gene Cloning and expression analysis of tea tree cinnamic acid 4-hydroxylase gene. *J. Tea Sci.* **2015**, *35*, 35–44.
16. Zhao, L.; Ma, L.; Yang, Z.; Feng, W.; Zheng, X. Cloning and expression analysis of the *C4H* gene of *Solanum vulgare*. *Acta Pharm Sin.* **2017**, *52*, 821–831.
17. Li, L.; Zhao, Y.; Ma, L. New advances in the study of key enzymes of the phenylalanine metabolic pathway: PAL, C4H, 4CL. *Chin J. Bioinform.* **2007**, *20*, 187–189.
18. Liang, L.; Han, X.; Zhang, Z.; Guo, Q.; Xu, Y.; Liu, J.; Lian, Y. Cloning and expression analysis of the cinnamic acid-4-hydroxylase (*C4H*) gene from *Cinnamomum album*. *Chin. J. Tradit. Chin. Med.* **2014**, *39*, 1767–1771.
19. Baek, M.H.; Chung, B.Y.; Kim, J.H.; Kim, J.S.; Lee, S.S.; An, B.C.; Lee, I.J.; Kim, T.H. cDNA cloning and expression pattern of cinnamate-4-hydroxylase in the Korean black raspberry. *BMB Rep.* **2008**, *41*, 529–536. [CrossRef]
20. Tao, S.; Wang, D.; Jin, C.; Sun, W.; Liu, X.; Zhang, S.; Gao, F.; Khanizadeh, S. Cinnamate-4-hydroxylase gene is involved in the step of lignin biosynthesis in Chinese white pear. *J. Am. Soc. Hortic. Sci.* **2015**, *140*, 573–579. [CrossRef]
21. Li, Y.; Ran, L.; Mo, T.; Liu, N.; Zeng, J.; Liang, A.; Wang, C.; Suo, Q.; Chen, Z.; Wang, Y.; et al. Yellow Petal locus GaYP promotes flavonol biosynthesis and yellow coloration in petals of Asiatic cotton (*Gossypium arboreum*). *TAG Theor. Appl. Genet.* **2023**, *136*, 98. [CrossRef]
22. Schwinn, K.E.; Boase, M.R.; Bradley, J.M.; Lewis, D.H.; Deroles, S.C.; Martin, C.R.; Davies, K.M. *MYB* and *bHLH* transcription factor transgenes increase anthocyanin pigmentation in petunia and lisianthus plants, and the petunia phenotypes are strongly enhanced under field conditions. *Front. Plant Sci.* **2014**, *5*, 603. [CrossRef]
23. Tanaka, Y. Cloning and expression of flavonol synthase from *Petunia hybrida*. *Plant J.* **2010**, *4*, 1003–1010.
24. Pelletier, M.K.; Shirley, B.W. Characterization of flavonol synthase and leucoanthocyanidin dioxygenase genes in Arabidopsis. Further evidence for differential regulation of "early" and "late" genes. *Plant Physiol.* **1997**, *113*, 1437. [CrossRef]
25. Nielsen, K.; Deroles, S.C.; Markham, K.R.; Bradley, M.J.; Podivinsky, E.; Manson, D. Antisense flavonol synthase alters copigmentation and flower color in lisianthus. *Mol. Breed.* **2002**, *9*, 217–229. [CrossRef]
26. Almeida, J.R.; D'Amico, E.; Preuss, A.; Carbone, F.; de Vos, C.H.; Deiml, B.; Mourgues, F.; Perrotta, G.; Fischer, T.C.; Bovy, A.G. et al. Characterization of major enzymes and genes involved in flavonoid and proanthocyanidin biosynthesis during fruit development in strawberry (*Fragaria xananassa*). *Arch. Biochem. Biophys.* **2007**, *465*, 61–71. [CrossRef] [PubMed]
27. Takahashi, R.; Githiri, S.M.; Hatayama, K.; Dubouzet, E.G.; Shimada, N.; Aoki, T.; Ayabe, S.; Iwashina, T.; Toda, K.; Matsumura H. A single-base deletion in soybean flavonol synthase gene is associated with magenta flower color. *Plant Mol. Biol.* **2007**, *63*, 125–135. [CrossRef]
28. Luo, P. Molecular Cloning and Functional Analysis of Fiavonoids Biosynthesis Related Genes from *Rosa rugosa*. Ph.D. Thesis, Huazhong Agricultural University, Wuhan, China, 2016; pp. 2–4.
29. Lukačin, R.; Wellmann, F.; Britsch, L.; Martens, S.; Matern, U. Flavonol synthase from Citrus unshiu is a bifunctional dioxygenase. *Phytochemistry* **2003**, *62*, 287–292. [CrossRef] [PubMed]
30. Li, Y.; Wei, C.M.; Li, X.Y.; Meng, D.C.; Gu, Z.J.; Qu, S.P.; Huang, M.J.; Huang, H.Q. De novo transcriptome sequencing of *Impatiens uliginosa* and the analysis of candidate genes related to spur development. *BMC Plant Biol.* **2022**, *22*, 553. [CrossRef]
31. Allan, A.C.; Hellens, R.P.; Laing, W.A. *MYB* transcription factors that colour our fruit. *Trends Plant Sci.* **2008**, *13*, 99–102. [CrossRef]
32. Nie, S.; Zhao, S.W.; Shi, T.L.; Zhao, W.; Zhang, R.G.; Tian, X.C.; Guo, J.F.; Yan, X.M.; Bao, Y.T.; Li, Z.C.; et al. Gapless genome assembly of azalea and multi-omics investigation into divergence between two species with distinct flower color. *Hortic. Res* **2023**, *10*, 241. [CrossRef]
33. Nie, S.; Ma, H.Y.; Shi, T.L.; Tian, X.C.; El-Kassaby, Y.A.; Porth, I.; Yang, F.S.; Mao, J.F. Progress in phylogenetics, multi-omics and flower coloration studies in *Rhododendron*. *Ornam. Plant Res.* **2024**, *4*, e003. [CrossRef]

4. Song, X.; Tian, Y.; Gao, K.; Li, J.; Li, Y.; Wang, J.; Deng, C.; Zhang, F.; Kong, D.; Fan, G.; et al. Genetic and QTL analysis of flower color and pigments in small-flowered chrysanthemum based on high-density genetic map. *Ornam. Plant Res.* **2023**, *3*, 17. [CrossRef]
5. Zhang, Q.; Yang, W.; Liu, J.; Liu, H.; Lv, Z.; Zhang, C.; Chen, D.; Jiao, Z. Postharvest UV-C irradiation increased the flavonoids and anthocyanins accumulation, phenylpropanoid pathway gene expression, and antioxidant activity in sweet cherries (*Prunus avium* L.). *Postharvest Biol. Technol.* **2021**, *175*, 111490. [CrossRef]
6. Peng, H.; Yang, T.; Whitaker, B.D.; Trouth, F.; Shangguan, L.; Dong, W.; Jurick, W.M. Characterization of spermidine hydroxycinnamoyl transferases from eggplant (*Solanum melongena* L.) and its wild relative *Solanum richardii* Dunal. *Hortic. Res.* **2016**, *3*, 16062. [CrossRef] [PubMed]
7. Zeng, X.; Zheng, R.; Luo, J.; Wang, C. Cloning and Characterization of Cinnamate 4-hydroxylase (*C4H*) Genes from *Osmanthus fragrans*. *J. Hortic.* **2016**, *43*, 525–537.
8. Liu, H.; Su, B.; Zhang, H.; Gong, J.; Zhang, B.; Liu, Y.; Du, L. Identification and Functional Analysis of a Flavonol Synthase Gene from *Grape Hyacinth*. *Molecules* **2019**, *24*, 1579. [CrossRef] [PubMed]
9. Luo, P.; Ning, G.; Wang, Z.; Shen, Y.; Jin, H.; Li, P.; Huang, S.; Zhao, J.; Bao, M. Disequilibrium of Flavonol Synthase and Dihydroflavonol-4-Reductase Expression Associated Tightly to White vs. Red Color Flower Formation in Plants. *Front. Plant Sci.* **2016**, *6*, 1257. [CrossRef]
0. Davies, K.M.; Schwinn, K.E.; Deroles, S.C.; Manson, D.G.; Lewis, D.H.; Bloor, S.J.; Bradley, J.M. Enhancing anthocyanin production by altering competition for substrate between flavonol synthase and dihydroflavonol 4-reductase. *Euphytica* **2003**, *131*, 259–268. [CrossRef]

horticulturae

MDPI

Article

Resistances and Physiological Responses of *Impatiens uliginosa* to Copper Stress

Jiapeng Zhu [1], Xinyi Li [2], Haiquan Huang [2,*] and Meijuan Huang [2,*]

[1] Faculty of Culture and Tourism, Qujing Normal University, Qujing 655011, China; zhuqjnu@163.com

[2] College of Landscape Architecture and Horticulture Sciences, Southwest Research Center for Engineering Technology of Landscape Architecture (State Forestry and Grassland Administration), Yunnan Engineering Research Center for Functional Flower Resources and Industrialization, Research and Development Center of Landscape Plants and Horticulture Flowers, Southwest Forestry University, Kunming 650224, China

* Correspondence: haiquanl@163.com (H.H.); xmhhq2001@163.com (M.H.)

Abstract: The phytoremediation of soil and water that has been significantly contaminated with metals has potential ecological and economical ramifications, as well as the advantages of high efficiency, and is an environmentally friendly method of ecological pollution control. This study aimed to examine the impact of varying concentrations of Copper (Cu^{2+}) (0, 5, 10, 15, 20, and 25 $mg{\cdot}L^{-1}$) on the growth, development, physiology, biochemistry, mineral elements, and features of Cu^{2+} enrichment of *Impatiens uliginosa*. This plant is endemic to Yunnan Province in China and is a wetland species. The results showed that the root lengths, stem diameters, plant height, and stem and leaf biomass of *I. uliginosa* showed a phenomenon of "low promotion and high inhibition," while the root biomass showed a trend of gradual decreasing. At the early stage of Cu^{2+} stress (day 6), the activities of peroxidase and catalase and the contents of malondialdehyde (MDA) of *I. uliginosa* were directly proportional to the concentration of Cu^{2+}. As the treatment time increased, the activation of a defense mechanism in vivo enabled *I. uliginosa* to adapt to the high Cu^{2+} environment, and the content of MDA gradually decreased. As the concentration of Cu^{2+} increased, its contents in the roots, stems, and leaves also gradually increased. In particular, when the concentration of Cu^{2+} reached 25 $mg{\cdot}L^{-1}$, its contents in the roots of *I. uliginosa* increased by 39.16-fold compared with that of the control group (CK). The concentration-dependent influence of the contents of iron (Fe) and zinc (Zn) in the roots and leaves were observed. Low concentrations of Cu^{2+} promoted iron content in roots and leaves, and vice versa, while Zn content decreased with the increasing concentration of Cu^{2+}. It was conclusively shown that *I. uliginosa* has the potential to remediate low concentrations of Cu^{2+} pollution in water and is a textbook ornamental plant to remediate bodies of water that are polluted with Cu^{2+}.

Keywords: *Impatiens uliginosa*; Cu^{2+} stress; physiological response; enrichment characteristics

1. Introduction

Heavy metal pollution is strongly destructive for the environment and organisms and is characterized by long-term and potential irreversibility. It can be enriched or amplified through the food chain, thereby harming animal, plant, and human health [1]. As the industrial and agricultural economy has rapidly expanded, the handling of mineral resource exploration and smelting waste has unfortunately fallen short of proper standards, leading to inadequate disposal methods such as careless discharges from factories and inappropriate sewage irrigation practices in agriculture. A large amount of heavy metals enters the environment through the activities of humans, which results in a series of problems of environmental pollution. The frequency of the occurrence of cases of metal toxicity has been increasing [2–4]. Over the past 20 years, owing to the needs of human production and life, copper has become one of the three metals with the largest output in the world [5], and in the past half century, copper has been the second most widely discharged metal into the natural environment in the world with a content of 9.39×10^5 t [6]. The control and remediation of Cu^{2+} pollution has become an important topic in today's society, and it is an important environmental problem that merits urgent remediation [7].

The research has shown that Cu^{2+} is among the eight crucial trace elements that are required for optimal plant growth and development, and it is involved in many physiological and biochemical processes [8,9]. Cu^{2+} can regulate various physiological and metabolic processes in plants [10], including photosynthesis and respiration and the metabolism of glucose, protein, and cell walls. However, excess Cu^{2+} in plants can lead to changes in DNA, cell membrane integrity, and enzyme activity, thereby affecting the yields of crops and plants [10–13]. The toxicity of Cu^{2+} is limited to a narrow range of concentrations, which is typically between 0.19 $mg{\cdot}L^{-1}$ and 6.4 $mg{\cdot}L^{-1}$ for most plant species [14]. A deficiency in Cu^{2+} primarily affects plant reproductive organs and young leaves and leads to tissue necrosis, leaf yellowing, discoloration, developmental retardation, and the inhibition of root growth. It promotes the production of reactive oxygen species (ROS) and also affects the development of pollen and plant embryos, viable pollen and seeds, fruit yield, and nutritional characteristics [15–17]. However, excessive Cu^{2+} inhibits the division and elongation of cells and the activity of many enzymes, which leads to a decrease in respiration and photosynthesis. It then inhibits seed germination, seedling growth, and the accumulation of biomass, which causes disorders of the physiological metabolic networks and eventually leads to significant toxicity to plants [18–20]. Dianchi Lake is a typical shallow plateau lake in China. The average concentration of Cu^{2+} in the water in Dianchi Lake has been found to be 105.32 $\mu g{\cdot}L^{-1}$, which is far higher than the average concentration of Cu^{2+} in non-polluted natural water (2 $\mu g{\cdot}L^{-1}$) [21]. The selection of plants resistant to Cu^{2+} is the key to the successful implementation of phytoremediation technology in controlling and repairing the water area of Dianchi Lake.

Impatiens uliginosa is an annual herb of the Balsaminaceae family. It is widely distributed in Guangxi and Yunnan Provinces. It grows under wet forests and next to gullies and streams. Its bright colors, long flowering period, large biomass, developed root system, and strong adaptability have led to its high value as an ornamental plant that is frequently grown in gardens. It is also valuable medicinally [22]. *I. uliginosa* grows luxuriously in Dianchi Lake where the Cu^{2+} concentration is 50-fold higher than that of non-polluted natural water [21], which indicates that the plant is somewhat tolerant to copper. Thus, it is a facultative metal plant, which makes it an ideal candidate for bioremediation. Currently, relatively little research has been conducted on *I. uliginosa*. What research has been conducted has focused on the characteristics of seed germination [22,23], the cloning of genes regulating flower development [24–26], and the determination of metal elements [27,28]. Studies on the physiological and biochemical characteristics of *I. uliginosa* have been limited to changes in the seeds, seedlings, and flower color [23,29,30]. Although the influence of Cu^{2+} on *I. uliginosa* in the natural environment is a dynamic and long-lasting process, to our knowledge, research on this topic has not yet been reported. This study examined the effects of Cu^{2+} stress on the morphology and development of growth, physiological and biochemical parameters, and the contents of metal ions of *I. uliginosa*. The characteristics of Cu^{2+} in response to stress are discussed, and technical advice for the scientific cultivation of *I. uliginosa* is provided.

2. Materials and Methods

2.1. Plant Materials and Cu^{2+} Treatments

To enhance the simulation of a natural habitat and improve the observation of morphological indicators in the stress experiment, a soil culture was first used, followed by a hydroponic setup. *I. uliginosa* seeds were collected in Laoyuhe Wetland Park, Kunming, Yunnan Province, China (102°46′7.15″ E, 24°49′22.79″ N). The heavy metal tested was $CuSO_4{\cdot}5H_2O$ (Damao Chemical Reagent Factory, Tianjin, China), which was dissolved in deionized water. The seeds were germinated in distilled water in Petri dishes. Two weeks later, the seedlings with uniform growth were selected and transplanted into the arboreal garden of Southwest Forestry University (Kunming, China). After 50 days, the seedlings with uniform biomass were collected, and the bases of the seedlings were washed with distilled water. The seedlings were affixed to a foam plastic tray with holes, which were used for planting cups. They were then placed in a 60 × 40 × 40 cm incubator for subsequent use. After acclimation with one-half Hoagland nutrient solution for two weeks, the seedlings were subjected to Cu^{2+} stress treatment on September 7, 2019. Five gradients of Cu^{2+} concentration were established, including 5 mg·L^{-1}, 10 mg·L^{-1}, 15 mg·L^{-1}, 20 mg·L^{-1}, and 25 mg·L^{-1}, and seedlings in one-half Hoagland nutrient solution were used as the control. Samples were taken at 0 d, 6 d, 12 d, 18 d, and 24 d after treatment. The roots, stems, and leaves were wrapped in tin foil and placed in tanks of liquid nitrogen for further use.

2.2. Biomass Determination

After the experimental treatment, *I. uliginosa* was separated into three parts, including the roots, stems, and leaves. The surface water was removed using filter paper, and the fresh weight (g) was measured. It was dried in an oven (EYELA, Tokyo, Japan) at 105 °C for 15 min and then dried at 80 °C until a constant weight was reached. The dry weight (g) was then obtained.

2.3. Plant Height, Root Length, and Stem Diameter

On days 0, 6, 12, 18, and 24 after the experiment had been started, the dimensions of *I. uliginosa*, including its elevation, root extension, and stem width, were meticulously gauged using a Vernier caliper and documented in centimeters. The plant height was measured as the height from stem base to the plant tip, stem diameter as the width of stem base, and root length was determined by averaging the longest and shortest root lengths for an overall representation.

2.4. Determination of Oxidative Stress Indicators and Antioxidant Responses

After Cu^{2+} stress, 0.5 g of fresh leaves of the treated groups were weighed and placed in a pre-cooled mortar. After 5 mL of sodium phosphate buffer (pH 7.8) (Tianjin Zhiyuan Chemical Reagent Co. Ltd., Tianjin, China) had been added several times, the leaves were ground in an ice bath. After the ground homogenate turned white, it was poured into a 10 mL centrifuge tube and centrifuged at 4000 rpm at 4 °C for 20 min. The supernatant (crude enzyme solution) was added to a test tube and stored at 4 °C for later use, as described by Dhindsa et al. [31]. The activity of superoxide dismutase (SOD) was determined using the method of inhibition of photochemical reduction of nitroblue tetrazolium chloride [32]. The samples were exposed to 4000 Lux for 20 min in the plant incubator (SANYO, Osaka, Japan), and the absorbance readings were recorded at 560 nm. The activity of peroxidase (POD) was analyzed through the guaiacol method, as outlined by Sakharov et al. [33], with absorbance readings taken at a wavelength of 470 nm. The activity of the enzyme catalase (CAT) was determined using the procedure outlined by Chaoui et al. [34], with absorbance readings taken at a wavelength of 240 nm. The measurement of malondialdehyde (MDA) content was conducted through the thiobarbituric acid method, as outlined by Tewari et al. [35], with absorbance readings taken at wavelengths of 600 nm, 532 nm, and 450 nm.

2.5. Determination of the Concentration of Metals

To analyze the content of the metals in dried plant parts (roots, stems, leaves), a 0.15 g sample was ground and passed through a 60-mesh sieve. The sample was then mixed with HNO_3 (Damao Chemical Reagent Factory, Tianjin, China) and HCl (Yanglin Industrial Development Zone Shandian Pharmaceutical Co. Ltd., Yunnan, China) (3:1) and put in ETHOS A (Milestone, Santa Clara, CA, USA) for microwave fermentation. The digestion tank was removed after the heating procedure was completed, and the temperature decreased to below 80 °C. The sample was transferred to a polyethylene tank in a fume hood (ShuangXu Electronics Co. Ltd., Shanghai, China) and placed on a C-MAG HS 10 digital (IKA, Staufen, Germany) magnetic floating agitator to remove the acid. The liquid was transferred to a 25 mL volume bottle after it had become transparent. The concentration of metal ions was determined by flame atomic absorption spectrometer (Shimadzu Corporation, Kyoto, Japan).

2.6. Transfer Coefficient, Enrichment Coefficient, and the Rate of Distribution of Cu^{2+} in Plants

The bioenrichment factor (BCF) and transport factor (TF) were used to evaluate the ability of plants to accumulate metals from the soil and transport them from the root to stem [36].

Bioconcentration factor (BCF) = Cu^{2+} concentration in tissue/Cu^{2+} concentration in solution

Translocation factor (TF) = Cu^{2+} concentration in tissues/Cu^{2+} concentration in roots

$$\text{Allocation rate} = \{C_i \times W_i / \sum(C_i \times W_i)\} \times 100\%$$

where i refers to different parts of *I. uliginosa* (roots, stems, and leaves); C_i is the concentration of Cu^{2+} ($mg \cdot kg^{-1}$) in the tissues of *I. uliginosa*, and W_i is the biomass dry weight (kg) of all the parts of *I. uliginosa*.

2.7. Statistical Analysis

Each experimental group was repeated three times and analyzed using SPSS 24.0 (IBM, Inc., Armonk, NY, USA). The variability of the samples was expressed as the mean ± standard deviation (SD). Statistical significance was determined at the 0.05 level using the least significant difference (LSD) test. The graphs and tables were created using Microsoft Excel 2010 (Redmond, WA, USA) and Adobe Photoshop CS6 (San Jose, CA, USA).

3. Results

3.1. Effects of Cu^{2+} Stress on the Growth and Development of I. uliginosa

As shown in Figure 1, the height and stem diameter of *I. uliginosa* gradually increased with the extension of treatment time under different levels of Cu^{2+} stress. However, these parameters varied under this type of treatment. Compared with the treatment on day 0, the height of *I. uliginosa* increased by 64.50% (CK), 81.74%, 82.66%, 63.15%, 49.94%, and 40.33% after treatment. Under treatment with the same concentrations, the height of *I. uliginosa* after treatment at 6 d was as follows: 28.80%, 34.70%, 20.55%, 14.41%, and 14.84%, which were all higher than the height of CK (8.38%). Compared with the 0 d treatment, the increase in height of *I. uliginosa* after 24 d was 62.73% (CK), 59.22%, 55.45%, 51.43%, 57.27%, and 50.42%. The height increased the most during the 0–6 d period, and the values were 25.45% (CK), 18.45%, 15.45%, 14.29%, 14.55% and 6.84%. The amount of increase decreased in the following order: CK > 5 $mg \cdot L^{-1}$ > 10 $mg \cdot L^{-1}$ > 15 $mg \cdot L^{-1}$ > 20 $mg \cdot L^{-1}$ > 25 $mg \cdot L^{-1}$.

Roots are the first natural barriers that enable plants to sense and respond to their environment. Under stress conditions, the root system will be affected to varying degrees and can even die and fall off. As shown in Figure 1, the length of *I. uliginosa* roots gradually decreased as the concentration of Cu^{2+} increased, and the treatment time was extended. These changes were significantly different from those of the control group (CK) ($p > 0.05$). The length of *I. uliginosa* roots decreased gradually as the treatment time increased despite being treated with the same concentration. This decrease was influenced by various factors, such as the external environment and root damage. However, when subjected to 5–10 $mg \cdot L^{-1}$ Cu^{2+} stress, the roots grew longer than those of the CK. However, the difference between them was not significant ($p > 0.05$). Under 15–25 $mg \cdot L^{-1}$ Cu^{2+} stress, the roots grew slowly owing to the high concentration of Cu^{2+} stress, which indirectly protected the artificial damage and other factors caused by the overly long roots during the measurements.

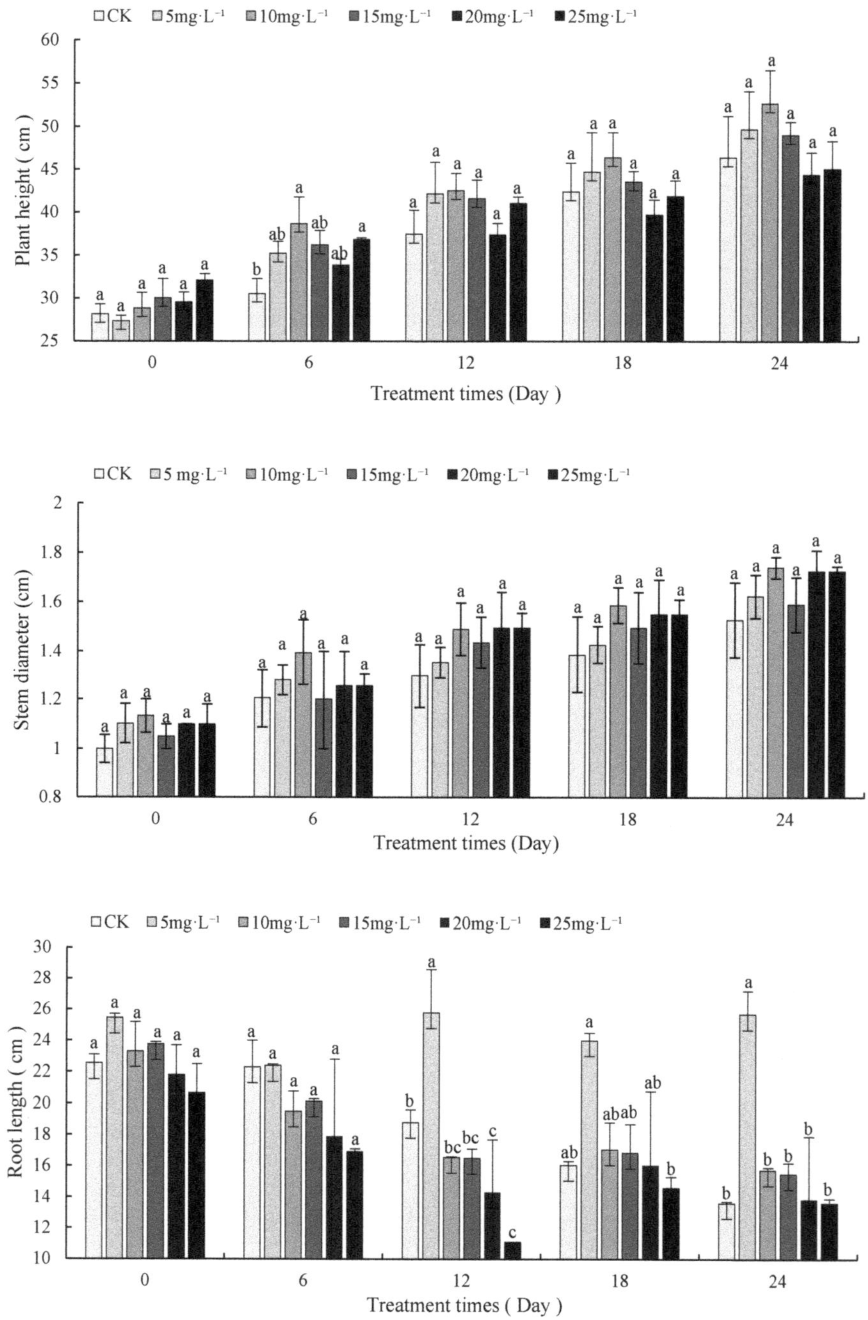

Figure 1. Effects of different levels of Cu^{2+} stress on the growth indices of *I. uliginosa*. Note: Values represent the mean ± standard error of three replicates. Different lowercase letters indicate significant differences in values within the same concentration ($p < 0.05$).

3.2. Effects of Cu^{2+} Stress on the Biomass of *I. uliginosa*

Table 1 shows a gradual decrease in the root biomass of *I. uliginosa* with an increase in the concentration of Cu^{2+}. Additionally, the root biomass was found to be lower than that of the control group (CK) at all ranges of concentration. At a concentration of Cu^{2+} of 25 mg·L^{-1}, the root fresh weight reached its lowest point at 9.87 g, which did not reveal any significant difference when compared with the control group (CK) ($p > 0.05$). However, the dry weight differed significantly when compared with the control group (CK) ($p < 0.05$) at 0.66 g. With the increase in concentration of Cu^{2+}, the biomass of stems and leaves of *I. uliginosa* increased first and then decreased. When the Cu^{2+} concentration was 20 mg·L^{-1}, the fresh weight of the stems and leaves reached their maximum values of 58.22 g and 13.40 g, respectively. This represents an increase of 89.21% and 52.79% compared with the control group (CK), although this difference was not statistically significant ($p > 0.05$).

Table 1. Effects of different levels of Cu^{2+} stress on the biomass of *I. uliginosa*.

Cu^{2+} Concentration (mg·L^{-1})	Fresh Weight (g)			Dry Weight (g)		
	Root	Stem	Leaf	Root	Stem	Leaf
0 (CK)	12.91 ± 1.91 a	30.77 ± 1.33 b	8.77 ± 1.36 a	1.17 ± 0.15 a	1.66 ± 0.16 a	1.05 ± 0.06 a
5	11.13 ± 0.38 a	40.90 ± 6.93 ab	11.68 ± 4.45 a	0.79 ± 0.05 ab	2.03 ± 0.57 a	1.32 ± 0.67 a
10	11.20 ± 5.90 a	45.93 ± 10.42 ab	12.87 ± 6.22 a	0.89 ± 0.37 ab	2.47 ± 1.17 a	1.48 ± 0.50 a
15	11.93 ± 0.51 a	39.83 ± 4.55 ab	12.45 ± 4.34 a	0.94 ± 0.15 ab	2.26 ± 0.16 a	1.20 ± 0.19 a
20	10.32 ± 3.92 a	58.22 ± 4.96 a	13.40 ± 2.31 a	0.83 ± 0.18 ab	3.16 ± 1.07 a	2.33 ± 0.78 a
25	9.87 ± 2.26 a	40.50 ± 7.97 ab	11.66 ± 5.43 a	0.66 ± 0.12 b	2.48 ± 0.93 a	1.48 ± 0.63 a

Note: Values represent the mean ± standard error of three replicates. Different lowercase letters within the same column indicate significant differences ($p < 0.05$).

3.3. Effects of Cu^{2+} Stress on the Activities of Antioxidant Enzymes and Content of Malondialdehyde in *I. uliginosa*

Oxidative stress was assessed by measuring MDA content in the leaves, which is an important biomarker of oxidative injury. In the range of low concentrations (0–15 mg·L^{-1}), the content of MDA of *I. uliginosa* tended to decrease first and then increased with the extension of treatment time, and the difference was significant when compared with the CK ($p < 0.05$) (Figure 2). In the range of high concentrations (20–25 mg·L^{-1}), the content of MDA tended to increase first and then decrease with the extension of treatment time, and the difference was significant compared with the CK ($p < 0.05$). From 6 to 18 days post-treatment, the content of MDA gradually increased as the concentration of Cu^{2+} increased. At a concentration of 25 mg·L^{-1}, the contents of MDA were 31.19, 24.39, and 15.91 μmol·L^{-1} FW, respectively, which were significantly different from those of the control group ($p < 0.05$). The increases were as follows: 190.41%, 301.15%, and 121.90%. After 24 d of stress with different concentrations of Cu^{2+}, the content of MDA fluctuated over a small range. When the concentration reached 25 mg·L^{-1}, the content of MDA was 19.10 μmol·L^{-1} FW, which was significantly different from that of Cu^{2+} (15.72 μmol·L^{-1} FW) ($p < 0.05$). The increase was 21.50%.

As shown in Figure 2, the activity of SOD from *I. uliginosa* tended to increase first and then decrease at different times after Cu^{2+} stress. On the sixth day after treatment, the activity increased slowly at first and then decreased rapidly with the increase in treatment concentration. When the concentration reached 25 mg·L^{-1}, the activity of SOD decreased to its lowest level (258.10 U·g^{-1}·FW·h^{-1}), which was significantly different from that of the CK (421.66 U·g^{-1}·FW·h^{-1}) ($p < 0.05$). Within the range of Cu^{2+} concentrations of 5–20 mg·L^{-1}, the activity first increased and then decreased with the increase in Cu^{2+} concentration. This difference was significant when compared with that of the CK ($p < 0.05$). However, at a concentration of 25 mg·L^{-1}, the activity of SOD increased gradually with the increase in treatment time.

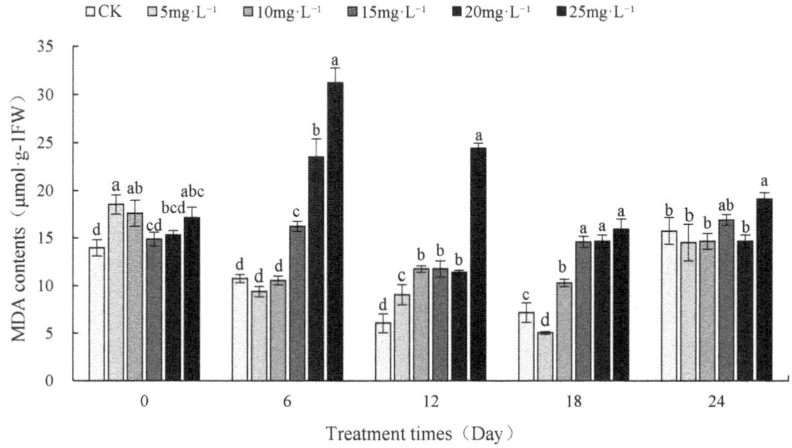

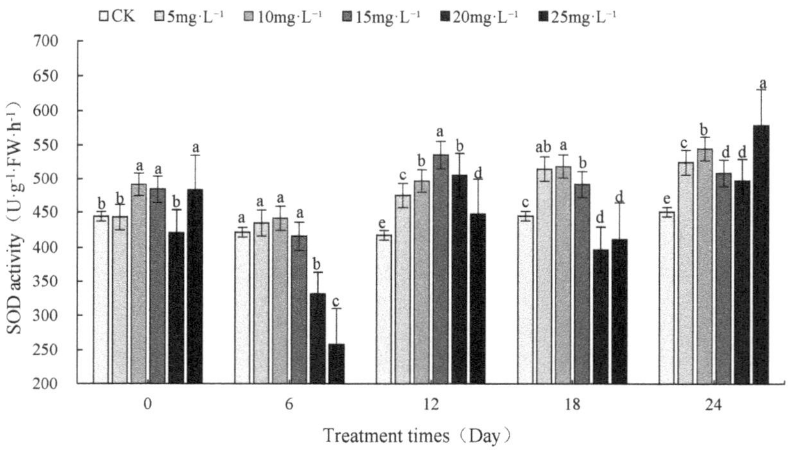

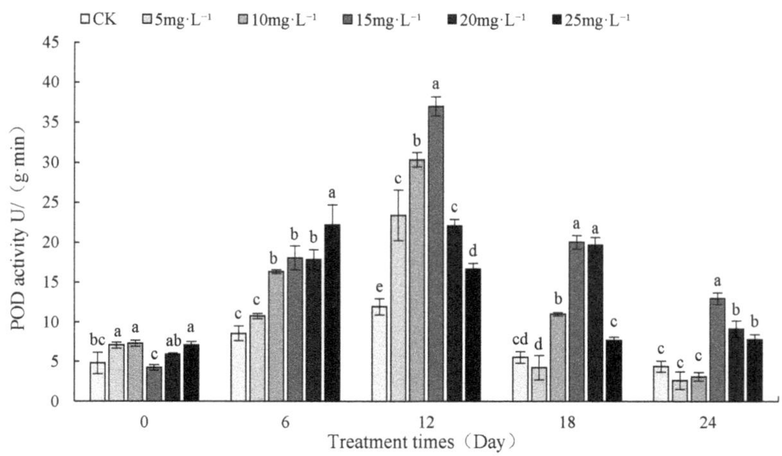

Figure 2. *Cont.*

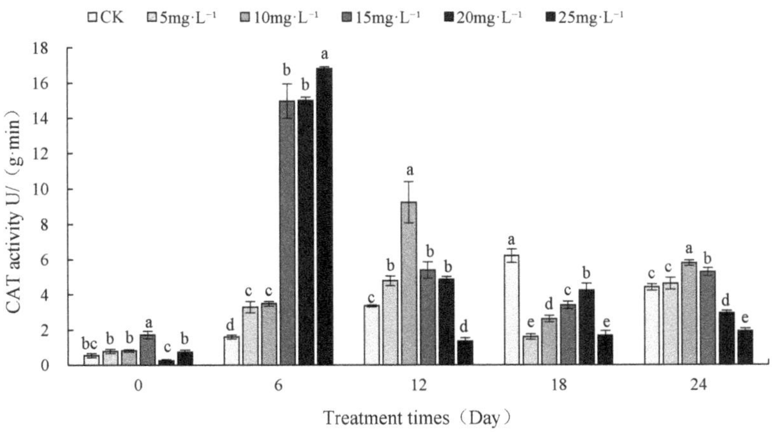

Figure 2. Effects of Cu^{2+} stress on the activities of antioxidant enzymes and the contents of malondialdehyde (MDA) of *I. uliginosa*. Note: Values represent the mean ± standard error of three replicates, and different lowercase letters indicate significant differences ($p < 0.05$).

Under different concentrations of Cu^{2+} stress, the activity of POD first increased and then decreased over time (Figure 2). It increased sharply from 0–12 d and reached its maximum on 12 d. The activity of POD decreased gradually from 12 to 24 days. Under the 6 d treatment, the activity gradually increased with increasing concentrations. When the concentration of Cu^{2+} reached 25 mg·L^{-1}, the activity of POD increased by 160.17% compared with the CK, and this difference was found to be statistically significant ($p < 0.05$). After 12–24 d of processing conditions, the POD activity of *I. uliginosa* exhibited a pattern of initially rising and then declining as the Cu^{2+} concentration increased. When the concentration was 15 mg·L^{-1}, the POD activity of *I. uliginosa* peaked on the days 12, 18 and 24, compared with that of the CK. The respective increases were 209.62%, 263.11%, and 193.41%, with a statistically significant difference ($p < 0.05$).

Over time, the CAT activity of *I. uliginosa* increased rapidly at first and then decreased and maintained a stable trend (Figure 2). In the treatment time of 0–6 d, the activity increased rapidly and reached its maximum at 6 d. In the treatment time of 6 d, the CAT activity increased gradually as the concentration of Cu^{2+} increased and reached its maximum at 25 mg·L^{-1} (16.81 U·g^{-1}·min^{-1}), which was significantly different from that of the CK (1.61 U·g^{-1}·min^{-1}) ($p < 0.05$). From days 12 to 24, the CAT activity gradually decreased and remained mostly stable. As the concentration of Cu^{2+} increased, the CAT activity first increased and then decreased.

3.4. Effect of Cu^{2+} Stress on the Mineral Contents in Different Parts of I. uliginosa

As shown in Table 2, the contents of Cu^{2+} in the roots, stems, and leaves of *I. uliginosa* gradually increased as the degree of Cu^{2+} stress increased, which was significantly different from that in the CK ($p < 0.05$). The degree of Cu^{2+} stress significantly increased the contents of Cu^{2+} in the roots. This increase was found to be 16.50-, 22.63-, 29.42-, 37.51-, and 38.16-fold higher than that of the control group (CK), and the difference was statistically significant ($p < 0.05$). The contents of Cu^{2+} in the stems of *I. uliginosa* increased by 0.91-, 1.94-, 3.36-, 6.26-, and 9.50-fold, respectively. The contents of Cu^{2+} in the leaves increased by 1.7-, 6.04-, 7.92-, 5.12-, and 4.2- fold, respectively. This study showed that the contents of iron (Fe) in the roots and leaves first increased and then decreased with the increase in degree of Cu stress. When the concentration of Cu^{2+} was 10 mg·L^{-1}, the highest contents of Fe in the roots and leaves were 2076.75 mg·kg^{-1} and 764.97 mg·kg^{-1}, respectively. This was a significant improvement compared with the CK (1634.40 mg·kg^{-1} and 582.29 mg·kg^{-1}, respectively) with a difference of 27.06% and 31.37%, respectively ($p < 0.05$). The content of

Fe in the stems decreased first and then increased with the increase in degree of Cu^{2+} stress. When the concentration reached 15 mg·L^{-1}, the content of Fe in stems was at its lowest, 157.55 mg·kg^{-1}, which was not significantly different from that of the CK (184.33 mg·kg^{-1}) ($p > 0.05$). Under the stress of Cu^{2+}, the contents of Zn in the roots, stems, and leaves of *I. uliginosa* generally decreased. Under different degrees of Cu^{2+} stress, the contents of Zn in different parts of the plant decreased compared with the CK. The changes in the roots were 28.34%, 22.62%, 49.33%, 26.18%, and 17.21%, stems (18.18%, 2.10%, 35.22%, 51.26%, and 82.22%), and leaves (67.35%, 16.63%, 30.39%, 44.94%, and −3.05%).

Table 2. Effects of Cu^{2+} stress on the content of metal elements in different parts of *I. uliginosa*.

Item	Cu^{2+} Concentration (mg·L^{-1})	Element						
		Cu mg·kg^{-1}	Fe mg·kg^{-1}	Zn mg·kg^{-1}	K g·kg^{-1}	Ca g·kg^{-1}	Mg mg·kg^{-1}	Na mg·kg^{-1}
Root	0 (CK)	61.37 ± 1.46 e	1634.40 ± 72.93 c	205.12 ± 66.06 a	2.00 ± 0.23 a	8.10 ± 1.14 a	554.55 ± 13.38 a	833.62 ± 9.34 c
	5	1074.31 ± 220.42 d	1797.06 ± 41.31 b	146.99 ± 26.97 ab	2.11 ± 0.28 a	6.68 ± 0.89 b	574.65 ± 29.78 a	1189.29 ± 74.39 a
	10	1450.28 ± 181.44 c	2076.75 ± 51.99 a	158.72 ± 29.72 ab	2.30 ± 0.18 a	6.84 ± 0.34 ab	581.58 ± 15.30 a	1259.09 ± 5.19 a
	15	1867.17 ± 365.34 b	1814.89 ± 74.28 b	103.96 ± 14.17 b	2.16 ± 0.16 a	5.48 ± 0.78 bc	560.34 ± 1.83 a	945.49 ± 53.20 b
	20	2363.90 ± 135.02 a	1303.88 ± 104.21 d	151.42 ± 9.79 ab	2.06 ± 0.05 a	5.08 ± 0.70 c	572.44 ± 16.87 a	971.24 ± 1.05 b
	25	2403.22 ± 111.52 a	1167.38 ± 53.84 e	169.81 ± 12.19 a	2.18 ± 0.02 a	6.46 ± 0.40 bc	551.28 ± 4.49 a	1258.16 ± 31.19 a
Shoot	0 (CK)	5.18 ± 1.24 e	184.38 ± 64.29 b	142.76 ± 12.56 a	2.30 ± 0.11 a	9.93 ± 0.38 a	611.71 ± 19.71 a	468.87 ± 41.59 ab
	5	9.90 ± 0.74 de	194.12 ± 17.73 b	116.80 ± 3.07 b	2.36 ± 0.11 a	10.41 ± 0.28 a	613.91 ± 18.59 a	399.32 ± 14.45 bc
	10	15.22 ± 3.17 cd	159.13 ± 9.95 b	139.76 ± 9.22 a	2.25 ± 0.09 a	10.57 ± 0.42 a	619.12 ± 4.31 a	427.61 ± 53.85 abc
	15	22.61 ± 1.95 c	157.55 ± 5.46 b	92.48 ± 5.18 c	2.17 ± 0.04 a	10.50 ± 0.19 a	628.87 ± 1.69 a	508.01 ± 81.60 a
	20	37.59 ± 5.09 b	188.49 ± 59.92 b	69.58 ± 3.72 d	1.92 ± 0.07 a	9.90 ± 0.51 a	617.56 ± 13.74 a	369.99 ± 22.06 c
	25	54.42 ± 10.92 a	284.56 ± 37.75 a	24.53 ± 13.53 e	1.76 ± 0.05 b	10.33 ± 0.04 a	610.94 ± 6.39 a	429.62 ± 8.13 abc
Leaf	0 (CK)	15.73 ± 1.03 c	582.29 ± 105.16 b	154.83 ± 61.73 a	1.64 ± 0.14 b	9.33 ± 0.24 b	591.93 ± 10.82 bc	419.22 ± 17.65 a
	5	42.50 ± 11.98 c	430.72 ± 38.95 c	50.55 ± 15.87 c	1.98 ± 0.02 a	8.88 ± 0.47 b	578.24 ± 0.12 c	395.68 ± 13.37 ab
	10	110.68 ± 20.55 b	764.97 ± 8.18 a	129.08 ± 36.24 ab	1.77 ± 0.08 b	10.02 ± 0.42 a	608.11 ± 9.63 ab	385.33 ± 39.15 ab
	15	140.24 ± 22.71 a	544.69 ± 51.38 b	107.78 ± 21.95 abc	1.98 ± 0.08 a	10.63 ± 0.25 a	624.74 ± 6.88 a	399.63 ± 18.88 ab
	20	96.19 ± 14.92 b	297.97 ± 47.42 d	85.29 ± 7.76 bc	1.92 ± 0.07 a	10.02 ± 0.37 a	594.21 ± 6.85 bc	359.64 ± 23.37 b
	25	81.72 ± 15.32 b	325.22 ± 43.20 d	159.55 ± 28.86 a	1.76 ± 0.05 b	10.28 ± 0.22 a	598.39 ± 3.19 b	410.32 ± 28.43 a

Note: Values represent the mean ± standard error of three replicates. Different lowercase letters indicate significant differences ($p < 0.05$).

The contents of potassium (K) in the roots and stems of *I. uliginosa* did not change significantly compared with that of the CK under Cu^{2+} stress, and the increase was between 5.65% and 15% with no significant difference ($p > 0.05$). The contents of K in the leaves tended to increase with the increase in concentration, which resulted in readings of 1.98, 1.77, 1.98, 1.92, and 1.76 g·kg^{-1}, respectively. Compared with the CK (1.64 g·kg^{-1}), the increase was in the following order: 20.73%, 7.92%, 20.73%, 17.07%, and 7.32%, and the differences were significant ($p < 0.05$). Under the conditions of Cu^{2+} stress, the contents of Ca in the roots of *I. uliginosa* decreased with the increase in concentration, and the values were 6.68, 6.84, 5.48, 5.08, and 6.46 g·kg^{-1}, respectively. Compared with the CK (8.10 g·kg^{-1}), the degree of decrease was as follows: 17.53%, 15.56%, 32.35%, 37.28%, and 20.24%, and these changes were significant ($p < 0.05$). The content of Ca in the stems and leaves tended to increase in parallel with the concentration of Cu^{2+} applied. With this increase in Cu^{2+}, the contents of Mg in the roots, stems, and leaves of *I. uliginosa* barely changed. In the treatment of 5–20 mg·L^{-1}, the content of Mg in the roots and stems of *I. uliginosa* was higher than that of the CK with an increase of 0.36–4.87%. When the concentration reached 25 mg·L^{-1}, the contents of Mg in the roots and stems were slightly lower than those of the CK with a decrease of 0.13–0.59%, and there was no significant difference between them ($p > 0.05$). At a concentration of 5 mg·L^{-1}, there was a 2.31% decrease compared with the control (591.93 mg·kg^{-1}), but the difference was not statistically significant ($p > 0.05$). In the treatment of 10–25 mg·L^{-1}, the content of Mg in the leaves of *I. uliginosa* was higher than that of the CK, and the amount of increase was between 0.39% and 5.54%.

Under varying degrees of Cu^{2+} stress, the contents of Na in the roots of *I. uliginosa* were 118.29, 1259.09, 945.49, 971.24, and 1258.16 mg·kg^{-1}, respectively, which were significantly different from that of the CK (833.62 mg·kg^{-1}). The increases were as follows: 42.67%, 51.04%, 13.42%, 16.51%, and 50.93%. The content of Na in the stem increased first and then decreased. When the concentration reached 20 mg·L^{-1}, the content of Na in the stem decreased to its lowest value, 369.99 mg·kg^{-1}, which was 21.09% lower than that of the CK (468.87), and the difference was significant ($p < 0.05$). Compared with the CK (419.22), the content of Na in leaves of *I. uliginosa* tended to decrease. However, this difference was not significant ($p > 0.05$). The percentage of decrease was between 2.12% and 14.21%.

3.5. Effect of Cu^{2+} Stress on the Enrichment and Transport of Cu^{2+} in I. uliginosa

As shown in Table 3, as the degree of Cu^{2+} stress increased, the enrichment coefficient of Cu^{2+} in each part of *I. uliginosa* decreased. The root showed the highest level of Cu^{2+} enrichment, followed by the leaf and then the stem. This suggests that *I. uliginosa* primarily stores Cu^{2+} in its roots and transports less Cu^{2+} to the tissues of its shoots. The characteristics of enrichment were significantly enhanced with the increase in degree of Cu^{2+} stress. However, in the Cu^{2+} stress treatment, the enrichment coefficients of the roots and leaves decreased gradually with the increase in the degree of treatment, and the difference was significant compared with the CK ($p < 0.05$). When the plants were treated with 10–25 mg·L^{-1} of Cu^{2+}, the enrichment coefficient of the stem of *I. uliginosa* tended to gradually increase, but it was always lower than 5 mg·L^{-1}. As indicated by the transport coefficient of Cu^{2+} to *I. uliginosa*, the ratio of middle leaf to root of *I. uliginosa* was higher than that of the stem to root under different concentrations of Cu^{2+}. In addition, it showed different changes with the increase in concentration. These ratios could reflect the ability of *I. uliginosa* to transport Cu^{2+} from the belowground parts to those aboveground. The results also indicated that under Cu^{2+} stress, *I. uliginosa* could control the Cu^{2+} in the belowground parts and effectively transport it to the aboveground parts, thus, reducing the amount of damage to the belowground parts. Therefore, *I. uliginosa* can tolerate Cu^{2+} and has the ability to enrich it.

Table 3. Cu^{2+} concentration and translocation during Cu^{2+} stress in *I. uliginosa*.

Cu^{2+} Concentration (mg·L^{-1})	Bioconcentration Factor			Translocation Factor	
	Root	Stem	Leaf	Stem/Root	Leaf/Root
5	190.071 ± 14.107 a	2.043 ± 0.145 a	9.868 ± 0.512 a	0.011 ± 0.002 cd	0.052 ± 0.001 bc
10	155.476 ± 1.871 b	1.366 ± 0.237 a	9.884 ± 0.197 a	0.009 ± 0.001 d	0.064 ± 0.002 ab
15	111.212 ± 11.424 c	1.581 ± 0.036 a	9.761 ± 1.889 a	0.014 ± 0.002 bc	0.089 ± 0.026 a
20	114.324 ± 1.121 c	1.764 ± 0.224 a	4.722 ± 1.033 b	0.015 ± 0.002 b	0.041 ± 0.009 bc
25	93.872 ± 3.040 c	2.015 ± 0.474 a	3.356 ± 0.838 b	0.025 ± 0.000 a	0.030 ± 0.001 c

The data are Note: Values represent the mean ± standard error of three replicates. Different lowercase letters indicate a significant difference ($p < 0.05$).

3.6. Effects of Cu^{2+} Stress on the Allocation of Cu^{2+} Elements in I. uliginosa

The rate of Cu^{2+} allocation is the percentage of amount of Cu^{2+} that accumulates in each part of the total plant based on the total amount of Cu^{2+} accumulated by the plant. As shown in Figure 3, under different degrees of Cu^{2+} treatments, the rates of Cu^{2+} allocation in the roots, stems, and leaves of *I. uliginosa* ranged from 86.08% to 90.58%, 2.23 to 5.13%, and 6.03 to 11.40%, respectively. As the amount of Cu^{2+} stress increased, the rate of allocation of Cu^{2+} in the roots, stems, and leaves of *I. uliginosa* tended to decrease compared with the CK. However, there were differences in the rate of distribution of Cu^{2+} between different concentrations of treatments and parts of *I. uliginosa*. In the range of

5–25 mg·L^{-1} Cu^{2+}, the rates of Cu^{2+} that were distributed to the roots of *I. uliginosa* were 90.58%, 86.09%, 88.50%, 89.62%, and 88.83%, respectively. Compared with the CK (61.41%), the rate of increase was between 40.19% and 47.50%. With the increase in amount of Cu^{2+} used to stress the plants, its distribution to the stems of *I. uliginosa* tended to gradually increase. When the concentration reached 25 mg·L^{-1}, the rate of distribution of Cu^{2+} to the stem reached its highest value, which was 5.13%. However, this value was much lower than that of the CK (13.27%). As the concentration increases, the distribution rate of Cu^{2+} in leaves initially increases, peaks, and then decreases. The highest allocation rate of Cu^{2+} to leaves was observed at a concentration of 10 mg·L^{-1} (11.40%), which was significantly lower than the control group (25.32%).

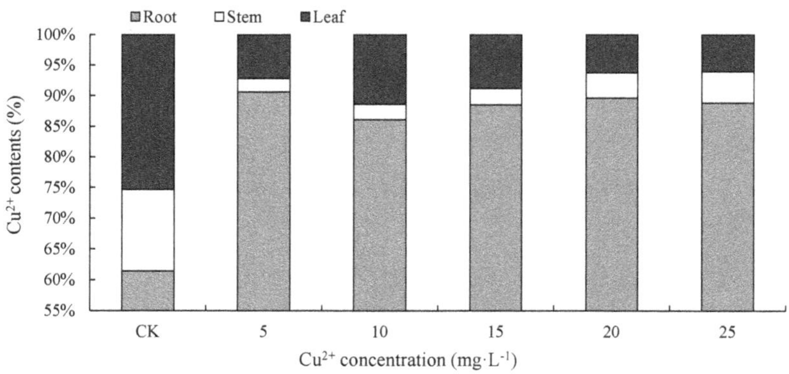

Figure 3. Distribution of Cu^{2+} in different organs of *I. uliginosa*. The data are the average of three experiments.

4. Discussion

4.1. Effects of Cu^{2+} Stress on the Morphological Characteristics of I. uliginosa

Following treatment with low concentration (5 mg·L^{-1}), the plant height and stem diameter of *I. uliginosa* were greater than those of the control, which showed the phenomenon of "low-level promotion and high-inhibition", which was consistent with the findings of other researchers [37–39]. The height and stem diameter of *I. uliginosa* increased continuously with the increase in time, and the increase reached its maximum on day 6 before gradually decreasing. There was a gradual decrease in root length with increasing time and concentration, and the root gradually turned from white to brown. Part of the root then decayed and fell off. Simultaneously, the germination of fibrous roots and lateral roots could be stimulated at low concentrations of Cu^{2+}, but a high concentration inhibited germination. Under the 24 d stress treatment with different concentrations of Cu^{2+}, the lengths of *I. uliginosa* roots were between 13.62 and 25.70 cm, and the presence of certain root lengths, root numbers, lateral roots, and fibrous roots can effectively ensure the normal nutritional requirements of the *I. uliginosa* shoots.

Different concentrations of Cu^{2+} stress also had significant differences on the biomass of *I. uliginosa*. The results indicated a decrease in both fresh weight and dry weight of the root system with the rise in Cu^{2+} stress levels. However, the biomass of stems and leaves was higher than that of the CK and reached its maximum value when the concentration was 20 mg·L^{-1}. The reason may be that the Cu^{2+} damages the root system and causes a decrease in plant root biomass under the condition of the different effects of varying concentrations of Cu^{2+} stress [40,41]. Alternatively, Cu^{2+} is an essential trace element for plants. Thus, at low concentrations, it can compensate for the demands of *I. uliginosa*. This could still result in the absorption of excess Cu^{2+} by the lateral roots as they germinate. This Cu^{2+} can be transferred to the shoot of the first and second branches and the leaves for transfer and fixation.

4.2. Effects of Cu^{2+} Stress on the Physiology and Biochemistry of I. uliginosa

Heavy metal pollution has a detrimental impact on plants and affects their growth, development, and physical characteristics. Additionally, it leads to an increase in the amounts of ROS, such as superoxide anion (O^{2-}), hydroxide (OH), and H_2O_2, in plants. The presence of these compounds can lead to the peroxidation of membrane lipids and the degeneration of biological macromolecules, such as proteins and nucleic acids, thus damaging the membrane structure and the plant itself [42]. MDA is one of the products of membrane lipid peroxidation, and the accumulation of MDA reflects the dynamics of free radical activities in plants to some extent [43]. Many studies have shown that MDA always accumulates in amounts proportional to the concentration of heavy metals [44–46]. The content of MDA in the sedge *Cyperus malaccensis* increased by 185.74% when subjected to 500 mg·L^{-1} of chromium stress [47]. Treatment with high amounts of Cu^{2+} significantly increased the contents of MDA in the leaves of the grape, and the increase in MDA contents gradually increased as the time of treatment was extended [48].The results of this study are consistent with those of Chen [49]. After 6 days of Cu^{2+} stress treatment, the content of MDA in *I. uliginosa* increased significantly with the increase in concentration and was significantly different from the control (CK). At a concentration of 25 mg L^{-1}, the content of MDA in each group reached its maximum under Cu^{2+} stress. However, when the treatment time was extended, the content of MDA in the leaves of *I. uliginosa* decreased gradually and tended to be stable, and it recovered to its original level before 24 d after treatment. The antioxidant system in *I.uliginosa* is activated when exposed to copper stress. This exposure led to a disruption in the homeostasis of scavenging reactive oxygen species (ROS), causing a rapid increase in malondialdehyde (MDA) levels within a short period. However, with the extension of the treatment time, *I. uliginosa* entered a period of adaptation to Cu^{2+} stress. Cu^{2+} activates enzymes in the antioxidant system in *I. uliginosa*, thus increasing its ability to scavenge superoxide free radicals in the plant.

To help manage the accumulation of ROS caused by Cu^{2+} stress, plants have formed a series of perfect defense systems to maintain the balance of an intracellular redox state during long-term natural evolution [50]. Plants can resist the damage caused by Cu^{2+} stress through a system composed of protective enzymes that scavenge ROS. POD, SOD, and CAT can coordinate with each other as the protective enzyme system of plants [51]. SOD catalyzes O_2^-·to generate H_2O_2 and O_2. CAT can degrade H_2O_2 and cooperates with SOD to minimize the formation of ·OH. POD can remove peroxides in cells, which can reduce the damage of ROS to plant cells to some extent [52]. The previous research has shown that the activities of POD, SOD, and CAT in grape (*Vitis vinifera*) roots initially increase and then decrease in response to varying concentrations of Cu (0.5, 1, 1.5, and 2 mmol L^{-1}). However, the scavenging mechanism of ROS in plants also varies with different species. Saleem et al. [53] found that ROS was removed from the plant by increasing the antioxidant activity, while a further increase in the concentration of Cu^{2+} resulted in a decrease in antioxidant activity. Contreras et al. [54] reported that the activities of SOD, POD, and CAT were enhanced under Cu treatment in Antarctic pearlwort (*Colobanthus quitensis*). The results of this study demonstrate that in a short period of time (6 days) under Cu^{2+} stress, the SOD activity of *I. uliginosa* increased first and then decreased with the increase in concentration. However, there was no significant difference in the low concentration (5–15 mg·L^{-1}), while the activities of POD and CAT gradually increased with the increase in the concentration. This resulted in a significant difference. With the extension of treatment time, the SOD activity reacted in the following manner. As the concentration of Cu^{2+} stress increased, the SOD activity of *I. uliginosa* increased first and then decreased, but at 18 d after treatment with 25 mg·L^{-1}, the SOD activity increased more than that of the CK. The reason may be that the change in antioxidant proteins (SOD, POD, and CAT) was related to the type, concentration, and treatment time with heavy metals. The POD activity reached its maximum in the whole treatment cycle at 12 days after treatment, which increased first and then decreased with the increase in concentration.

It then gradually decreased over time. On the sixth day after treatment, CAT activity first increased and then decreased with the increase in concentration.

4.3. Effects of Cu^{2+} Stress on the Uptake of Mineral Elements in I. uliginosa

The balance of mineral nutrient elements is the basis for the normal growth and development of plants [55]. Simultaneously, the uptake of minerals by plants is fundamentally affected by various environmental factors, such as salinity, drought, oxygen content, and heavy metals [56,57]. An appropriate amount of Cu^{2+} can promote the growth of plants, but excess Cu^{2+} causes a change in the absorption of nutrients and characteristics of transport of plants, which has have a toxic effect on them [58]. Studies have shown that the absorption and translocation of mineral elements by plants generally occurs through the interaction with other mineral elements [59]. Cu^{2+} stress decreased the contents of Zn, N, and K in leaves and the contents of K, Ca, P, and Mg in roots. The application of Cu^{2+} increases the contents of Ca and Mg in leaves [38,60]. Zhang et al. found that with the increase in concentration of copper, the contents of N, P, K, Ca, and Mg in the roots and leaves of sugar beet (*Beta vulgaris* subsp. *vulgaris* Altissima Group) tended to increase [60]. Chen et al. [61] found that treatment with a low concentration of Cu^{2+} increased the contents of K, Mg, Ca, and Zn in the stems and leaves of willow (*Salix* sp.), while a high concentration of Cu^{2+} reduced the contents of K, Ca, Mg, and Zn in the stems and leaves of weeping willow (*Salix babylonica*), and *Salix* 172 increased the content of Fe. He [62] found that the contents of Na, K, Mg, and Fe in the roots of cockscomb (*Celosia cristata*) decreased significantly under Cu^{2+} stress. The results of this study showed that the contents of K and Mg and the absorption of Ca, Na, and Zn in the stems, stems and leaves, and roots and leaves were not affected by Cu^{2+} stress treatment. However, the content of Ca in the roots decreased significantly with the increase in concentration, while the contents of Ca^{2+} in the leaves increased significantly. The reason may be that Ca^{2+} channels on the plasma membrane of roots of plants under Cu^{2+} stress are blocked, which results in a decrease in the net uptake of Ca^{2+} by root tip cells [63]. In addition, the contents of Ca in the roots are transported to the shoot, which results in the accumulation of Ca^{2+} in the leaves. The content of Na in the roots increased significantly, which may be related to the fact that Na, as an osmotic substance in plants, can regulate their levels of osmotic pressure [64]. With the increase in Cu^{2+} stress, the permeability of cell membranes gradually increased. Owing to the increase in membrane peroxidation, the permeability of the cell membrane increased, which led to an increase in the absorption of metal elements by the cells. The content of Fe was consistent with the results of Dong [65], and there was no significant change in the stem. The contents of Fe in the roots and leaves showed a phenomenon of "low promotion and high inhibition" with the change in concentration, which may be owing to the competition between copper ions and Fe, which resulted in a lack of Fe, thus affecting the formation of chlorophyll [66].

The Cu^{2+} content in the roots, stems, and leaves of *I. uliginosa* increased gradually with the increase in concentration, and the content of Cu^{2+} in the roots increased more apparently, which may indicate that there is a strong correlation between the content of Cu^{2+} in the environment and that in plants. The characteristics of Cu^{2+} enrichment of *I. uliginosa* were as follows: root > leaf > stem. Excessive Cu^{2+} would destroy the structure of the cell membrane and lead to a rapid increase in the content of Cu^{2+} in the plant. This may be a self-protective mechanism for *I. uliginosa* to manage Cu^{2+} stress and also indicates that it is a common phenomenon that the content of heavy metals in plant roots is higher than that in the shoots, which is related to the effective precipitation or inactivation of heavy metals by the roots [62]. The transfer coefficient showed that the middle leaf/root was higher than that of the stem/root, indicating that the stem primarily played a role in transport and could not store the excess Cu^{2+} content transported from the root. Thus, the content of excess Cu^{2+} would eventually be transferred upward to the leaves for storage in addition to meeting the normal nutritional requirements. Under normal conditions, the rate of distribution of Cu^{2+} from the roots, stems and, leaves of *I. uliginosa* is as follows: 61.41%, 13.27%, and 25.32%. As the concentration of stress increased, the distribution of Cu^{2+} in the root takes place in a water rate of up to 86.09–90.58%. This a self-protective mechanism of *I. uliginosa* when it has accumulated too much Cu^{2+} in its system for redistribution to relieve the concentration of Cu^{2+} in its stems and leaves. After that, the content of Cu^{2+} in the shoots is preferentially transported to the leaves for isolation to protect the stem tissues and maintain the normal supply of plant nutrients. With the increase in concentration, the leaf partitioning rate of Cu^{2+} increased first and then decreased. When the concentration reached 10 mg·L^{-1}, the leaf partitioning reached its highest rate, which was 11.40%. When the leaf to Cu^{2+} partitioning rate reached saturation, the stem to Cu^{2+} partitioning rate showed a gradual tendency to increase.

5. Conclusions

In a certain range of concentration (5–15 mg·L^{-1}), the morphology of growth and physiological and biochemical characteristics of *I. uliginosa* showed a phenomenon of "low promotion and high inhibition" to Cu^{2+} stress, which was accompanied by a prolonged time of Cu^{2+} stress. The activation of antioxidant enzyme systems in the plant could alleviate the Cu^{2+} stress and gradually adapt to the adversity by producing enzymes, such as SOD, POD, and CAT. However, beyond a certain range, the effect would be toxic. *I. uliginosa* has minimal impact on the absorption of mineral nutrients such as Na, Mg, Zn, and K. However, there is interference in the absorption of Ca, Fe, and Cu elements, including the absorption of copper, which increases gradually with the increase in the concentration of processing. It tends to gradually increase and accumulate excess copper elements in the plant through the root–leaf–stem sequence. In conclusion, *I. uliginosa* can adapt to the environment of Cu^{2+} stress through the redistribution of metal elements in its tissues and the improvement in antioxidant enzyme activities. It has a certain potential to repair the low concentration of Cu^{2+} pollution in water. In summary, *I. uliginosa* is a textbook alternative ornamental plant that can be used for the remediation of water that is contaminated with copper.

Author Contributions: J.Z. and M.H. were responsible for the experimental design. J.Z. carried out sample collection, experiments, data analysis, and article writing. H.H. and X.L. participated in the experiment. All authors have read and agreed to the published version of the manuscript.

Funding: This study was financially supported by Key Project of Yunnan Provincial Agricultural Joint Special Program (202101BD070001-018), the National Natural Science Foundation of China (32060364, 32060366), Doctoral Tutor Team for Genetic Improvement and High-efficient Propagation of Landscape Plants in Yunnan Province, and Yunnan Province Local Undergraduate Universities Basic Research Joint Special Project-Youth Project.(202101BA070001-023).

Data Availability Statement: The original contributions presented in the study are included in the article, further inquiries can be directed to the corresponding authors.

Conflicts of Interest: The authors declare no conflicts of interest.

References

1. Hou, D.; O'Connor, D.; Igalavithana, A.D.; Alessi, D.S.; Luo, J.; Tsang, D.C.W.; Sparks, D.L.; Yamauchi, Y.; Rinklebe, J.; Ok, Y.S. Metal contamination and bioremediation of agricultural soils for food safety and sustainability. *Nat. Rev. Earth Environ.* **2020**, *1*, 366–381. [CrossRef]
2. Farid, M.; Ali, S.; Zubair, M.; Saeed, R.; Rizwan, M.; Sallah-Ud-Din, R.; Azam, A.; Ashraf, R.; Ashraf, W. Glutamic acid assisted phyto-management of silver-contaminated soils through sunflower; physiological and biochemical response. *Environ. Sci. Pollut. Res. Int.* **2018**, *25*, 25390–25400. [CrossRef] [PubMed]
3. Kumar, S.; Lai, L.; Kumar, P.; Feliciano, Y.M.V.; Battaglia, M.L.; Hong, C.O.; Owens, V.N.; Fike, J.; Farris, R.; Galbraith, J. Impacts of nitrogen rate and landscape position on soils and switchgrass root growth parameters. *Agron. J.* **2019**, *111*, 1046–1059. [CrossRef]
4. Parveen, A.; Saleem, M.H.; Kamran, M.; Haider, M.Z.; Chen, J.T.; Malik, Z.; Rana, M.S.; Hassan, A.; Hur, G.; Javed, M.T.; et al. Effect of Citric Acid on Growth, Ecophysiology, Chloroplast Ultrastructure, and Phytoremediation Potential of Jute (*Corchorus capsularis* L.) Seedlings Exposed to Copper Stress. *Biomolecules* **2020**, *10*, 592. [CrossRef] [PubMed]
5. Fu, Z.; Guo, W.; Dang, Z.; Hu, Q.; Wu, F.; Feng, C.; Zhao, X.; Meng, W.; Xing, B.; Giesy, J.P. Refocusing on Nonpriority Toxic Metals in the Aquatic Environment in China. *Environ. Sci. Technol.* **2017**, *51*, 3117–3118. [CrossRef] [PubMed]

6. Singh, O.V.; Labana, S.; Pandey, G.; Budhiraja, R.; Jain, R.K. Phytoremediation: An overview of metallic ion decontamination from soil. *Appl. Microbiol. Biotechnol.* **2003**, *61*, 405–412. [CrossRef] [PubMed]
7. Covre, W.P.; Ramos, S.J.; Pereira, W.V.D.S.; Souza, E.S.; Martins, G.C.; Teixeira, O.M.M.; Amarante, C.B.D.; Dias, Y.N.; Fernandes, A.R. Impact of copper mining wastes in the Amazon: Properties and risks to environment and human health. *J. Hazard. Mater.* **2022**, *421*, 126688. [CrossRef] [PubMed]
8. Nazir, F.; Hussain, A.; Fariduddin, Q. Hydrogen peroxide modulate photosynthesis and antioxidant systems in tomato (*Solanum lycopersicum* L.) plants under copper stress. *Chemosphere* **2019**, *230*, 544–558. [CrossRef] [PubMed]
9. Garcia, L.; Welchen, E.; Gonzalez, D.H. Mitochondria and copper homeostasis in plants. *Mitochondrion* **2014**, *19 Pt B*, 269–274. [CrossRef]
10. Ravet, K.; Pilon, M. Copper and iron homeostasis in plants: The challenges of oxidative stress. *Antioxid. Redox Signal* **2013**, *19*, 919–932. [CrossRef]
11. Kohli, S.K.; Handa, N.; Bali, S.; Arora, S.; Sharma, A.; Kaur, R.; Bhardwaj, R. Modulation of antioxidative defense expression and osmolyte content by co-application of 24-epibrassinolide and salicylic acid in Pb exposed Indian mustard plants. *Ecotoxicol. Environ. Saf.* **2018**, *147*, 382–393. [CrossRef] [PubMed]
12. Lu, Y.; Luo, D.; Lai, A.; Liu, G.; Liu, L.; Long, J.; Zhang, H.; Chen, Y. Leaching characteristics of EDTA-enhanced phytoextraction of Cd and Pb by *Zea mays* L. in different particle-size fractions of soil aggregates exposed to artificial rain. *Environ. Sci. Pollut. Res. Int.* **2017**, *24*, 1845–1853. [CrossRef]
13. Rehman, M.; Liu, L.; Wang, Q.; Saleem, M.H.; Bashir, S.; Ullah, S.; Peng, D. Copper environmental toxicology, recent advances, and future outlook: A review. *Environ. Sci. Pollut. Res. Int.* **2019**, *26*, 18003–18016. [CrossRef] [PubMed]
14. Yruela, I. Copper in plants: Acquisition, transport and interactions. *Funct. Plant Biol.* **2009**, *36*, 409–430. [CrossRef] [PubMed]
15. Hunter, C.; Ware, M.A.; Gleason, S.M.; Pilon-Smits, E.; Pilon, M. Recovery after deficiency: Systemic copper prioritization and partitioning in the leaves and stems of hybrid poplar. *Tree Physiol.* **2022**, *42*, 1776–1785. [CrossRef] [PubMed]
16. Burkhead, J.L.; Gogolin-Reynolds, K.A.; Abdel-Ghany, S.E.; Cohu, C.M.; Pilon, M. Copper homeostasis. *New Phytol.* **2009**, *182*, 799–816. [CrossRef]
17. Lafuente, M.T.; Sampedro, R.; Vélez, D.; Romero, P. Deficient copper availability on organoleptic and nutritional quality of tomato fruit. *Plant Sci.* **2023**, *326*, 111537. [CrossRef] [PubMed]
18. Saleem, M.H.; Ali, S.; Seleiman, M.F.; Rizwan, M.; Rehman, M.; Akram, N.A.; Liu, L.; Alotaibi, M.; Al-Ashkar, I.; Mubushar, M. Assessing the Correlations between Different Traits in Copper-Sensitive and Copper-Resistant Varieties of Jute (*Corchorus capsularis* L.). *Plants* **2019**, *8*, 545. [CrossRef] [PubMed]
19. Kadri, O.; Karmous, I.; Kharbech, O.; Arfaoui, H.; Chaoui, A. Cu and CuO Nanoparticles Affected the Germination and the Growth of Barley (*Hordeum vulgare* L.) Seedling. *Bull. Environ. Contam. Toxicol.* **2022**, *108*, 585–593. [CrossRef]
20. Gong, Q.; Wang, L.; Dai, T.; Zhou, J.; Kang, Q.; Chen, H.; Li, K.; Li, Z. Effects of copper on the growth, antioxidant enzymes and photosynthesis of spinach seedlings. *Ecotoxicol. Environ. Saf.* **2019**, *171*, 771–780. [CrossRef]
21. Li, B.; Dao, J.R.; Zhu, R.Y.; He, H.; Meng, X.Q.; Han, F.X. Distribution, accumulation and risk assessment of heavy metal pollution in Dianchi Lake. *Environ. Chem.* **2021**, *40*, 1808–1818. (In Chinese)
22. Zhu, J.P.; Wen, Y.H.; Ji, W.L.; Wang, Q. Exudates of *Impatiens uliginosa* Franch. on Its own seed germination. *Seed* **2020**, *39*, 41–46. (In Chinese)
23. Zhu, J.P.; Luo, C.; Li, Y.; Li, Q.M.; Zhao, Q.Y.; Wang, Q.; Huang, M.J.; Huang, H.Q. Effects of copper stress on seed germination and seedling growth of *Impatiens uliginosa*. *J. Biol.* **2023**, *40*, 64–68. (In Chinese)
24. Li, Y.; Wei, C.M.; Li, X.Y.; Meng, D.C.; Gu, Z.J.; Qu, S.P.; Huang, M.J.; Huang, H.Q. De novo transcriptome sequencing of *Impatiens uliginosa* and the analysis of candidate genes related to spur development. *BMC Plant Biol.* **2022**, *22*, 553. [CrossRef] [PubMed]
25. Wei, C.M.; Li, Y.; Meng, D.C.; Li, Z.F.; Li, Y.; Huang, M.J.; Huang, H.Q. Cloning and expression analysis of TIP genes related to spur development in *Impatiens uliginosa* Franch. *Plant Physiol. J.* **2022**, *58*, 1757–1765. (In Chinese)
26. Tan, Y.; Zhang, X.; Li, Q.M.; Li, X.Y.; Luo, L.; He, H.H.; Liang, G.R.; Huang, H.Q.; Huang, M.J. Transcriptomic Analysis of Flower Color Changes in *Impatiens uliginosa* in Response to Copper Stress. *Horticulturae* **2024**, *10*, 412. [CrossRef]
27. Huang, Q.; Guo, J.W.; Wang, Q.; Wen, Y.H.; Huang, M.J.; Huang, H.Q. Determination and analysis of metal elements content in four different colors of *Impatiens uliginosas*. *Heilongjiang Agric. Sci.* **2019**, *295*, 86–89. (In Chinese)
28. Shao, L.Y.; Guo, J.W.; Wang, Q. Determination of total nitrogen and total phosphorus in *Impatiens uliginosa* in the Dianchi Lake Watershed. *Environ. Sci. Surv.* **2021**, *40*, 36–37. (In Chinese)
29. Li, Q.M.; Li, W.X.; Cao, M.H.; Liu, S.; Zhang, T.Y.; Wang, Q.; Huang, M.J.; Huang, H.Q. Physiological and biochemical correlations between color of *Impatiens uliginosa* flower and nutrient supply on copper. *Fujian J. Agric. Sci.* **2021**, *36*, 1323–1329. (In Chinese)
30. Li, Q.M.; Li, W.X.; Li, X.Y.; Li, Y.; Qu, S.P.; Huang, M.J.; Huang, H.Q. Effects of copper stress on flower development and physiological and biochemical characteristics of *Impatiens uliginosa*. *Shandong Agric. Sci.* **2022**, *54*, 74–79. (In Chinese)
31. Dhindsa, R.S.; Plumb-dhindsa, P.; Thorpe, T.A. Leaf Senescence: Correlated with Increased Levels of Membrane Permeability and Lipid Peroxidation, and Decreased Levels of Superoxide Dismutase and Catalase. *J. Exp. Bot.* **1981**, *32*, 93–101. [CrossRef]
32. Giannopolitis, C.N.; Ries, S.K. Superoxide Dismutases I. Occurrence in higher plants. *Plant Physiol.* **1977**, *59*, 309–314. [CrossRef] [PubMed]
33. Sakharov, I.Y.; Ardila, G.B. Variations of peroxidase activity in cocoa (*Theobroma cacao* L.) beans during their ripening, fermentation and drying. *Food Chem.* **1999**, *65*, 51–54. [CrossRef]

4. Chaoui, A.; Mazhoudi, S.; Ghorbal, M.H.; Ferjani, E.E. Cadmium and Zinc Induction of Lipid Peroxidation and Effects on Antioxidant Enzyme Activities in Bean (*Phaseolus vulgaris* L.). *Plant Sci.* **1997**, *127*, 139–147. [CrossRef]
5. Tewari, R.K.; Kumar, P.; Sharma, P.N.; Bisht, S.S. Modulation of oxidative stress responsive enzymes by excess cobalt. *Plant Sci.* **2002**, *162*, 381–388. [CrossRef]
6. Majid, N.M.; Islam, M.M.; Justin, V.; Abdu, A.; Ahmadpour, P. Evaluation of heavy metal uptake and translocation by *Acacia mangium* as a phytoremediator of copper contaminated soil. *Afr. J. Biotechnol.* **2011**, *10*, 8373–8379. [CrossRef]
7. Xiang, Y.; Wang, X.; Song, W.; Du, J.; Yin, X. Integrative Omics Analyses Reveal the Effects of Copper Ions on Salvianolic Acid Biosynthesis. *Front. Plant Sci.* **2021**, *12*, 746117. [CrossRef]
8. Hossain, M.S.; Abdelrahman, M.; Tran, C.D.; Nguyen, K.H.; Chu, H.D.; Watanabe, Y.; Hasanuzzaman, M.; Mohsin, S.M.; Fujita, M.; Tran, L.P. Insights into acetate-mediated copper homeostasis and antioxidant defense in lentil under excessive copper stress. *Environ. Pollut.* **2020**, *258*, 113544. [CrossRef] [PubMed]
9. Hayat, K.; Khan, A.; Bibi, F.; Salahuddin; Murad, W.; Fu, Y.; Batiha, G.E.; Alqarni, M.; Khan, A.; Al-Harrasi, A. Effect of Cadmium and Copper Exposure on Growth, Physio-Chemicals and Medicinal Properties of *Cajanus cajan* L. (*Pigeon Pea*). *Metabolites* **2021**, *11*, 769. [CrossRef]
0. Yue, C.; Wang, Z.; Yang, P. Review: The effect of light on the key pigment compounds of photosensitive etiolated tea plant. *Bot. Stud.* **2021**, *62*, 21. [CrossRef]
1. Zunaidi, A.A.; Lim, L.H.; Metali, F. Comparative assessment of the heavy metal phytoextraction potential of vegetables from agricultural soils: A field experiment. *Heliyon* **2023**, *9*, e13547. [CrossRef]
2. Zhao, H.; Guan, J.; Liang, Q.; Zhang, X.; Hu, H.; Zhang, J. Effects of cadmium stress on growth and physiological characteristics of sassafras seedlings. *Sci. Rep.* **2021**, *11*, 9913. [CrossRef] [PubMed]
3. Zhang, L.; Gao, B. Effect of Isosteviol on Wheat Seed Germination and Seedling Growth under Cadmium Stress. *Plants* **2021**, *10*, 1779. [CrossRef] [PubMed]
4. Priyanka, N.; Geetha, N.; Manish, T.; Sahi, S.V.; Venkatachalam, P. Zinc oxide nanocatalyst mediates cadmium and lead toxicity tolerance mechanism by differential regulation of photosynthetic machinery and antioxidant enzymes level in cotton seedlings. *Toxicol. Rep.* **2021**, *8*, 295–302. [CrossRef]
5. Georgiadou, E.C.; Kowalska, E.; Patla, K.; Kulbat, K.; Smolińska, B.; Leszczyńska, J.; Fotopoulos, V. Influence of Heavy Metals (Ni, Cu, and Zn) on Nitro-Oxidative Stress Responses, Proteome Regulation and Allergen Production in Basil (*Ocimum basilicum* L.) Plants. *Front. Plant Sci.* **2018**, *9*, 862. [CrossRef]
6. Çelekli, A.; Kapı, E.; Soysal, Ç.; Arslanargun, H.; Bozkurt, H. Evaluating biochemical response of filamentous algae integrated with different water bodies. *Ecotoxicol. Environ. Saf.* **2017**, *142*, 171–180. [CrossRef] [PubMed]
7. Saud, S.; Wang, D.; Fahad, S.; Javed, T.; Jaremko, M.; Abdelsalam, N.R.; Ghareeb, R.Y. The impact of chromium ion stress on plant growth, developmental physiology, and molecular regulation. *Front. Plant Sci.* **2022**, *13*, 994785. [CrossRef] [PubMed]
8. Shangguan, L.; Fang, X.; Chen, L.; Cui, L.; Fang, J. Genome-wide analysis of autophagy-related genes (ARGs) in grapevine and plant tolerance to copper stress. *Planta* **2018**, *247*, 1449–1463. [CrossRef] [PubMed]
9. Chen, W. Effects of Heavy Metal Stress on Growth and Physiological Characteristics of Turf Plants. Ph.D. Thesis, Gansu Agricultural University, Lanzhou, China, 2014. (In Chinese).
0. Peláez-Vico, M.Á.; Fichman, Y.; Zandalinas, S.I.; Van Breusegem, F.; Karpiński, S.M.; Mittler, R. ROS and redox regulation of cell-to-cell and systemic signaling in plants during stress. *Free Radic. Biol. Med.* **2022**, *193*, 354–362. [CrossRef]
1. Salinitro, M.; Hoogerwerf, S.; Casolari, S.; Zappi, A.; Melucci, D.; Tassoni, A. Production of Antioxidant Molecules in *Polygonum aviculare* (L.) and *Senecio vulgaris* (L.) under Metal Stress: A Possible Tool in the Evaluation of Plant Metal Tolerance. *Int. J. Mol. Sci.* **2020**, *21*, 7317. [CrossRef]
2. Bellini, E.; Maresca, V.; Betti, C.; Castiglione, M.R.; Fontanini, D.; Capocchi, A.; Sorce, C.; Borsò, M.; Bruno, L.; Sorbo, S.; et al. The Moss *Leptodictyum riparium* Counteracts Severe Cadmium Stress by Activation of Glutathione Transferase and Phytochelatin Synthase, but Slightly by Phytochelatins. *Int. J. Mol. Sci.* **2020**, *21*, 1583. [CrossRef] [PubMed]
3. Saleem, M.H.; Fahad, S.; Rehman, M.; Saud, S.; Jamal, Y.; Khan, S.; Liu, L. Morpho-physiological traits, biochemical response and phytoextraction potential of short-term copper stress on kenaf (*Hibiscus cannabinus* L.) seedlings. *Peer J.* **2020**, *8*, e8321. [CrossRef] [PubMed]
4. Contreras, R.A.; Pizarro, M.; Köhler, H.; Sáez, C.A.; Zúñiga, G.E. Copper stress induces antioxidant responses and accumulation of sugars and phytochelatins in Antarctic *Colobanthus quitensis* (Kunth) Bartl. *Biol. Res.* **2018**, *51*, 48. [CrossRef] [PubMed]
5. Mahender, A.; Ali, J.; Prahalada, G.D.; Sevilla, M.A.L.; Balachiranjeevi, C.H.; Md, J.; Maqsood, U.; Li, Z. Genetic dissection of developmental responses of agro-morphological traits under different doses of nutrient fertilizers using high-density SNP markers. *PLoS ONE* **2019**, *14*, e0220066. [CrossRef] [PubMed]
6. Ma, X.; Fan, L.; Mao, F.; Zhao, Y.; Yan, Y.; Tian, H.; Xu, R.; Peng, Y.; Sui, H. Discrimination of three Ephedra species and their geographical origins based on multi-element fingerprinting by inductively coupled plasma mass spectrometry. *Sci. Rep.* **2018**, *8*, 10271. [CrossRef] [PubMed]
7. Uddin, M.M.; Zakeel, M.C.M.; Zavahir, J.S.; Marikar, F.M.M.T.; Jahan, I. Heavy Metal Accumulation in Rice and Aquatic Plants Used as Human Food: A General Review. *Toxics* **2021**, *9*, 360. [CrossRef] [PubMed]
8. Liu, Z.; Bai, Y.; Luo, L.; Wan, J.; Wang, W.; Zhao, G. Effects of high dose copper on plant growth and mineral nutrient (Zn, Fe, Mg, K, Ca) uptake in spinach. *Environ. Sci. Pollut. Res. Int.* **2021**, *28*, 37471–37481. [CrossRef] [PubMed]

59. Yang, Y.; Xie, J.; Li, J.; Zhang, J.; Zhang, X.; Yao, Y.; Wang, C.; Niu, T.; Bakpa, E.P. Trehalose alleviates salt tolerance by improving photosynthetic performance and maintaining mineral ion homeostasis in tomato plants. *Front. Plant Sci.* **2022**, *13*, 974507. [CrossRef] [PubMed]
60. Zhang, F.S.; Liu, N.X.; Wu, Y.M. Effect of Excessive Cu on the Nutrients Absorption of Sugar Beet. *Chin. Agric. Sci. Bull.* **2015**, *31*, 160–165. (In Chinese)
61. Chen, C.H.; Liu, Z.K.; Chen, G.C.; Zhang, J.F. Effects of copper stress on uptake of mineral elements by *Salix jiangsuensis* CL J^{-172} and *Salix babylonica* Linn. *J. Cent. South Univ. For. Technol.* **2011**, *31*, 28–32. (In Chinese)
62. He, T.T. Study on Physiological Response and Enrichment Ability of *Celosia argentea* and *Celosia cristata* to Copper Stress. Master's Thesis, University of Chinese Academy of sciences, Wuhan, China, 2013. (In Chinese).
63. Xu, G.D.; Ge, S.F.; Zhang, Y.; Wu, Y.H.; Mei, X.M.; Liu, P. Effect of salicylic acid on growth and nutrient uptake in Hydroponic Tobacco under cu stress. *Acta Agron. Sin.* **2015**, *41*, 956–962. (In Chinese) [CrossRef]
64. Lu, J.L. *Plant Nutrition*; China Agricultural University Press: Beijing, China, 2003. (In Chinese)
65. Dong, C.L. Physiological reaction of *Paeonia ostii* and *Paeonia suffruticosa* and the transcriptome analyses of *Paeonia ostii* in response to copper stress. Master's Thesis, Nanjing Agricultural University, Nanjing, China, 2013. (In Chinese).
66. Pätsikkä, E.; Kairavuo, M.; Sersen, F.; Aro, E.M.; Tyystjärvi, E. Excess copper predisposes photosystem II to photoinhibition in vivo by outcompeting iron and causing decrease in leaf chlorophyll. *Plant Physiol.* **2002**, *129*, 1359–1367. [CrossRef] [PubMed]

horticulturae

MDPI

Article

Variations in C:N:P Stoichiometry and Non-Structural Carbohydrates in Different Parts of Pomelo (*Citrus maxima*) Flowers at Three Phenophases

Jiali Liao [1], Shiyao Hu [1], Yiming Kong [1], Haohao Pan [1], Maoyuan Zhu [1], Ting Yu [2], Hongling Hu [1], Guoqing Zhuang [3,*] and Shun Gao [1,4,*]

[1] Department of Forestry, Faculty of Forestry, Sichuan Agricultural University, Chengdu 611130, China; liaojiali@stu.sicau.edu.cn (J.L.); 202201344@stu.sicau.edu.cn (S.H.); 202302070@stu.sicau.edu.cn (Y.K.); panhaohao@stu.sicau.edu.cn (H.P.); zhumaoyuan@stu.sicau.edu.cn (M.Z.); huhongling@sicau.edu.cn (H.H.)

[2] School of Life Science and Engineering, Southwest Jiaotong University, Chengdu 611756, China; yuting212916@my.swjtu.edu.cn

[3] Ecological Conservation, Restoration and Resource Utilization on Forest and Wetland Key Laboratory of Sichuan Province, Sichuan Academy of Forestry, Chengdu 610081, China

[4] Forest Ecology and Conservation in the Upper Reaches of the Yangtze River Key Laboratory of Sichuan Province & Sichuan Mt. Emei Forest Ecosystem National Observation and Research Station, Sichuan Agricultural University, Chengdu 611130, China

* Correspondence: zgq@sclky.com (G.Z.); shungao@sicau.edu.cn (S.G.)

Abstract: Carbon (C), nitrogen (N) and phosphorus (P), and non-structural carbohydrates (NSCs) are basic nutrients and energy sources for flower development. In this study, the morphological traits, C, N, P, and NSC concentrations, and C:N:P ratios in pistil, stamen, and petal of *C. maxima* flower at three phenophases (BBCH54, BBCH59, and BBCH61) were comparatively analyzed. Morphology diverged markedly among the three phenophases, whereas relative water contents were stable. C, N, P, and NSC showed larger variations at three phenophases and parts in *C. maxima* flower. Maximal C:N, C:P, and N:P occurred in pistils, pistils, and petals at BBCH61, respectively. C:N:P stoichiometry was the most responsive to ontogeny, indicating development-specific elemental storage and biomass partitioning of *C. maxima* flowers. NSC contents (glucose, fructose, sucrose, starch) differed significantly among organs and phenophases, and peak NSC appeared in the pistils at the three phenophases. High correlations between NSCs and C:N:P ratios suggested coordinated resource allocation. Correlation analysis showed that significant differences occurred at three phenophases for the accumulation and allocation of C, N, P, and NSCs. Principal component analysis (PCA) ordinated samples along PC-1 (44.2%) and PC-2 (24.4%), cumulatively explaining 68.6% of variance, corroborating development- and organ-dependent divergence. These data elucidated the intricate regulatory dynamics of nutrient contents among the three parts during the flower development of *C. maxima*, providing a robust quantitative framework for targeted nutrient management strategies.

Keywords: *Citrus maxima* flower; phenophases; non-structural carbohydrates; C:N:P stoichiometry; principal component analysis

1. Introduction

Carbon (C) and nitrogen (N) can provide carbon skeletons, amino- and nucleic-acid backbones, and phosphorus (P) is required for nucleic acids, ATP, and membrane lipids [1,2]. Their absolute contents and relative proportions respond plastically to genotype, ontogeny, and environmental conditions (temperature, rainfall, soil nutrients, etc.), which drive plant architecture, metabolism, and reproduction. C, N, and P stoichiometry can govern resource-acquisition strategies, growth rates, and ecosystem productivity [3,4]. During flower development, they are tightly coupled to dynamic re-allocation of C, N, and P among floral organs and to the concomitant remobilization of non-structural carbohydrates (NSCs) [5,6]. Recent investigations in *Hibiscus rosa-sinensis, Camellia sinensis, Juglans sigillata, Michelia maudiae* 'Rubicunda', and *Houpoea officinalis* flowers pronounced stage-dependent shifts in C:N:P ratios. These changes are more tightly correlated with developmental stages than those of organs [5,7–9]. Moreover, the C:N:P stoichiometric ratios are tightly coupled to the dynamics of NSC accumulation and partitioning, thereby exerting decisive control during the floral development [8,9]. These studies implicated nutrients' stoichiometric flexibility as an internal clock that synchronizes floral organ expansion, anther dehiscence, nectar secretion, and post-anthesis senescence [10,11].

NSCs, including fructose, glucose, sucrose, and starch, serve as the primary source of energy and carbon skeletons for plant metabolism and growth [12,13]. Beyond energetic roles, NSCs function as osmotic regulators and signaling molecules and modulate gene expression, hormone crosstalk, and developmental transitions in plants [14,15]. Studies have demonstrated that flower formation is highly dependent on the metabolism and availability of NSCs in flowering plants [16,17]. Carbohydrates drive cell division, expansion, and differentiation; provide C skeletons for the biosynthesis of secondary metabolites; and create the osmotic gradients necessary for petal expansion [8,12,18]. In Rosa hybrida petals, glucose and fructose levels rise dramatically during the opening phase, comprising up to 50% of the final dry mass and lowering osmotic potential [19]. Similar patterns of glucose and fructose content have been observed in various species during flower development, including *Asiatic lily, Michelia maudiae* 'Rubicunda', and *Houpoea officinalis*. In these species, stage-specific fluctuations in hexose levels are closely associated with inflorescence differentiation, pigment deposition, and the accumulation of secondary metabolites [8,9,20]. In woody perennials, starch reserves act as a buffer to reconcile the asynchronous supply and demand of carbohydrates. These reserves are remobilized to support flowering when the current photosynthate is insufficient [21]. Quantitative analysis of pistillate tissues at the time of pollination indicates that starch and soluble carbohydrates are strong predictors of subsequent fruit set and seed viability [7]. Collectively, these findings highlighted that the partitioning, interconversion, and utilization of NSCs must be precisely coordinated to meet the dynamically changing energy and C requirements throughout the entire process of flower development.

Flowers are the reproductive epicenters of angiosperms, driving propagation, genetic diversification, and evolutionary progress [19,22]. Structurally, they are reservoirs of carbohydrates, phenolic acids, flavonoids, mineral elements, and pigments, which support organogenesis, color development, and stress tolerance [19]. Flower ontogeny proceeds through tightly regulated phases: bud initiation, floral organ differentiation, anthesis, and senescence. A unified gene-signaling network controls the timing and progression of these stages [23]. This network integrates endogenous signals, including specifically phytohormone levels, carbohydrate status, transcription-factor activity, and external environmental signals, namely temperature, photoperiod, and water availability, ensuring that each developmental transition is synchronized with both internal physiology and external conditions [24,25]. Throughout these developmental stages, significant changes in biomass, morphometrics, coloration, mineral and NSC allocation, and secondary metabolites have been documented in *Michelia maudiae* 'Rubicunda,' *Houpoea officinalis Zingiber mioga*, and *Rosa damascena* [8,9,19,26]. Understanding their temporal distribution during flower development is imperative for defining nutrient-use efficiency and optimizing fertilization practices [1,25,27]. Thus, a mechanistic understanding of these processes enables growers to adjust cultural practices, thereby extending post-harvest longevity and enhancing the ornamental and commercial quality of these flowers.

Citrus maxima (Burm.) Merr. (*C. maxima*), commonly referred to as pomelo or "You-zi" in China, is an evergreen fruit tree that has been cultivated for over 3 000 years. It is now the world's most widely planted and economically important horticultural plant [28]. Pomelo is highly valued for its large, crisp, mildly acidic fruit and long post-harvest life, and is predominantly grown across subtropical and tropical Asia, which is the largest producer in terms of both harvested area and production volume [29]. The fruit is a rich source of sugars, organic acids, vitamin C, flavonoids, carotenoids, coumarins, and dietary fiber. These components endow the fruit with antioxidant, anti-inflammatory, hypoglycemic, and hypolipidemic properties [30,31]. Clinical and animal studies have consistently associated these bioactives with reducing risks of cardiovascular disease, type-2 diabetes, renal calculi, osteoporosis, and certain cancers [32]. Despite the extensive literature on pomelo fruit, relatively little attention has been given to pomelo flowers. *C. maxima* produces globose flowers arranged in terminal panicles, with annual anthesis occurring from late March to early April, accompanied by a pronounced floral fragrance [33]. Traditionally in China, C. maxima flowers are used to make fragrant tea drinks or are incorporated into cakes, herbal drinks, and casseroles. Recent analyses have revealed that *C. maxima* flowers contain substantial concentrations of phenolics, essential oils, vitamins, amino acids, and mineral elements with potent antioxidant and radical-scavenging activities [34]. Studies on *C. maxima* flowers have been extensively conducted within the food science domain, generating comprehensive datasets on their phytochemical composition, bioactivity, and safety profiles. These efforts are designed to attract both academic investigators and industry stakeholders [33]. Recently, an increasing number of researchers have paid attention to the dynamic changes in C, N, P, and element composition during flower development in *Rosa damascena*, *Michelia maudiae* 'Rubicunda', and *Houpoea officinalis* plants [8,9,19]. However, detailed analysis of the spatiotemporal dynamics of C, N, P, and NSC partitioning among the discrete floral organs of *C. maxima* throughout development remains notably scarce. Herein, the goals of this study were to illustrate the variations of morphometric traits, C:N:P stoichiometry, and NSC contents in pistils, stamens, and petals of *C. maxima* flowers at three phenophases (BBCH 54, BBCH59, and BBCH61). The present findings may provide critical insights into the developmental dynamics of C:N:P stoichiometry, together with NSC fluxes, across the discrete floral organs of *C. maxima*. This information also offered evidence-based insights for flowering management protocols, particularly precision fertilization strategies across the three discrete phenophases.

2. Materials and Methods

2.1. Study Site

The study site is located in Sichuan Agricultural University, District Wenjiang, Chengdu, China, and it is a flat alluvial plain with elevations ranging from 511.3 to 647.4 m above sea level. The site has a typical subtropical humid monsoon climate, with an average monthly temperature of 16.4 °C and a daily temperature range of 13–17 °C. The highest average temperatures occur in July, at 25.6 °C, while the lowest average temperature is in January, at 5.5 °C. The annual total precipitation is 985.1 mm, primarily concentrated in the summer months, accompanied by an average relative humidity of 81–84%. Solar radiation is moderate, with an annual sunshine duration of 1 104.5 h. The weather data were obtained from the China Meteorological Administration.

2.2. Flower Collection and Measurements

Five uniformly cultivated, 10-year-old *C. maxima* trees were meticulously selected and tagged for floral sampling. For each individual tree, the onset and cessation of anthesis were meticulously recorded daily. Observations conducted in 2024 indicated that the flowering period of *C. maxima* spans about 40 days, extending from early March to mid-April. The *C. maxima* flower consists of a pistil, stamens, and five petals. Development was classified according to the Biologische Bundesanstalt, Bundessortenamt und Chemische Industrie (BBCH) scale into three phenophases [35,36], including young bud stage (BBCH 54), first bloom stage (BBCH 59), and full bloom stage (BBCH 61). Fresh flowers were harvested from 09:00 to 11:00 h. Twenty flowers per phenophase were collected from each tree ($n = 60$ per stage). Ten flowers were randomly subsampled to determine morphometric indices, fresh mass (FM), dry mass (DM), and relative water content (RWC). After photographic documentation, each flower was dissected into pistil, stamens, and petals (Figure 1). Three parts were oven-dried at 65 °C for 24 h to constant mass, reweighed, and pulverized to pass through a 0.15 mm mesh prior to further analyses.

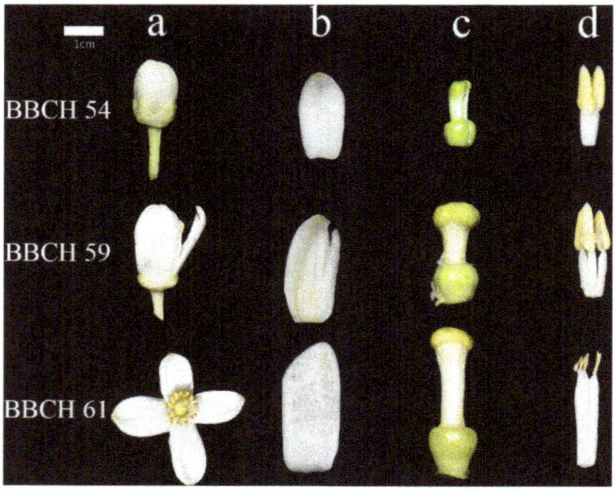

Figure 1. Morphology changes in three parts in the *C. maxima* flower at three phenophases. (**a**) Whole flower, (**b**) petals, (**c**) pistil, (**d**) stamen. BBCH54, young bud stage. BBCH59, first bloom stage. BBCH61, full bloom stage. Scale bar = 1 cm.

2.3. Determinations of C, N, and P Contents

Approximately 0.5 g of finely ground sample was precisely weighed into polytetrafluoroethylene (PTFE, Teflon®, Houston, TX, USA) digestion vessels. Subsequently, 10 mL concentrated HNO_3 and 2 mL of 30% (v/v) H_2O_2 were added to the vessels. The mixtures were then subjected to complete microwave-assisted digestion. After cooling to room temperature, the digest was transferred to a 25 mL volumetric flask and diluted to the volume with ultrapure (18.2 MΩ·cm) Milli-Q deionized water. C content was quantified by wet dichromate oxidation ($H_2SO_4/K_2Cr_2O_7$), followed by titration with $FeSO_4$. N content was determined using the Kjeldahl method. P content was measured using the molybdenum-antimony colorimetric procedure [5,9]. C, N, and P contents were expressed as mg per g dry mass (mg/g DM). C:N, N:P, and C:P ratios were calculated on a mass basis from the corresponding concentrations. Element accumulation (mg/per flower) in each part of *C. maxima* flower was calculated as the product of organ-specific elemental concentration (mg/g DM) and the corresponding dry mass (g). The proportional allocation of elements among reproductive organs was then expressed as Allocation ratio (%) = [Element accumulation in pistil, stamen, or petal (mg)/∑Element accumulation across all organs (mg)] ×100.

2.4. Determinations of NSC Contents

NSC contents were quantified with the anthrone–H_2SO_4 colorimetric protocol [8,9]. Approximately 0.2 g of material was extracted three times with 10 mL of 80% (*v*/*v*) ethanol. The suspensions were heated in a boiling-water bath for 30 min and then centrifuged at 5000× *g* for 10 min. The pooled supernatants were retained for the quantification of glucose, fructose, and sucrose. For glucose analysis, 0.1 mL of extract was reacted with 5 mL anthrone (0.2% *w*/*v* anthrone in 80% H_2SO_4) at 90 °C for 15 min, and absorbance was recorded at 620 nm after cooling to room temperature. For fructose analysis, identical volumes of extract and anthrone reagent were incubated at 25 °C for 90 min, and absorbance was recorded at 620 nm. For sucrose analysis, 0.1 mL of extract was hydrolyzed with 0.1 mL of 7.6 M KOH at 100 °C for 10 min. After cooling, 5 mL of anthrone was added, and the mixtures were heated at 90 °C for 15 min, and absorbance was then measured at 620 nm. For starch analysis, the extracted residues were suspended in 10 mL of 30% (*v*/*v*) perchloric acid and incubated at 80 °C for 10 min. The slurry was held overnight at 4 °C, then centrifuged at 4000× *g* for 10 min at 4 °C. The supernatants were collected and diluted to 50 mL with ultrapure water. Starch content was determined using the above anthrone procedure. The contents of glucose, fructose, sucrose, and starch were calculated from standard curves (Table S1) and expressed as mg g^{-1} dry mass (mg g^{-1} DM). NSC accumulation (mg/per flower) in each part of *C. maxima* flower was calculated as the product of organ-specific NSC concentration (mg g^{-1} DM) and the corresponding dry mass (g). The proportional allocation of NSCs among reproductive organs was calculated as follows: Allocation ratio (%) = [NSC accumulation in pistil, stamen, or petal (mg)/∑NSC accumulation across all organs (mg)] ×100.

2.5. Statistical Analysis

All data are presented as mean ± standard deviation (SD). Differences among phenophases and flower parts were assessed by one-way analysis of variance (ANOVA), followed by Duncan's multiple-range test at $p \leq 0.05$. Regression and Pearson correlation analyses were executed in SPSS 22.0 (SPSS Inc., Chicago, IL, USA). Additional data handling was performed in SPSS 22.0 and Microsoft Excel 2021 (Microsoft Corp., Redmond, WA, USA). Multivariate exploration of the dataset was conducted by principal component analysis (PCA) using the averaged values of each measured parameter across floral phenophases. Statistical significance was accepted at $p \leq 0.05$ for all tests. Figures were generated, and PCA was executed in Origin 2021 (OriginLab Corp., Northampton, MA, USA).

3. Results

3.1. Variations of Morphological Indicators and Biomass

Morphometric indices and biomass allocation across three parts of *C. maxima* flower were monitored at three distinct phenophases, as summarized in Tables 1 and S2. Pistil length (PL), stamen length (SL), petal length (PL), and petal width (PW) exhibited a progressive increase up to anthesis (BBCH61), reaching their respective maxima of 22.74 ± 0.37 mm, 19.38 ± 0.25 mm, 27.78 ± 0.38 mm, and 11.28 ± 0.97 mm (Table S1). The FM of pistil, stamen, and petal rose monotonically from BBCH54 to BBCH61, with values ranging from 0.34 ± 0.02 to 0.71 ± 0.10 g, 0.25 ± 0.02 to 0.43 ± 0.04 g, and 0.50 ± 0.09 to 1.27 ± 0.11 g, respectively (Table 1). The DM followed a similar trend, peaking at 0.19 ± 0.02 g, 0.08 ± 0.01 g, and 0.19 ± 0.02 g, respectively. The RWC in pistil and stamen declined steadily from

BBCH54 to BBCH61, reaching 72.04% and 81.31%, respectively. In contrast, petal RWC peaked at BBCH59 (86.63%) before a modest decrease at BBCH61.

Table 1. Growth indexes of three parts of *C. maxima* flower at three phenophases.

Organs		BBCH54	BBCH59	BBCH61
Pistil	Fresh mass (g)	0.34 ± 0.02 b	0.62 ± 0.04 a	0.71 ± 0.10 a
	Dry mass (g)	0.11 ± 0.01 c	0.16 ± 0.003 b	0.19 ± 0.02 a
	RWC (%)	74.37 ± 1.38 a	73.26 ± 0.41 a	72.04 ± 1.82 a
Stamen	Fresh mass (g)	0.25 ± 0.02 b	0.43 ± 0.01 a	0.43 ± 0.04 a
	Dry mass (g)	0.04 ± 0.03 b	0.08 ± 0.03 a	0.08 ± 0.01 a
	RWC (%)	83.49 ± 1.15 a	81.65 ± 1.27 a	81.31 ± 0.97 a
Petal	Fresh mass (g)	0.50 ± 0.09 c	0.83 ± 0.10 b	1.27 ± 0.11 a
	Dry mass (g)	0.08 ± 0.02 b	0.11 ± 0.01 b	0.19 ± 0.02 a
	RWC (%)	83.14 ± 0.54 b	86.63 ± 1.05 a	85.62 ± 0.99 a

Data are expressed as the means ± standard deviation (n = 3). Lowercase letters represent significant differences with a significance level of 0.05.

3.2. Variations in C, N, and P Contents

Figure 2 illustrates that both phenophases and floral parts significantly influence the C, N, and P contents in *C. maxima* flowers. In pistils and stamens, C contents ranged from 378.0 to 533.6 mg g^{-1} DM and 417.5 to 448.7 mg g^{-1} DM, respectively, with maxima recorded at BBCH61 (pistil) and BBCH54 (stamen). In contrast, C contents in the petals were relatively stable, ranging narrowly from 457.3 to 470.3 mg g^{-1} DM, peaking at BBCH59 (Figure 2a). As exhibited in Figure 2b, stamens exhibited the highest N concentrations (32.5–37.1 mg g^{-1} DM), followed by petals (27.1–30.9 mg g^{-1} DM) and pistils (25.1–29.7 mg g^{-1} DM). The maximal N contents were observed at BBCH57 (stamen), BBCH61 (petal), and BBCH61 (pistil), respectively. Figure 2c revealed that P contents are the highest in the stamens (0.889–1.18 mg g^{-1} DM), intermediate in the petals (0.768–0.843 mg g^{-1} DM), and the lowest in pistils (0.731–0.907 mg g^{-1} DM). Across three floral parts, P contents increased from BBCH54 to BBCH57 and then declined at BBCH61. Collectively, these findings indicated that C, N, and P contents in *C. maxima* flowers are closely associated with both developmental progression and organs.

3.3. Variations in C, N, and P Accumulation and Allocation Proportion

Figure 3 depicts the dynamic patterns of C, N, and P allocation fractions and accumulation within the pistil, stamens, and petals of *C. maxima* flowers across three phenophases (BBCH54, BBCH57, and BBCH61). Figure 3a shows that the proportion of C allocated to the pistil increased progressively from 29.2% (BBCH54) to 37.9% (BBCH61). Conversely, relative C allocation to stamens declined to 29.6%, while petal C allocation decreased to 32.5% by BBCH61. Figure 3b shows that N allocation in the pistil rose steadily from 28.2% to 31.9%, and N allocation in the petal increased from 30.4% to 33.2%. In contrast, N allocation in the stamen dropped significantly, from 41.4% (BBCH54) to 34.9% (BBCH61). In Figure 3c, P allocation in the pistil exhibits a pronounced increase, from 28.4% (BBCH54) to 32.8% (BBCH61). P allocation in the stamen declined from 41.7% to 35.2%. P allocation in the petal displayed a transient decrease at BBCH57 (29.9%) before recovering to 32.0% at BBCH61. As shown in Figure 3d–f, irrespective of the floral parts, C, N, and P accumulation in the flower escalated continuously with flower development, showing highly significant differences among phenophases. C, N, and P contents in whole flowers expanded from 311.8 to 639.4 mg/per flower, 21.5 to 42.2 mg/per flower, and 0.62 to 1.15 mg/per flower, respectively, underscoring pronounced stage-dependent nutrient accumulation.

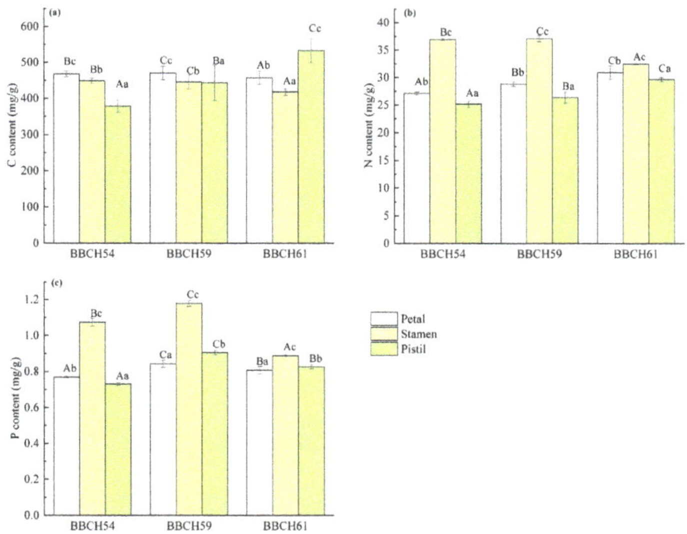

Figure 2. Variations in C (**a**), N (**b**), and P (**c**) contents of three parts in *C. maxima* flowers at three phenophases. Data are expressed as the means ± standard deviation (n = 3). Lowercase letters represent significant differences, with a significance level of 0.05. Lowercase letters represent significant differences at different stages, and capital letters represent significant differences in the three parts, with a significance level of 0.05.

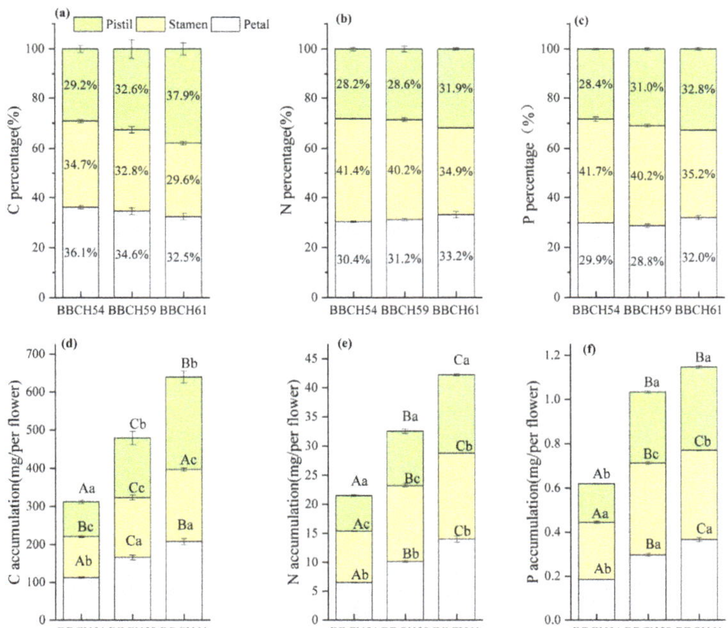

Figure 3. Variations in C, N, and P allocation ratios and accumulations of pistil, stamens, and petals in *C. maxima* flower at three phenophases. (**a**) C allocation ratios. (**b**) N allocation ratios. (**c**) P allocation ratio. (**d**) C accumulation. (**e**) N accumulation. (**f**) P accumulation. Data are expressed as the means ± standard deviation (n = 3). Lowercase letters represent significant differences, with a significance level of 0.05. Lowercase letters represent significant differences at different stages, and capital letters represent significant differences in the three parts, with a significance level of 0.05.

3.4. Variations in C, N, and P Stoichiometry

Figure 4 demonstrates that both phenophases and floral parts significantly influence the stoichiometric ratios of C, N, and P in *C. maxima* flowers. Across the three phenophases, C:N values ranged from 14.8 to 17.3 in the petals, 12.0 to 12.9 in the stamens, and 15.0 to 18.0 in pistils, with maxima recorded at BBCH54, BBCH61, and BBCH61, respectively (Figure 4a). Pistils consistently exhibited the highest C:N ratios relative to those of the stamens and petals. C:P ratios varied markedly among stages and organs: 558.7–608.8 (petals), 489.0–645.7 (pistils), and 377.9–469.8 (stamens). Petals displayed the highest C:P values at each phenophase (Figure 4b). N:P ratios ranged from 34.1 to 38.3 (petals), 29.1 to 35.9 (pistils), and 31.5 to 36.5 (stamens) across three phenophases (Figure 4c), peaking at BBCH61 (petals and pistils) and BBCH54 (stamens). Collectively, these findings indicate that the C:N, C:P, and N:P ratios in *C. maxima* flowers are dynamically regulated by development stage and organ-specific nutrient demands.

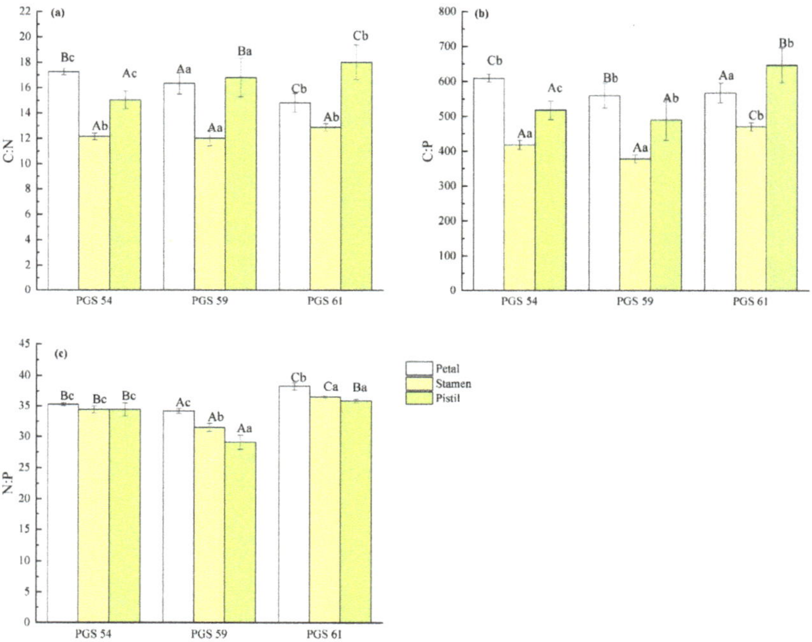

Figure 4. Variations in C:N (**a**), C:P (**b**), and N:P (**c**) ratios of pistil, stamens, and petals in *C. maxima* flower at three phenophases. Data are expressed as the means ± standard deviation ($n = 3$). Lowercase letters represent significant differences, with a significance level of 0.05. Lowercase letters represent significant differences at different stages, and capital letters represent significant differences in the three parts, with a significance level of 0.05.

3.5. Non-Structural Sugars (NSC) Contents

As depicted in Figure 5, the spatiotemporal distribution of NSCs in the pistil, stamens, and petals of *C. maxima* flowers exhibited significant divergence across the three phenophases. In Figure 5a, glucose fluctuated from 163.3 mg g^{-1} (BBCH 61) to 184.6 mg g^{-1} (BBCH 59) in the stamens, 136.9 mg g^{-1} (BBCH 61) to 214.3 mg g^{-1} (BBCH 54) in the petals, and 293.5 mg g^{-1} (BBCH 61) to 375.8 mg g^{-1} (BBCH 59) in the pistil. Figure 5b shows that fructose content reached maxima of 19.40 mg g^{-1} (stamens, BBCH 59), 29.82 mg g^{-1} (petals, BBCH 54), and 56.30 mg g^{-1} (pistil, BBCH 54). Figure 5c shows that sucrose contents ranged from 92.1 mg g^{-1} (BBCH 54) to 139.5 mg g^{-1} (BBCH 59) in the stamens, 68.6 mg g^{-1} (BBCH 61) to 114.8 mg g^{-1} (BBCH 59) in the petals, and 116.2 mg g^{-1} (BBCH 54) to 68.9 mg g^{-1}

(BBCH 61) in the pistil. In Figure 5d, starch contents peaked at 33.2 mg g^{-1} (stamens, BBCH 54), 30.4 mg g^{-1} (pistil, BBCH 61), and 37.2 mg g^{-1} (petals, BBCH 59). These patterns indicate a dynamic interplay of NSC partitioning, implying coordinated inter-conversion and translocation of soluble sugars among floral parts during *C. maxima* flower development.

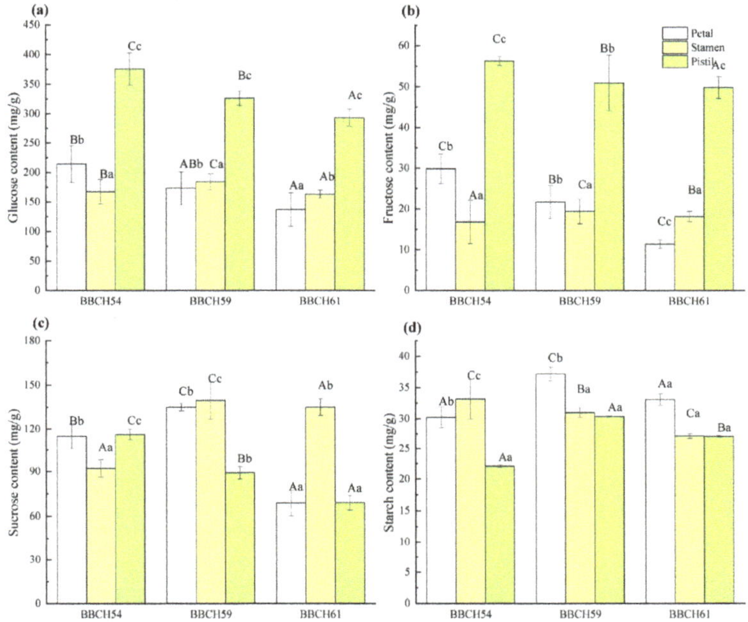

Figure 5. Variations in glucose (**a**), fructose (**b**), sucrose (**c**), and starch (**d**) contents of pistil, stamens, and petals in *C. maxima* flower at three phenophases. Data are expressed as the means ± standard deviation (n = 3). Lowercase letters represent significant differences, with a significance level of 0.05. Lowercase letters represent significant differences at different stages, and capital letters represent significant differences in the three parts, with a significance level of 0.05.

3.6. NSC Accumulation and Allocation Proportion

Figure 6 shows the development-dependent re-partitioning of NSCs within *C. maxima* flowers across the three phenophases. As shown in Figure 6a, stamens and petals received 22.1–27.5% and 23.1–28.3% of total glucose, respectively, with maxima at BBCH 61 and BBCH 54. The pistil retained a stable share (about 49%), indicating constitutive allocation. For fructose, allocation in the stamens rose progressively from 16.3% (BBCH 54) to 22.8% (BBCH 61), whereas petals exhibited a continuous decline from 29.0% to 14.4%. The pistil was the predominant sink for fructose partitioning, with a peak allocation of 62.8% at BBCH 61, which was significantly higher than 54.7% and 55.4% observed at BBCH 54 and BBCH 59, respectively (Figure 6b). As shown in Figure 6c, the sucrose allocation ratios of pistil, stamen, and petal fluctuated from 24.5 to 36.0%, 28.5 to 49.5%, and 25.2 to 37.1%, respectively, with maxima at BBCH 54, BBCH 61, and BBCH 59. As shown in Figure 6d, the starch allocation ratios in the pistil, stamen, and petal exhibited minimal change trends, ranging from 25.8% to 31.0%, 31.1% to 38.8%, and 35.4% to 37.9% respectively, with maxima appearing at BBCH54/BBCH59, BBCH54, and BBCH59/BBCH61, respectively. Total glucose, fructose, and starch increased monotonically from 182.5 to 269.5 mg g^{-1} (Figure 6e), 24.8 to 36.0 mg g^{-1} (Figure 6f), and 20.6 to 39.6 mg g^{-1} (Figure 6g), respectively. However, the accumulation of sucrose (128.3 mg/g, Figure 6h) in flowers at BBCH59 was significantly higher than that at BBCH54 and BBCH61.

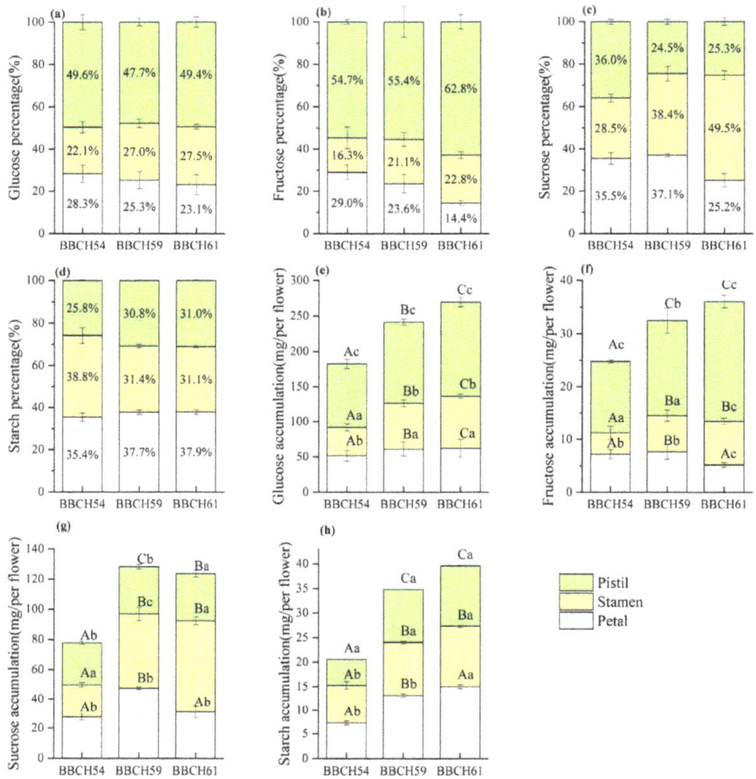

Figure 6. Variations in glucose, fructose, sucrose, and starch allocation ratios (**a–d**) and accumulations (**e–h**) of pistil, stamens, and petals in *C. maxima* flower at three phenophases. Data are expressed as the means ± standard deviation ($n = 3$). Lowercase letters represent significant differences, with a significance level of 0.05. Lowercase letters represent significant differences at different stages, and capital letters represent significant differences in the three parts, with a significance level of 0.05.

3.7. Correlation Analysis

As shown in Figure 7a, correlation analysis showed the associations between C:N:P stoichiometric ratios and NSC content in the stamens, petals, and pistils of *C. maxima* flowers across three phenophases. Specifically, glucose exhibited strong positive correlations with C contents and C:N ratio ($r \approx 0.87$), indicating that a high C availability coupled with nitrogen (N) limitation significantly promotes glucose accumulation. In contrast, N and P contents, as well as N:P and C:P ratios, showed pronounced negative correlations with fructose, sucrose, and starch ($r \approx -0.6$ to -0.8), suggesting that N and P sufficiency may inhibit synthesis or accelerate consumption of these NSCs. The negative impact of P is particularly notable, implying that P may constrain the interconversion between starch and soluble sugars by modulating phosphorylase activity. Collectively, C:N:P stoichiometry might act as a critical regulator of sugar metabolism, with a higher C:N ratio potentially enhancing glucose storage, while relative excesses of N and P may mitigate this process.

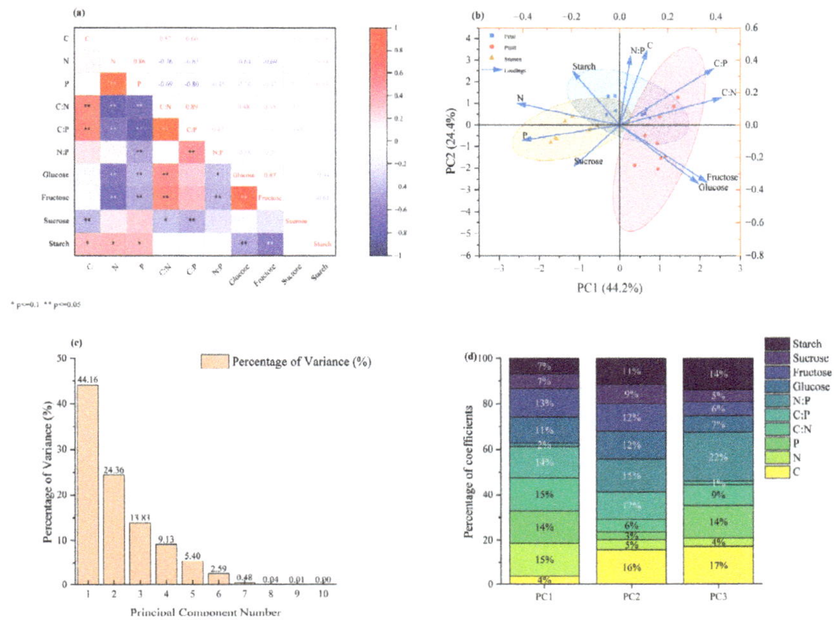

Figure 7. Correlation analysis and PCA of tested parameters in *C. maxima* flowers at three phenophases. (**a**) Correlation analysis of the test parameters. Correlation matrix showing significant *p*-values (<0.05; <0.01) of different tested parameters, where their color indicates the correlation slope (red Pearson's correlation coefficient = 1.0 and blue one = −1.0). Asterisks indicate significant differences: * $p < 0.05$, ** $p < 0.01$. (**b**) Distribution plots of PC1 and PC2. (**c**) The histogram from PC1 to PC10. (**d**) The proportion of the contribution of the tested parameters to PC1, PC2, and PC3. The locations of the variables in PC1 and PC2 are indicated by the direction and strength of the vector lines. The percentage of variation explained by each component is given next to the axis. The location of the trait in the diagram closest to the intersection of 0 on the X-axis (PC1) and Y-axis (PC2) shows similarity.

As shown in Figure 7b–d, PCA showed significant temporal dynamics and part-specific differences in tested indicators in *C. maxima* flowers across three phenophases. The PCA biplot delineated that the variables (C, N, P, NSC contents, and C:N:P ratios) were partitioned into four principal clusters. C content, C:N, C:P, and N:P ratios were positioned in the upper right quadrant (Figure 7b), signifying their positive correlation with development stages. Retaining the first three PCs, which collectively explained 82.4% of total inertia (PC1 = 44.2%, PC2 = 24.4%, PC3 = 13.8%), the ordination resolved discrete organ-specific chemotypes. Petals were positioned along positive PC1, reflecting a C-dominant signature enriched in glucose and fructose, coupled with elevated C:N and C:P ratios. Pistils were positioned in the quadrant defined by positive PC1 and negative PC2, indicating an intermediate C status and a significant N:P-driven reallocation toward starch accumulation. In contrast, stamens were projected onto the quadrant defined by negative PC1 and PC2, indicating a stoichiometric profile characterized by high N and P content, but low C content. As shown in Table S3, loading trajectories revealed a hierarchical and three-tier control of sugar–starch partitioning along the C:N:P stoichiometric gradient. Firstly, high positive loadings for C:N (0.423), C:P (0.391), glucose (0.327), and fructose (0.362), juxtaposed with negative loadings for N (−0.424) and P (−0.403), indicated a C surplus that may promote the accumulation of soluble sugars in *C. maxima* flower. Second, the coordinated positive loadings of N:P (0.427) and starch (0.331), contrasted with negative loadings for glucose (−0.360) and fructose (−0.351), suggested a stoichiometrically regulated transformation of soluble sugars into starch. Moreover, the antagonism between carbon (0.464) and N:P (−0.582) promoted overflow-induced starch sequestration under conditions of extreme stoichiometric imbalance. These findings underscored that C, N, P contents, C:N:P stoichiometric ratios, and NSCs in the three parts of *C. maxima* flowers were highly correlated with phenophases.

4. Discussion

Flower development constitutes an extremely intricate and multifaceted process, encompassing a wide range of physiological and biochemical alterations in plants [19,25]. Elucidating the diverse processes that maintain flower development is of paramount importance for enhancing the visual quality and longevity of flowers. *C. maxima*, a highly regarded ornamental and economically significant citrus species, exhibited intricate flower development processes that are intricately linked to its morphological traits and underlying biochemical dynamics [30,33]. Understanding the complex interrelationships among flower development, morphological traits, and C, N, and P dynamics is crucial for optimizing cultivation practices and enhancing the ornamental value of this plant. The present study documented the variations in morphometric traits, C:N:P stoichiometry, and NSC contents in the pistils, stamens, and petals of *C. maxima* flowers across three phenophases. Correlation analysis and PCA revealed significant temporal dynamics and organ-specific differences in the tested parameters across the three phenophases (Figure 7 and Table S3). This study provides crucial insights into the developmental dynamics of morphological parameters and C:N:P stoichiometry, along with NSC fluxes, across the discrete floral organs of *C. maxima*. Variations in morphological parameters, C:N:P stoichiometry, and NSC contents of flowers have been reported in *Styrax japonicus* and other ornamental species during the process of flower development [8,9,19,37]. These findings not only facilitate a deeper understanding of the variations in nutritional composition in *C. maxima* flowers during flower development but also offer evidence-based insights for flowering management protocols, particularly precision fertilization strategies across the three discrete phenophases.

C, N, and P, along with their stoichiometric ratios, are pivotal to modulating a wide array of physiological and biochemical processes and exert a substantial influence on flower development. These elements serve as crucial indicators of the nutrient status of plants and their ability to sequester resources, thereby affecting the allocation of resources between vegetative and reproductive growth [2,9]. The intricate interplay among C, N, and P in floral tissues is a fundamental aspect of plant reproductive biology [3,4]. Extensive studies have documented the variability in C, N, and P contents during flower development in several species, such as *Juglans sigillata*, *Cercis chinensis*, and olive [5,7,24]. The transition from vegetative to reproductive growth is a critical juncture where the balance of these nutrients is particularly important [9]. An elevated C:N ratio is often indicative of a transition from vegetative growth, which requires significant N inputs, to reproductive growth, where C storage becomes the predominant process for flower bud differentiation. This shift is essential for the successful differentiation of flower buds and the subsequent development of flowers [7]. Yang et al. (2024) [9] showed that the C:P ratio in different floral parts of *Houpoa officinalis* increased progressively with flower development. This suggests that as flowers progress from bud stages to flowering, the relative availability of P decreases compared to C, potentially affecting the energy and nutrient balance essential for reproductive success. The present study demonstrated that the dynamics of C, N, and P in *C. maxima* flowers are closely linked to developmental progression and organs (Figures 2–4). The variation in C, N, and P contents and stoichiometric ratios reflects the nutrient requirements and dynamic balances between development stages and floral parts, as demonstrated by the strong positive correlations with NSC contents (Figure 7a). These differential C, N, and P contents and allocation patterns observed in the pistils, stamens, and petals of *C. maxima* flowers have several direct and interrelated ramifications for flower development. One potential explanation is the progressive increase in absolute C, N, and P accumulation in the flower, thereby creating an increasingly larger metabolic sink (Figure 3d–f). Moreover, the fact that the share of C allocation to the stamen decreases while the share of C allocated to the pistil increases simultaneously (reaching 37.9% by BBCH61) suggests that the import of photoassimilates into the floral sinks speeds up, which occurs at the cost of maintaining the stamen. Another possible explanation is that by BBCH61, there is a steep decrease in the allocation of N (by 34.9%) and P (by 35.2%) to the stamen, while the shares of N and P in the pistil increase. This indicates the more energetically demanding processes of ovule development and seed set. In contrast, the stamens exhibit a decline in nutrient allocation, which may reflect a reduction in metabolic activity as the focus shifts to pistil development. The concept of metabolic sinks is particularly relevant in understanding these allocation patterns. As flowers develop, they become increasingly demanding metabolic sinks, attracting photoassimilates and nutrients from other parts of the plant [4,8,9]. Furthermore, the strong positive correlations between C, N, and P contents and NSC contents (Figure 7a) suggest that these elements play a crucial role in maintaining the energy balance within floral tissues. This interdependence highlights the importance of a balanced nutrient supply in supporting the energetic demands of flower development [22,23]. These findings highlighted that C, N, and P contents and their stoichiometric ratios are influenced by phenophases and floral parts, emphasizing the allocation and transport of these nutrients across compartments based on different physiological needs [38,39]. However, the molecular mechanisms that govern nutrient uptake, allocation, and transport during flower development in *C. maxima* still need to be elucidated.

NSCs serve as energy sources, osmotic regulators, and metabolic precursors, significantly contributing to flower bud differentiation and development. The deficiency of NSCs can have detrimental effects on flower development, leading to undersized petals or complete cessation of development [25,40,41]. In flowering plants, there are considerable variations in NSC content and type from floral bud development to blooming and senescence, which is accompanied by changes in color, morphological indices, and physiological and biochemical processes [12,16,19]. For example, in rose petals, the decrease in osmotic potential is primarily attributed to increased soluble carbohydrate content [42]. Numerous studies have shown that changes in NSC content are closely related to developmental stages and organ-specific requirements in flowers. In *Houpoa officinalis*, the concentrations of glucose, fructose, starch, and sucrose varied significantly across different floral parts (stamen, pistil, petal) and developmental stages [9]. Flower development is highly dependent on carbohydrate metabolism, as documented in species such as *Rosa damascena*, *Dendrobium crumenatum*, *Borago officinalis*, and *Centaurea cyanus* [19,22,43]. These reports highlight the role of NSCs in maintaining cellular turgor and facilitating processes such as petal expansion and scent emission. In our study, we observed gradual increases in soluble sugar and starch content across the three parts of *C. maxima* flowers (Figure 5). The accumulation and proportion patterns of NSCs varied significantly during the development process, and the allocation patterns among the three parts also showed significant differences (Figure 6). These variations may be due to several factors. One potential reason is that high petal glucose (28.3%, Figure 6e) and sucrose (37.1%, Figure 6g) levels, coupled with rising starch (35.4–37.9%, Figure 6h), generate the turgor and volatile precursors required for rapid corolla expansion and scent emission under variable light conditions, thereby maximizing pollinator encounter rates. Another potential reason is that a sharp rise in pistil fructose allocation (62.8%, Figure 6f) and continued starch accumulation in the pistil (30.4%, Figure 6h) create a buffered sink that sustains early endosperm proliferation even when current photosynthate is diverted to competing sinks. This may ensure that reproductive processes are not compromised by transient fluctuations in carbohydrate availability in *C. maxima* flowers. During Gladiolus flower development, total soluble sugars gradually increase, peaking after full bloom [23]. In *Centaurea cyanus*, glucose and fructose contents increase with flower development, while sucrose content does not show significant variability [22]. These changes in NSC contents and/or accumulation during flower development reflect the complex interplay between nutrient availability, developmental stage, and species-specific requirements. Moreover, highly significant correlations were observed between NSC variables and C:N:P stoichiometric variables in *C. maxima* flowers at three phenophases (Figure 7a). This is likely due to the dynamic balances of NSC storage, migration, conversion, and availability. These findings reveal a dynamic nutrient allocation trade-off between reproductive (pistils, stamens) and attractive (petals) parts, offering a comprehensive understanding of the NSC requirements and allocation patterns that underpin flower development of *C. maxima*.

5. Conclusions

In summary, the present work has demonstrated significant variations in C:N:P stoichiometry and NSC contents in *C. maxima* flowers, which are dependent on both the developmental stage and the floral parts. Elevated soluble-sugar contents seem to be essential for coordinating floral development, guaranteeing synchronized growth, differentiation, and development through optimized nutrient reallocation in *C. maxima* flowers. Correlation analysis and PCA further confirmed that nutrient stoichiometry and NSCs are closely associated with developmental stage and floral parts, highlighting the strong connection among phenophases, organ specificity, and nutrient dynamics. These findings not only enhance our understanding of nutrient allocation during flower development but also provide a robust theoretical framework for the estimation of nutrient status in *C. maxima* flowers. Given that flower development is controlled by a complex regulatory network, there are still uncertainties regarding the impacts of phenophases and/or floral parts on the variable C, N, P content and their ratios, as well as NSC contents in *C. maxima* flowers. These uncertainties call for further investigation, which drives morphogenesis, resource partitioning, and effective nutrient management during flower development in *C. maxima* plant via integrating physiological and biochemical analysis.

Supplementary Materials: The following supporting information can be downloaded at: https://www.mdpi.com/article/10.3390/horticulturae11091053/s1, Table S1: Standard curves of glucose, fructose, sucrose, and starch in this study. Table S2: Morphological indexes of three parts of *C. maxima* flower at three phenophases. Table S3: The first three principal component load and contribution rate of percentage of variance (%) in *C. maxima* flowers.

Author Contributions: Conceptualization, G.Z. and S.G.; methodology, J.L.; software, J.L.; validation, J.L., S.H. and Y.K.; formal analysis, J.L., S.H. and Y.K.; investigation, J.L.; resources, J.L.; data curation, J.L.; writing—original draft preparation, J.L., G.Z. and S.G.; writing—review and editing, J.L., S.H., Y.K., H.P., M.Z., T.Y., H.H., G.Z. and S.G. visualization, J.L., S.H. and Y.K.; supervision, G.Z. and S.G.; project administration, G.Z. and S.G.; funding acquisition, G.Z. and S.G. All authors have read and agreed to the published version of the manuscript.

Funding: This research received no external funding.

Data Availability Statement: The data presented in this study are available upon request from the corresponding author.

Acknowledgments: We are grateful to all of the group members and workers for their assistance in the field experiment.

Conflicts of Interest: The authors declare no conflicts of interest.

References

1. Baslam, M.; Mitsui, T.; Sueyoshi, K.; Ohyama, T. Recent Advances in Carbon and Nitrogen Metabolism in C3 Plants. *Int. J. Mol. Sci.* **2019**, *22*, 318. [CrossRef]
2. Krouk, G.; Kiba, T. Nitrogen and Phosphorus Interactions in Plants: From Agronomic to Physiological and Molecular Insights. *Curr. Opin. Plant Biol.* **2020**, *57*, 104–109. [CrossRef]
3. Ågren, G.I. Stoichiometry and Nutrition of Plant Growth in Natural Communities. *Annu. Rev. Ecol. Evol. Syst.* **2008**, *39*, 153–170. [CrossRef]
4. Sardans, J.; Rivas-Ubach, A.; Peñuelas, J. The C:N:P Stoichiometry of Organisms and Ecosystems in a Changing World: A Review and Perspectives. *Perspect. Plant Ecol. Evol. Syst.* **2012**, *14*, 33–47. [CrossRef]
5. Ren, H.Y.; Qian, W.Z.; Yi, L.; Ye, Y.L.; Gu, T.; Gao, S.; Cao, G.X. Nutrient Composition and Antioxidant Activity of *Cercis chinensis* Flower in Response to Different Development Stages. *Horticulturae* **2023**, *9*, 961. [CrossRef]
6. Jia, S.; Wang, Y.; Hu, J.; Ding, Z.; Liang, Q.; Zhang, Y.; Wang, H. Mineral and Metabolic Profiles in Tea Leaves and Flowers During Flower Development. *Plant Physiol. Biochem.* **2016**, *106*, 316–326. [CrossRef]
7. Zhang, W.; Li, J.; Zhang, W.; Njie, A.; Pan, X. The Changes in C/N, Carbohydrate, and Amino Acid Content in Leaves During Female Flower Bud Differentiation of *Juglans sigillata*. *Acta Physiol. Plant.* **2022**, *44*, 19. [CrossRef]
8. Yu, T.; Yang, Y.; Wang, H.; Qian, W.; Hu, Y.; Gao, S.; Liao, H. The Variations of C/N/P Stoichiometry, Endogenous Hormones, and Non-Structural Carbohydrate Contents in *Michelia maudiae* 'Rubicunda' Flower at Five Development Stages. *Horticulturae* **2023**, *9*, 1198. [CrossRef]
9. Yang, Y.; Qiu, Y.; Cheng, Y.; Yu, T.; Zhu, M.; Qian, W.; Gao, S.; Zhuang, G. Non-Structural Carbohydrate Content and C:N:P Stoichiometry in *Houpoea officinalis* Flowers in Response to Development Stages. *Horticulturae* **2024**, *10*, 784. [CrossRef]
10. Verlinden, S. Changes in Mineral Nutrient Concentrations in *Petunia Corollas* during Development and Senescence. *HortScience* **2003**, *38*, 71–74. [CrossRef]
11. Wang, T.-W.; Tan, J.; Li, L.-Y.; Yang, Y.; Zhang, X.-M.; Wang, J.-R. Combined Analysis of Inorganic Elements and Flavonoid Metabolites Reveals the Relationship between Flower Quality and Maturity of *Sophora japonica* L. *Front. Plant Sci.* **2023**, *14*, 1255637. [CrossRef]
12. Rolland, F.; Baena-Gonzalez, E.; Sheen, J. Sugar Sensing and Signaling in Plants: Conserved and Novel Mechanisms. *Ann. Rev. Plant Biol.* **2006**, *57*, 675–709. [CrossRef]
13. Mishra, B.S.; Sharma, M.; Laxmi, A. Role of Sugar and Auxin Crosstalk in Plant Growth and Development. *Physiol. Plant.* **2022**, *174*, e13546. [CrossRef]
14. Therby-Vale, R.; Lacombe, B.; Rhee, S.Y.; Nussaume, L.; Rouached, H. Mineral Nutrient Signaling Controls Photosynthesis: Focus on Iron Deficiency-Induced Chlorosis. *Trends Plant Sci.* **2022**, *27*, 502–509. [CrossRef]
15. Wang, Y.; Funnell, K.A.; Eason, J.R.; Morgan, E.R.; Woolley, D.J. Osmotic Regulation via Carbohydrate Metabolism Drives Petal Expansion and Floret Opening in Gentian 'Showtime Spotlight.' *Sci. Hortic.* **2016**, *211*, 19–25. [CrossRef]
16. Cho, L.H.; Pasriga, R.; Yoon, J.; Jeon, J.S.; An, G. Roles of Sugars in Controlling Flowering Time. *J. Plant Biol.* **2018**, *61*, 121–130. [CrossRef]
17. Norikoshi, R.; Imanishi, H.; Ichimura, K. Changes in Cell Number, Osmotic Potential and Concentrations of Carbohydrates and Inorganic Ions in Tweedia Caerulea during Flower Opening. *J. Jpn. Soc. Hortic. Sci.* **2013**, *82*, 51–56. [CrossRef]
18. Xu, J.; Li, Q.; Yang, L.; Li, X.; Wang, Z.; Zhang, Y. Changes in Carbohydrate Metabolism and Endogenous Hormone Regulation During Bulblet Initiation and Development in *Lycoris radiata*. *BMC Plant Biol.* **2020**, *20*, 180. [CrossRef]
19. Önder, S.; Tonguç, M.; Önder, D.; Erbaş, S.; Mutlucan, M. Flower Color and Carbohydrate Metabolism Changes During the Floral Development of Rosa Damascena. *S. Afr. J. Bot.* **2023**, *156*, 234–243. [CrossRef]
20. Roderick, B.; John, E.; Julian, H.; Allan, W. Flower opening in Asiatic lily is a rapid process controlled by dark-light cycling. *Ann. Bot.* **2000**, *86*, 1169–1174. [CrossRef]
21. Hedhly, A.; Vogler, H.; Schmid, M.W.; Pazmino, D.; Gagliardini, V.; Santelia, D.; Grossniklaus, U. Starch Turnover and Metabolism During Flower and Early Embryo Development. *Plant Physiol.* **2016**, *172*, 2388–2402. [CrossRef]
22. Fernandes, L.; Pereira, J.A.; Saraiva, J.A.; Ramalhosa, E.; Casal, S. Phytochemical Characterization of *Borago officinalis* L. and *Centaurea cyanus* L. during Flower Development. *Food Res. Int.* **2019**, *123*, 771–778. [CrossRef]
23. Jhanji, S.; Kaur, G.; Kaur, R.; Dhatt, U.K. Physiological and Biochemical Changes during Flower Development and Senescence in Chrysanthemum and Gladiolus. *Acta Physiol. Plant.* **2023**, *45*, 14. [CrossRef]

24. Ulger, S.; Sonmez, S.; Karkacier, M.; Ertoy, N.; Akdesir, O.; Aksu, M. Determination of endogenous hormones, sugars and mineral nutrition levels during the induction, initiation and differentiation stage and their effects on flower formation in olive. *Plant Growth Regul.* **2004**, *42*, 89–95. [CrossRef]
25. Borghi, M. Roles of Sugar Metabolism and Transport in Flower Development. *Curr. Opin. Plant Biol.* **2025**, *85*, 102722. [CrossRef]
26. Wei, S.; Liu, H.; Li, J.; Ren, T.; Xie, J. Metabolite Variations of Sugars, Organic Acids, Fatty Acids and Amino Acids in Flower Buds of *Zingiber mioga* Roscoe at Different Developmental Stages. *J. Food Compos. Anal.* **2023**, *116*, 105050. [CrossRef]
27. Önder, S.; Tonguç, M.; Erbaş, S.; Önder, D.; Mutlucan, M. Investigation of Phenological, Primary and Secondary Metabolites Changes during Flower Developmental of *Rosa damascena*. *Plant Physiol. Biochem.* **2022**, *192*, 20–34. [CrossRef]
28. Xiang, N.; Zhao, Y.; Zhang, B.; Gu, Q.; Chen, W.; Guo, X. Volatiles Accumulation during Young Pomelo (*Citrus maxima* (Burm.) Merr.) Fruits Development. *Int. J. Mol. Sci.* **2022**, *23*, 5665. [CrossRef] [PubMed]
29. Ding, S.; Wang, P.; Pang, X.; Zhang, L.; Qian, L.; Jia, X.; Chen, W.; Ruan, S.; Sun, L. The New Exploration of Pure Total Flavonoids Extracted from *Citrus maxima* (Burm.) Merr. as a New Therapeutic Agent to Bring Health Benefits for People. *Front. Nutr.* **2022**, *9*, 958329. [CrossRef] [PubMed]
30. Huang, S.; Dong, T.; Xiong, B.; Qiu, X.; Sun, G.; Liao, L.; Fan, N.; Wang, X.; Deng, H.; He, S.; et al. Variation in The Content and Composition of Limonoids in Fruits of Four Pomelo Varieties During Fruit Development: The Natural Debittering Process in Pomelo Fruits. *J. Food Compos. Anal.* **2021**, *100*, 103928. [CrossRef]
31. Zhao, Y.; Yang, X.; Hu, Y.; Gu, Q.; Chen, W.; Li, J.; Guo, X.; Liu, Y. Evaluation of Carotenoids Accumulation and Biosynthesis in Two Genotypes of Pomelo (*Citrus maxima*) during Early Fruit Development. *Molecules* **2021**, *26*, 5054. [CrossRef]
32. Sapkota, B.; Devkota, H.P.; Poudel, P. *Citrus maxima* (Brum.) Merr. (Rutaceae): Bioactive Chemical Constituents and Pharmacological Activities. *Evid. Based Complement. Altern. Med.* **2022**, *2022*, 8741669. [CrossRef]
33. Agustí, M.; Reig, C.; Martínez-Fuentes, A.; Mesejo, C. Advances in Citrus Flowering: A Review. *Front. Plant Sci.* **2022**, *13*, 868831. [CrossRef]
34. Zhang, X.; Zhuang, X.; Xiong, T.; Huang, S.; Qian, X.; Guo, T.; Chen, M.; Xie, W.; Huang, Y. Extraction From Different Parts of *Citrus maxima* Flowers Using Ultrasound as an Aid and Study of Their Composition and Function. *Ultrason. Sonochem.* **2023**, *100*, 106632. [CrossRef]
35. Fadón, E.; Herrero, M.; Rodrigo, J. Flower development in sweet cherry framed in the BBCH scale. *Sci. Hortic.* **2015**, *192*, 141–147. [CrossRef]
36. Qian, W.; Hu, Y.; Lin, X.; Yu, D.; Jia, S.; Ye, Y.; Mao, Y.; Yi, L.; Gao, S. Phenological Growth Stages of *Abelmoschus manihot*: Codification and Description According to the BBCH Scale. *Agronomy* **2023**, *13*, 1328. [CrossRef]
37. Chen, C.; Chen, H.; Ni, M.; Yu, F. A Study on Petal Morphological and Physiological Characteristics of *Styrax Japonicus* during the Flowering Period. *Agronomy* **2021**, *11*, 1498. [CrossRef]
38. Önder, D. Variation in antioxidant capacity, antioxidant activity and mineral composition during flower development of oil-bearing rose (*Rosa damascena* Mill.). *Sci. Rep.* **2023**, *13*, 17255. [CrossRef]
39. Simko, I.; Zhao, R. Phenotypic Characterization, Plant Growth and Development, Genome Methylation, and Mineral Elements Composition of Neotetraploid Lettuce (*Lactuca sativa* L.). *Front. Plant Sci.* **2023**, *14*, 1296660. [CrossRef] [PubMed]
40. Goetz, M.; Rabinovich, M.; Smith, H.M. The Role of Auxin and Sugar Signaling in Dominance Inhibition of Inflorescence Growth by Fruit Load. *Plant Physiol.* **2021**, *187*, 1189–1201. [CrossRef] [PubMed]
41. Shah, K.; Zhu, X.; Zhang, T.; Chen, J.; Chen, J.; Qin, Y. Transcriptome Analysis Reveals Sugar and Hormone Signaling Pathways Mediating Flower Induction in Pitaya (*Hylocereus polyrhizus*). *Int. J. Mol. Sci.* **2025**, *26*, 1250. [CrossRef] [PubMed]
42. Yamada, K.; Norikoshi, R.; Suzuki, K.; Imanishi, H.; Ichimura, K. Determination of Subcellular Concentrations of Soluble Carbohydrates in Rose Petals during Opening by Nonaqueous Fractionation Method Combined with Infiltration–Centrifugation Method. *Planta* **2009**, *230*, 1115–1127. [CrossRef] [PubMed]
43. Yap, Y.M.; Loh, C.S.; Ong, B.L. Regulation of Flower Development in *Dendrobium crumenatum* by Changes in Carbohydrate Contents, Water Status and Cell Wall Metabolism. *Sci. Hortic.* **2008**, *119*, 59–66. [CrossRef]

horticulturae

MDPI

Article

Genome-Wide Identification and Pollen-Specific Promoter Analysis of the *DIR* Gene Family in *Rosa chinensis*

Qijing Dong, Qian Yang, Zitong Wang, Yuan Zhao, Sixu Guo, Yifang Peng, Qi Li and Yu Han *

National Engineering Research Center for Floriculture, State Key Laboratory of Efficient Production of Forest Resources, Beijing Key Laboratory of Ornamental Plants Germplasm Innovation & Molecular Breeding, Beijing Laboratory of Urban and Rural Ecological Environment, Key Laboratory of Genetics and Breeding in Forest Trees and Ornamental Plants, Ministry of Education, School of Landscape Architecture, Beijing Forestry University, Beijing 100083, China; irene3230502@bjfu.edu.cn (Q.D.); yangqian@bjfu.edu.cn (Q.Y.); wangzitong467@bjfu.edu.cn (Z.W.); zhaoyuan1012@bjfu.edu.cn (Y.Z.); guosixu02@bjfu.edu.cn (S.G.); pengyifang@bjfu.edu.cn (Y.P.); liqiecho0819@bjfu.edu.cn (Q.L.)

* Correspondence: hanyubjfu@bjfu.edu.cn

Abstract: Dirigent proteins (DIRs) are pivotal regulators of lignin/lignan biosynthesis and play multifaceted roles in plant development and stress adaptation. Despite their functional significance, *DIR* genes remain unexplored in *Rosa chinensis*, a globally important woody ornamental species. This study identified 33 *RcDIRs* through whole-genome analysis, including their chromosomal distribution, phylogenetic relationships, collinearity, protein and gene structure, conserved motifs, and cis-acting element distribution, and classified them into three phylogenetically independent subgroups (DIR-a, DIR-b/d, and DIR-e). Notably, the DIR-e subgroup includes an exclusive tandem cluster comprising RcDIR7-RcDIR12, representing the largest lineage-specific RcDIR expansion in *R. chinensis*. Structural characterization revealed that most *RcDIRs* exhibit a conserved single-exon architecture. Promoter *cis*-element analysis uncovered abundant stress-/hormone-responsive elements and three pollen-specific motifs (AAATGA, POLLEN1LELAT52, GTGANTG10), with *RcDIR12* from the DIR-e cluster showing high pollen-specific regulatory potential. Experimental validation included cloning the *RcDIR12* promoter from *R. chinensis* 'Old Blush', constructing proRcDIR12::*GUS* vectors, and conducting histochemical GUS assays with pollen viability/DAPI staining in transgenic *Arabidopsis*. Histochemical assays demonstrated GUS activity localization in mature trinucleate pollen grains, marking the first experimental evidence of pollen-specific DIRs in rose. Our findings not only elucidate the *DIR* family's genomic organization and evolutionary innovations in *R. chinensis* but also establish proRcDIR12 as a molecular tool for manipulating pollen development in plants.

Keywords: dirigent (DIR) gene family; *Rosa chinensis*; cis-regulatory elements; pollen-specific promoter; tandem gene cluster

1. Introduction

Rosa chinensis, a perennial woody plant of the genus *Rosa*, family *Rosaceae*, is known for its unique ornamental characteristics and significant economic value [1]. The modern *R. chinensis* is an interspecific hybrid developed in recent years, combining rich ornamental traits and agronomic significance, and is widely cultivated as an ornamental plant worldwide [2]. In plant sexual reproduction, normal pollen development is critical for achieving successful pollination and fruit set [3], while pollen viability and maturity directly determine the success rate of *R. chinensis* hybrid breeding. Studies demonstrate that pollen viability is a key factor influencing crossbreeding efficiency and fruiting rates in *R. chinensis*, with hybridization success rates significantly improving as pollen viability increases [4,5]. Although the importance of pollen in *R. chinensis* crossbreeding is widely acknowledged, research on pollen-specific promoters in this species remains scarce. Transcriptional regulation, a major mechanism of gene expression control, is coordinated by cis-acting elements (e.g., core promoters and enhancers) and trans-acting factors (e.g., transcription factors and co-regulators) [6]. In plants, pollen-specific promoters are essential for studying pollen development. For example, in *Triticum aestivum*, the pollen-specific promoter of Pollen-Specific Gene 076 (*PSG076*) was isolated and shown to drive GUS expression exclusively in late bicellular pollen to mature pollen stages and pollen tubes in transgenic tobacco, with no activity detected in other tissues [7]. In *Oryza sativa*, two late-stage pollen-specific promoters, pollen late-stage promoter 1 (PLP1) and pollen late-stage promoter 2 (PLP2), identified via a stable transformation system, activated GUS expression solely during late pollen development. Their activity and specificity were governed by specific promoter sequences and enhancer motifs such as 'AGAAA' and 'CAAT' [8].

Lignans are primarily found in conifers, while lignin is widely distributed in vascular plants. Dirigent proteins (DIRs) play a key role in the biosynthesis of lignin and lignans in plant cell walls, although their specific functions vary among different plants. DIRs were first identified in *Forsythia intermedia*, where they stereoselectively direct the coupling of two pinoresinol radical intermediates to form (+) or (−) pinoresinol, a key precursor for lignan biosynthesis [9,10]. To date, DIRs have been studied in various plants. For instance, in *Zea mays*, *ZmDIR11* is upregulated under drought stress, and its silencing or mutation significantly reduces drought tolerance in maize, indicating its positive regulatory role in the drought stress response [11]. In *Vitis vinifera*, *VvDIR4* enhances resistance to anthracnose in both *Arabidopsis* and grape by activating the SA/JA signaling pathway and promoting pathogen-induced lignin biosynthesis [12]. In *Setaria italica*, 38 *SiDIR* genes have been identified, with high expression in root tissues and responses to salt, heavy metals, and osmotic stress, suggesting their involvement in stress resistance and root development [13]. A total of 420 *DIR* genes have been identified in different rice species (cultivated and wild), with a significant expansion of members in cultivated rice. Their expression is responsive to various biotic and abiotic stresses and is enriched in roots, implying their potential role in environmental adaptation [14].

Studies have shown that dirigent proteins play an irreplaceable role in plant growth and development, cell structure maintenance, stress resistance enhancement, and secondary metabolism, especially in the lignin biosynthetic pathway [15]. Pollen walls are mainly composed of spore proteins and lignin; lignin may enhance the structural stability of pollen walls through interactions with spore proteins [16,17]. Although the functions of dirigent proteins have been explored in many studies, research on their promoters is still relatively scarce. In particular, the functions of DIRs in rose pollen development have not yet been revealed. In rose, pollen-specific promoters enable the precise genetic manipulation of male fertility traits without disrupting floral morphology—an essential feature for maintaining ornamental value. Furthermore, these promoters facilitate the study of the lineage-specific pollen development mechanisms underlying recurrent sterility in rose breeding populations. Screening for *RcDIRs* that are specifically expressed in pollen may provide important clues for uncovering their potential functions. This study aims to identify DIR genes in the rose genome and analyze their phylogeny, gene structure, and cis-acting elements, with the goal of thoroughly elucidating the rose *DIR* gene family and focusing on predicting the pollen-specific elements of *RcDIR* promoters. Based on this, a key *RcDIR*, named RcDIR12, was successfully predicted. Using the ancient rose cultivar 'Old Blush' as the material, the promoter sequence of *RcDIR12* was cloned, and a pBI121-proRcDIR12::*GUS* expression vector was constructed. Through the genetic transformation of *Arabidopsis thaliana*, transgenic lines were obtained, and the specific expression sites of proRcDIR12 were explored. The above studies provide important reference information for further revealing the function of the *RcDIR12* gene in rose pollen development and the application of the proRcDIR12 promoter.

2. Materials and Methods

2.1. Materials and Reagents

R. chinensis 'Old Blush' was grown in a greenhouse at Beijing Forestry University (Beijing, China, E 116°20′, N 40°0′) under the conditions of 25 °C day/18 °C dark, 12 h light/12 h dark, and 8000 lx light intensity. Wild-type *A. thaliana* (Columbia-0 ecotype) was grown in a growth chamber at 22 ± 2 °C with a photoperiod of 16 h light/8 h dark.

2.2. Identification and Chromosomal Localization of DIR Genes in R. chinensis

Twenty-five *A. thaliana* DIR sequences were downloaded from The *Arabidopsis* Information Resource (TAIR) website (http://www.arabidopsis.org/, accessed on 19 December 2024). Protein files of *R. chinensis* were obtained from the *R. chinensis* genome browser (https://lipm-browsers.toulouse.inra.fr/pub/RchiOBHm-V2/, accessed on 4 December 2024). The conserved domain of the DIR family protein (PF03018) was retrieved from the Pfam database (http://pfam-legacy.xfam.org/, accessed on 4 December 2024), and Hidden Markov Model (HMM) analysis and Protein Basic Local Alignment Search Tool (BLASTP) screening were performed. Specifically, HMMER search was conducted with an *E*-value threshold of 1×10^{-5} and a similarity threshold of >50%, while BLASTP was performed with an *E*-value threshold of 1×10^{-5} and an identity threshold of 50%. Initially, 34 *RcDIRs* were identified. Combining domain analysis and manual verification to remove proteins with incomplete domains, a total of 33 RcDIR gene family members were ultimately identified. The physicochemical properties of RcDIRs, including their number of amino acids, molecular weight, theoretical isoelectric point, instability index, hydropathy index, and aliphatic index, were calculated using the ExPASy tool (https://web.expasy.org/protparam/, accessed on 30 December 2024). Additionally, the subcellular localization of RcDIRs was predicted using the WoLF PSORT website (https://wolfpsort.hgc.jp/, accessed on 30 December 2024), and the chromosomal locations of *RcDIRs* were analyzed and visualized using TBtools-II (v2.210) software [18–20].

2.3. Phylogenetic Tree Construction and Collinearity Analysis of RcDIRs

The multiple sequence alignment of RcDIR and AtDIR sequences was performed using MEGA software (v11.0.13) [21]. Based on the alignment results, a phylogenetic tree of the DIR gene family was constructed using the Maximum Likelihood (ML) method, with parameters set to perform 1000 bootstrap replicates to assess the reliability of the tree topology. Subsequently, the phylogenetic tree was refined and annotated using iTOL (v7) (https://itol.embl.de/, accessed on 11 February 2025) [22]. The collinearity regions between the *A. thaliana* and *R. chinensis* genomes were analyzed and visualized using TBtools-II (v2.210) software.

2.4. Analysis of Gene Structure, Conserved Domains, Protein Structure, and Promoter Cis-Acting Elements of RcDIRs

The sequences of RcDIRs were aligned using the Jalview tool (v2.11.4.0) with the Muscle method [23]. Conserved motifs were analyzed using the MEME online tool (https://meme-suite.org/, accessed on 20 February 2025), with the number of motifs set to 10 and other parameters kept at the default values [24]. The gene structure features of *RcDIR* genes, including exon and intron distribution, were analyzed using the GSDS platform (v2.0) (http://gsds.cbi.pku.edu.cn/, accessed on 26 December 2024) [25]. The conserved motifs and gene structures of RcDIRs were visualized using TBtools-II (v2.210) software. We input the complete amino acid sequence of each target RcDIR. We use the default prediction mode of AlphaFold3 (https://alphafoldserver.com/, accessed on 24 March 2025) [26], which integrates sequence information and multiple sequence alignment (MSA) data. The accuracy of the structure is assessed using the pLDDT (predicted Local Distance Difference Test) and pTM (predicted Template Modeling score) provided by AlphaFold3, specifically determined as follows: very high (plDDT > 90), confident (90 > plDDT > 70), low (70 > pLDDT > 50), and very low (pLDDT < 50). A pTM score greater than 0.7 indicates higher confidence. After prediction, the structures were optimized and visualized using PyMOL (v2.6.0) software to clearly illustrate the overall structural features and important functional regions of the proteins [27]. To comprehensively analyze the promoter sequences of the *RcDIR* gene family, two different lengths of promoter regions were extracted from the genomic sequences. First, the nucleotide sequences 2000 bp upstream of each gene were extracted as promoter regions for the analysis of cis-acting regulatory elements. These promoter sequences were analyzed for the types, quantities, and functions of cis-acting elements using the PlantCARE website (https://bioinformatics.psb.ugent.be/webtools/plantcare/html, accessed on 19 February 2025) [28], and the results were graphically presented using R (v4.2.1).

To further investigate the regulatory elements associated with pollen-specific expression, promoter sequences 1200 bp upstream of the transcription start site were extracted from the genomic sequences of *RcDIR* gene family members. Based on the genomic annotation information, the orientation (sense or antisense strand) of each promoter sequence was determined. Subsequently, pollen-specific regulatory elements such as AGAAA and GTGA were specifically identified within these promoter sequences. The number of pollen-specific elements and their distribution on the sense and antisense strands were counted for each promoter sequence, and the results were compiled into a table.

2.5. Promoter Cloning and Expression Vector Construction

Genomic DNA was extracted from young leaves of *R. chinensis* 'Old Blush' using a plant DNA extraction kit (D6943-01, Omega Bio-Tek, Norcross, GA, USA). Based on the sequence information of the RcDIR12 gene (RchiOBHmChr2g0159671) provided by the *R. chinensis* genome database (https://lipm-browsers.toulouse.inra.fr/pub/RchiOBHm-V2/, accessed on 25 February 2025), primers for promoter cloning were designed (proRcDIR12-U: TGATTACGCCAAGCTTCTGACATATTAGCATTTGCTG; proRcDIR12-L: GACCAC-CCGGGGATCCCCAATTCAAAACAACATTGTA) to amplify the proRcDIR12 sequence. The specific promoter sequence was detailed in Table S1. The PCR system was performed according to the instructions of TOYOBO KOD-Plus (KOD-401, TOYOBO, Osaka, Japan). The PCR program was as follows: 94 °C for 5 min, 94 °C for 30 s, 56 °C for 30 s, and 72 °C for 1 min 30 s, for 32 cycles, and there was a 72 °C extension for 2 min. The PCR product was detected by 1% agarose gel electrophoresis, and the target band was recovered and purified using an Agarose Gel DNA Recovery Kit (DP209-02, Tiangen, Beijing, China) according to the instructions. The recovered product was ligated with pEASY®-Blunt (CB101-01, TransGen, Beijing, China), transformed into Trans1-T1 competent cells (CD501-02, TransGen, Beijing, China), and identified by PCR. The positive recombinant plasmid was sequenced by SinoGenoMax Company Limited (Beijing, China) and named pEASY-proRcDIR12. *Hind* III (R0104, NEB, lpswich, MA, USA) and *Bam*H I (R0136, NEB, MA, USA) restriction enzymes were used to digest the recombinant plasmid pEASY-proRcDIR12 and the target vector pBI121 (stored in our laboratory). The digestion products were recovered and purified using the Agarose Gel DNA Recovery Kit (DP209-02, Tiangen, Beijing, China). The promoter fragment was ligated with the target vector using T4 DNA ligase (EL0011, Thermo

Scientific™, Waltham, MA, USA) according to the instructions. The ligation product was transformed into Trans1-T1 *E. coli* competent cells, and positive clones were sequenced to obtain the recombinant expression vector pBI121-proRcDIR12::*GUS*.

2.6. Agrobacterium-Mediated Stable Genetic Transformation in A. thaliana

The recombinant vector pBI121-proRcDIR12::*GUS* was introduced into Agrobacterium tumefaciens AGL0 competent cells (AC1060, Weidibio, Shanghai, China). Positive colonies were identified by colony PCR, and selected colonies were cultured to an optical density at 600 nm (OD_{600}) of 0.6 to 0.8. The recombinant Agrobacterium suspension was used to infect wild-type *A. thaliana* (Columbia, Col) using the floral dip method [29]. Seeds from the T_0 generation were sown on MS medium containing kanamycin (Kan, 50 mg·L^{-1}). After germination for 1 week, seedlings were transplanted to soil and grown for 4 weeks. Leaves were collected and genomic DNA was extracted using the Omega Tissue DNA Kit (D3396-02, Omega Bio-Tek, GA, USA). RT-PCR was performed to detect the presence of the transgene using the following primers: U: ATTTCACCCTTCAGATGG, and L: GTGTAGTTTGGGCTGGTT. The PCR system was conducted according to the instructions of TB Green® Premix Ex Taq™ II (RR420Q, Takara, Tokyo, Japan). The PCR program was 94 °C for 5 min, 94 °C for 30 s, 51 °C for 30 s, and 72 °C for 1 min, for 28 cycles, and 72 °C for 2 min. Seeds from positive plants were harvested and propagated to the T_3 generation. Promoter GUS staining analysis was performed on T_3 generation seedlings.

2.7. GUS Staining of the Transgenic A. thaliana

The experimental materials included 45-day-old wild-type *A. thaliana* and transgenic *A. thaliana* expressing the pBI121-proRcDIR12::GUS construct at 15, 25, and 45 days old. The GUS staining solution was prepared according to the instructions provided with the GUS staining kit (G3060, Solarbio, Beijing, China). The plant materials to be tested were immersed in the GUS staining solution and incubated at 37 °C in the dark for 24 h. Subsequently, the samples were decolorized with a gradient ethanol solution for 48 h, and the expression of GUS in the plants was observed and recorded.

2.8. Pollen Viability Staining (I-KI Method)

Anthers from stage 13 flowers of wild-type and transgenic *A. thaliana* were placed on a slide, and one drop of ddH_2O was added [30]. The anthers were gently crushed with forceps to release pollen. One to two drops of I-KI staining solution (SL72602, Coolaber, Beijing, China) were added to completely submerge the pollen in the staining solution, as described in the product manual. The staining of pollen was observed under a Leica MZ10 F stereomicroscope (Leica Microsystems GmbH, Wetzlar, Germany). Pollen grains that appeared blue or black indicated strong viability; pollen grains that appeared brownish-yellow indicated some viability; and pollen grains that were colorless indicated no viability or sterility.

2.9. DAPI Staining

This section was conducted according to reference [31]. The DAPI working solution was prepared by mixing the stock solution (D9542, Sigma-Aldrich, Saint Louis, MO, USA) with buffer solution at a ratio of 1:1000 to 1:10,000. The DAPI working solution was placed on a slide, and flowers at stage 13 from both wild-type and transgenic *A. thaliana* were directly and evenly dipped into the staining solution. After the pollen grains fell into the staining solution, a cover slip was placed on top, and the slide was kept in the dark for 2 min. The development of pollen nuclei was observed using a Leica 2245 inverted fluorescence microscope (CKX941, Leica Microsystems GmbH, Wetzlar, Germany). DAPI staining allowed for a clear observation of the two sperm nuclei and one vegetative nucleus in normally mature pollen grains, thereby determining the developmental stage of the pollen nuclei.

3. Results

3.1. Identification and Physicochemical Property Analysis of the RcDIR Gene Family

A total of 33 *RcDIR* genes were identified in this study and were named *RcDIR1* to *RcDIR33* based on their chromosomal locations (Table 1). The length of amino acids in RcDIRs varied significantly, ranging from 167 to 413 with an average length of 203.55. The relative molecular weight (expressed in kD) of RcDIRs ranged from 18.67 to 42.57, with an average of 22.10 kD. Isoelectric point (pI) analysis showed that the pI values of RcDIRs ranged from 4.37 to 9.79, with an average of 7.66. Among them, 18 were basic proteins (pI > 7), and 15 were acidic proteins (pI < 7). The instability index (Instability Index Analysis, IIA) of RcDIRs ranged from 18.78 to 48.94, with an average value of 32.77. Only 4 RcDIRs (RcDIR1, RcDIR12, RcDIR17, and RcDIR21) were predicted to be unstable proteins, while the remaining 29 were stable proteins. The grand average of the hydropathicity (GRAVY) values of 25 RcDIRs were all greater than zero, indicating hydrophobic properties. Subcellular localization prediction results showed that more than 50% of RcDIRs were likely to be localized in the chloroplasts (Table 1). CDS sequences of the 33 *RcDIRs* identified in this study were presented in Table S1.

3.2. Chromosomal Distribution and Gene Cluster Formation of RcDIRs

Chromosomal localization analysis revealed that the 33 *RcDIR* genes are distributed across seven chromosomes of *R. chinensis* (Figure 1). The highest number of *RcDIRs*, a total of eight, are found on chromosome 2. Chromosomes 1, 3, and 5 each harbor four *RcDIRs*, while chromosome 4 contains two *RcDIRs*. Additionally, 29 of the *RcDIRs* (87.9% of the total) are organized into gene clusters on the chromosomes. Chromosome 1 carries two gene clusters, and chromosome 6 contains the largest tandem repeat cluster, which comprises six members (*RcDIR7–RcDIR12*). Chromosome 4 includes only one gene cluster, whereas chromosomes 2, 3, 5, and 7 each possess gene clusters with three to five members.

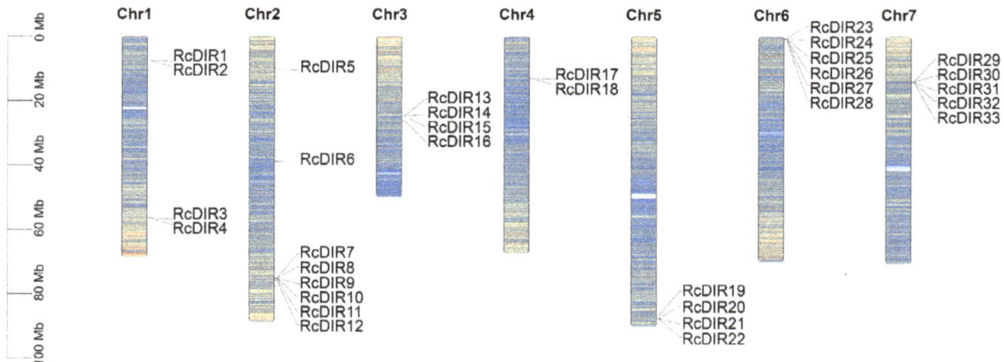

Figure 1. Chromosomal localization map shows the positions of *RcDIRs*. The labeled genes are arranged from top to bottom on the *R. chinensis* chromosomes, with chromosome numbers indicated at the top of each chromosome. The colors in the figure represent gene density, with lighter colors indicating low gene density and darker colors indicating high gene density.

Table 1. Identification and physicochemical properties of the 33 RcDIRs in *R. chinensis*.

Gene Name	Gene ID	Number of Amino Acid	Molecular Weight (kDa)	Theoretical pI	Instability	Aliphatic Index	Grand Average of Hydropathicity	Subcellular Localization
RcDIR1	RchiOBHmChr1g0319651	184	20.19	8.76	42.01	83.26	0.127	extracellular
RcDIR2	RchiOBHmChr1g0319661	187	20.76	5.92	34.56	88.07	0.159	extracellular
RcDIR3	RchiOBHmChr1g0365391	413	42.57	4.37	39.71	87.36	−0.012	vacuole
RcDIR4	RchiOBHmChr1g0365401	293	30.83	5.05	32.08	82.87	0.024	chloroplast
RcDIR5	RchiOBHmChr2g0097361	189	20.37	9.52	20.92	94.92	0.072	chloroplast
RcDIR6	RchiOBHmChr2g0125291	309	32.04	5.08	27.57	82.69	−0.058	chloroplast
RcDIR7	RchiOBHmChr2g0158831	177	19.25	9.33	39.47	95.31	0.119	chloroplast
RcDIR8	RchiOBHmChr2g0158841	131	14.36	9.52	28.73	93.74	0.051	chloroplast
RcDIR9	RchiOBHmChr2g0158861	205	22.21	8.85	30.27	94.63	0.307	chloroplast
RcDIR10	RchiOBHmChr2g0158891	179	19.58	9.13	24.79	94.69	0.134	chloroplast
RcDIR11	RchiOBHmChr2g0158911	178	19.28	6.10	32.49	96.97	0.206	extracellular
RcDIR12	RchiOBHmChr2g0159671	180	19.54	9.12	37.30	85.00	0.093	chloroplast
RcDIR13	RchiOBHmChr3g0478991	194	21.01	9.75	25.25	97.01	0.134	chloroplast
RcDIR14	RchiOBHmChr3g0479011	170	18.45	9.79	48.94	76.82	−0.106	nucleus
RcDIR15	RchiOBHmChr3g0479031	197	21.40	9.59	41.76	78.68	0.043	cytoplasm
RcDIR16	RchiOBHmChr3g0480301	184	20.83	6.95	33.55	86.36	0.024	chloroplast
RcDIR17	RchiOBHmChr4g0396611	250	25.81	5.81	41.60	100.24	0.274	chloroplast
RcDIR18	RchiOBHmChr4g0396621	249	25.58	5.01	36.07	89.00	0.117	extracellular
RcDIR19	RchiOBHmChr5g0080641	175	19.66	9.39	34.10	96.34	−0.025	chloroplast
RcDIR20	RchiOBHmChr5g0081441	196	21.70	7.96	36.46	93.01	0.048	chloroplast
RcDIR21	RchiOBHmChr5g0081451	196	21.72	6.97	40.35	92.55	0.033	vacuole
RcDIR22	RchiOBHmChr5g0081461	194	21.52	6.59	30.45	92.94	0.059	vacuole
RcDIR23	RchiOBHmChr6g0244551	191	21.50	8.93	37.37	87.7	0.108	extracellular
RcDIR24	RchiOBHmChr6g0244561	186	19.80	5.43	21.10	86.51	0.114	extracellular
RcDIR25	RchiOBHmChr6g0244571	233	25.34	6.65	38.33	82.45	−0.072	chloroplast
RcDIR26	RchiOBHmChr6g0244581	187	19.99	5.55	24.13	85.61	0.082	chloroplast
RcDIR27	RchiOBHmChr6g0244591	191	21.10	9.46	31.66	76.60	−0.165	chloroplast
RcDIR28	RchiOBHmChr6g0244601	192	21.29	9.78	29.25	81.30	−0.140	chloroplast
RcDIR29	RchiOBHmChr7g0195991	184	20.98	5.91	34.24	76.3	−0.037	extracellular
RcDIR30	RchiOBHmChr7g0196001	180	20.22	6.17	32.34	83.5	0.163	chloroplast
RcDIR31	RchiOBHmChr7g0196011	185	20.64	8.49	22	89.14	0.105	chloroplast
RcDIR32	RchiOBHmChr7g0196031	191	21.26	9.21	18.78	81.2	0.005	chloroplast
RcDIR33	RchiOBHmChr7g0196041	167	18.67	8.56	33.72	81.26	0.043	chloroplast

3.3. Phylogenetic Clustering, Structural Prediction, and Collinearity Analysis of the RcDIR Family

A phylogenetic tree was constructed using DIRs from both *A. thaliana* and *R. chinensis*. The analysis revealed that the 33 RcDIRs and 25 AtDIRs were clustered into three distinct subgroups: DIR-a, DIR-b/d, and DIR-e (Figure 2A). Different subgroups of dirigent proteins are known to have unique biological functions. Studies have shown that the DIR-a subgroup, which includes FiDIR1 from *F. intermedia* [9] and AtDIR5/AtDIR6 from *A. thaliana* [32], plays a key role in lignan biosynthesis through the stereoselective 8-8′ coupling of coniferyl alcohol. The DIR-b/d subgroup is involved in the enantioselective bimolecular coupling of coniferyl alcohol radicals and enantioselective oxidative coupling of sesquiterpene phenols such as (+)- or (−)-gossypol [33,34].

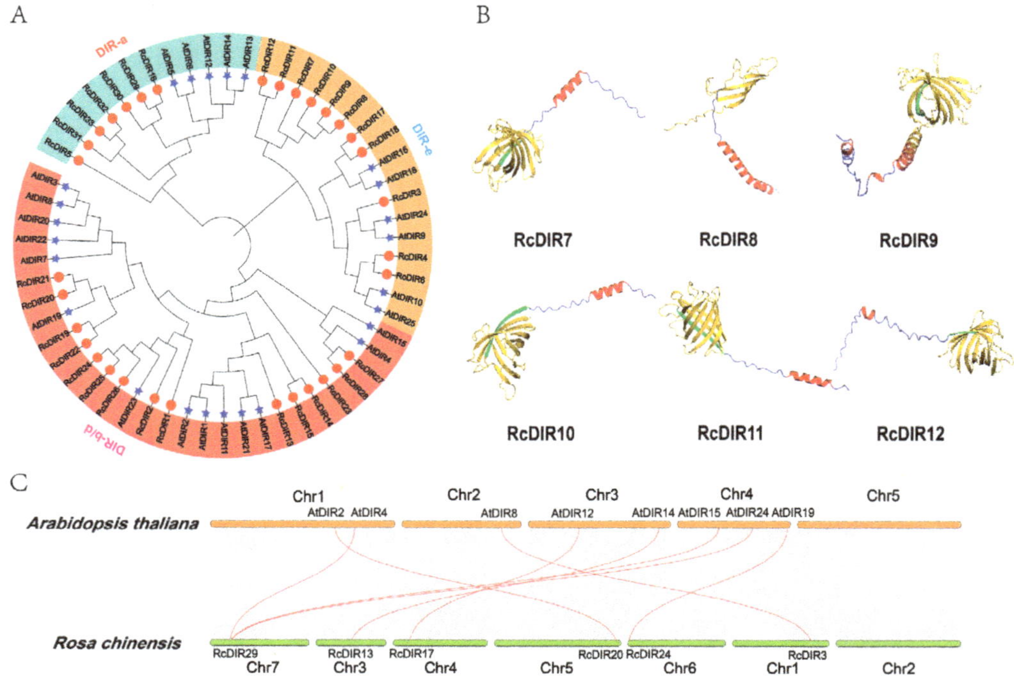

Figure 2. Phylogenetic evolution, protein structure, and collinearity analysis of the DIR family genes in *R. chinensis*. (**A**) Phylogenetic tree of the DIR family genes from *R. chinensis* and *A. thaliana*. Subgroups DIR-a, DIR-b/d, and DIR-e are indicated with green, orange-red, and yellow backgrounds, respectively. Specific colors are used to distinguish between different species: AtDIRs are marked with blue pentagons, and RcDIRs are marked with red circles. (**B**) Three-dimensional structural models of the *RcDIR* family members and their DIR domains. Red represents α-helices, green represents β-sheets, and gold represents the DIR domain. (**C**) Collinearity analysis of *DIR* genes between *R. chinensis* and *A. thaliana*. The gray lines in the background represent collinear pairs in the genomes of *R. chinensis* and *A. thaliana*; red lines represent collinear *DIR* gene pairs. "Chr1-7" denotes chromosome numbers.

The DIR-a subgroup comprises seven *R. chinensis* members and five *A. thaliana* members, with some RcDIRs forming conserved clusters with the key lignin biosynthesis genes AtDIR5 and AtDIR6 from *A. thaliana*. The DIR-b/d subgroup is the largest, containing 15 RcDIRs and 14 AtDIRs, accounting for 50.0% of all members. The DIR-e subgroup includes 11 RcDIRs and six AtDIRs, with a significantly higher proportion of *R. chinensis* genes compared to *A. thaliana*. Subgroups DIR-c, DIR-f, and DIR-g were not detected in this study, with DIR-c being a subgroup specific to monocots [35].

It is worth noting that the DIR-e family is divided into two groups, one of which is RcDIR7-RcDIR12, and these genes are also located in the same gene cluster according to the chromosomal localization results. Figure 2B shows the secondary structure elements of the RcDIR7-12 proteins and their spatial distribution. Each protein is composed of a single polypeptide chain, with a highly conserved core folding pattern. All proteins exhibit a mixed secondary structure composed of α-helices (red), β-sheets (green), and random coils. The DIR domain (gold) is located in the N-terminal to the central region, forming a continuous gold block.

The collinearity analysis results indicate (Figure 2C) that there are 8 collinear pairs between the 33 DIRs of *R. chinensis* and the 25 DIRs of *A. thaliana*. Among them, *RcDIR13* is collinear with *AtDIR14*, *RcDIR17* with *AtDIR12*, *RcDIR20* with *AtDIR2*, *RcDIR24* with *AtDIR19*, and *RcDIR13* with *AtDIR8*. Notably, *RcDIR29* from *R. chinensis* shows collinearity with three different genes in *A. thaliana* (*AtDIR4*, *AtDIR15*, and *AtDIR24*).

3.4. Conservation of Domains and Gene Structure Analysis of RcDIRs

As shown in Figure 3A, the motif analysis of the *RcDIR* gene family reveals that members generally contain four to six motifs, whose distribution patterns are closely related to the phylogenetic grouping. Based on the topology of the phylogenetic tree, grouping is performed through the first major bifurcation, dividing the RcDIRs into four main groups, within which genes exhibit similar motif patterns. For instance, in the first group, all members except RcDIR19 contain motifs 1 through 5. Motifs 1, 2, and 3 are found in all four main groups, indicating that these motifs are widely distributed and highly conserved in RcDIR sequences. Multiple sequence alignments further reveal the conservation patterns of these motifs: motifs 1, 2, and 3 are universally present in all members (Figure 3B). However, certain motifs such as motif 8 and motif 9 are only found in specific groups, that is, the DIR-b/d subgroup; motif 7 is only present in the DIR-a subgroup; and motifs 6 and 10 are specifically found in the DIR-e subgroup. Additionally, apart from a few exceptions like *RcDIR19* and *RcDIR20*, which contain two exons and one intron, the other *RcDIRs* only include one exon and lack introns, exhibiting a typical DIR gene structure (Figure 3A) [36].

3.5. Classification and Distribution of Cis-Acting Elements in the Promoters of RcDIRs

In this study, a total of 57 functional elements were identified and categorized into four major functional modules: light response (28 types), abiotic stress response (13 types), biotic stress response (9 types), and plant growth and development regulation (7 types) (Figure 4). The heatmap analysis results (left side of Figure 4) show that within the light response module, core regulatory elements such as G-Box, Box4, and I-box are widely distributed among *RcDIR* family members. Abiotic stress response elements mainly include Abscisic Acid Response Elements (ABREs), Antioxidant Response Elements (AREs), and MYB Transcription Factor Binding Sites (MBS I). In contrast, the distribution of elements in the biotic stress response and plant growth and development modules is relatively sparse, with no significant clustering observed. The element count statistics (right side of Figure 4) further confirm this distribution pattern, with the highest proportion of light response elements found in genes such as *RcDIR21*, *RcDIR13*, and *RcDIR2*, followed by abiotic stress, and the lowest proportion in biotic stress and growth and development.

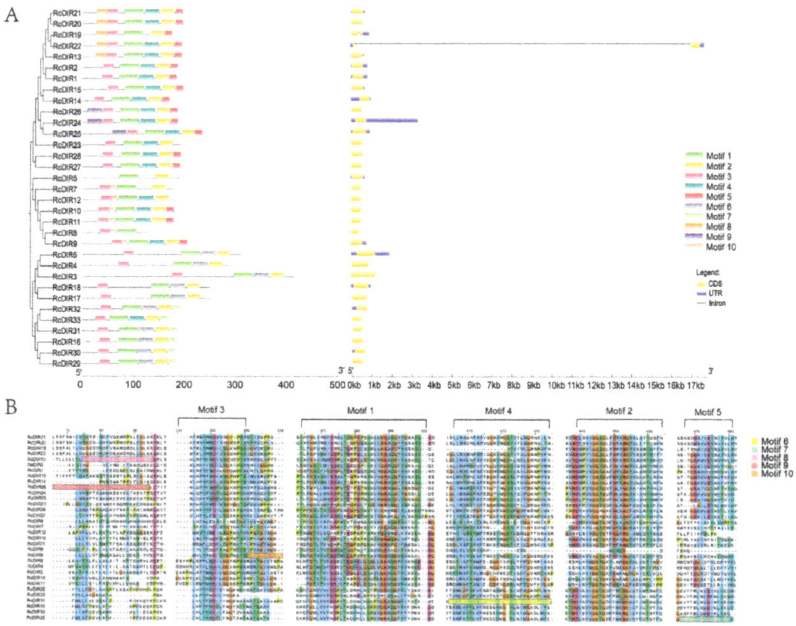

Figure 3. Structural characterization analysis of the RcDIR gene family in *R. chinensis*. (**A**) Phylogenetic tree, conserved motifs, and exon–intron structure of RcDIRs. Each motif (motif 1 to motif 10) is represented by a distinct color. A scale bar is shown at the bottom of the figure, and gene names are labeled on the left side, arranged according to their subgroup affiliations. Yellow bars represent exons, lines represent introns, and blue bars represent non-coding regions. (**B**) Amino acid sequence alignment and conservation analysis of DIR family members in *R. chinensis*. Each row in the figure represents the amino acid sequence of a family member, with sequence names listed on the left. Letters in different colors indicate amino acids of different properties: red for hydrophobic amino acids, blue for hydrophilic amino acids, green for acidic amino acids, and yellow for basic amino acids. Multiple conserved motifs (motif 1 to motif 10) are highlighted with colored boxes, which are highly conserved among different family members.

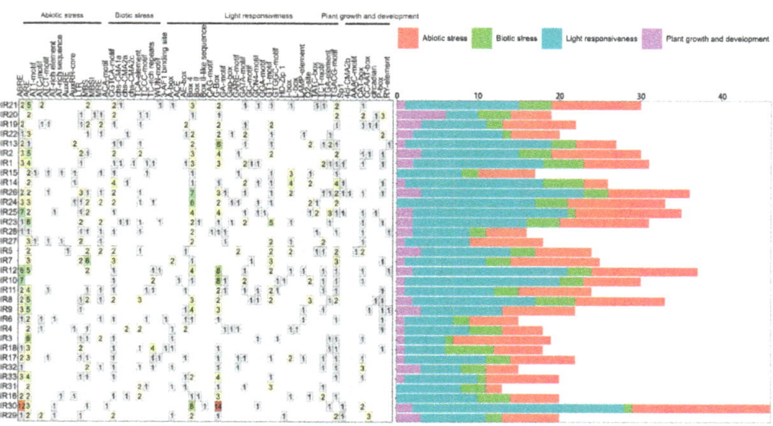

Figure 4. Analysis of cis-acting elements in the promoter regions of *RcDIR* gene family members. The table on the left side lists the presence of specific cis-acting elements in the promoter regions of each gene family member, with the shade of color indicating the frequency of occurrence of each element. The types of elements include abiotic stress (red), biotic stress (green), light responsiveness (cyan), and plant growth and development elements (purple).

3.6. Identification of Three Pollen-Specific Elements in the Promoters of the DIR Gene Family in R. chinensis

A systematic analysis was conducted to explore the distribution characteristics and potential functional associations of pollen-specific cis-acting elements in the promoter regions of *RcDIR* gene family members. Pollen-specificity was defined by the exclusive ability of these elements to drive expression in pollen tissues with undetectable activity in all non-pollen tissues. The results revealed the presence of three pollen-specific cis-acting elements in the promoter regions of the *RcDIR* gene family: AAATGA, POLLEN1LELAT52, and GTGANTG10 [37–39] (Table S2). The distribution of these three elements among the promoter regions of *RcDIR* gene family members showed significant variation. Taking the promoter of *RcDIR12* as an example, it was identified to contain one AAATGA element, six POLLEN1LELAT52 elements, and nine GTGANTG10 elements. All *RcDIR* gene family members contained the POLLEN1LELAT52 element, and all but *RcDIR13* and *RcDIR14* contained the GTGANTG10 element. However, only a few members contained the AAATGA element, and they generally included only one of either AAATGA or TCATTT. Notably, the promoter of *RcDIR24* contained the highest number of AAATGA elements, totaling seven; it also contained the highest number of GTGANTG10 elements, with sixteen. *RcDIR6* and *RcDIR17* had the highest number of POLLEN1LELAT52 elements, with 13 each. These differences suggest that the composition of pollen-specific cis-acting elements in the promoter regions of different members may be closely related to their roles in pollen development and function.

3.7. Specific Expression of proRcDIR12::GUS in Pollen of Transgenic A. thaliana

A recombinant expression vector, pBI121-proRcDIR12::*GUS*, was constructed (Figures S1 and S2). *A. thaliana* was infected using the floral dip method, and T_3 transgenic plants were identified as positive by PCR (Figure S3). GUS histochemical staining results showed that, compared to wild-type *A. thaliana*, 45-day-old transgenic plants only stained at the flower bud site (Figure 5A,B), and only 25-day-old transgenic flower buds displayed a blue color (Figure 5C,D); other parts did not stain.

Microscopic analysis was carried out to localize the specific staining sites within the flower buds. Further refined GUS staining experiments on flowers from wild-type and proRcDIR12::*GUS* transgenic *A. thaliana* revealed that the proRcDIR12 promoter drove GUS expression in the anthers of 45-day-old transgenic plants (Figure 6A,B). Specifically, immature anthers in the bud did not turn blue (Figure 6C), anthers in semi-open flowers showed a slight blue color (Figure 6D), anthers in fully open flowers turned dark blue, and scattered pollen also appeared blue (Figure 6E). Additionally, when the siliques began to develop, the tips of the young siliques of proRcDIR12::*GUS* transgenic plants, which are the location of the original stigma, also turned blue due to pollen adhesion (Figure 6F,G).

3.8. Specific Expression of proRcDIR12::GUS in Mature Trinucleate Pollen

To investigate the expression of proRcDIR12::*GUS* in pollen, pollen viability staining and DAPI staining were performed on pollen at different developmental stages of transgenic *A. thaliana*. The results showed that most pollen grains of the transgenic *A. thaliana* were viable (Figure 7A–C), and pollen at the mononuclear, binuclear, and trinucleate stages could be distinguished (Figure 7D–F). GUS staining revealed no expression in mononuclear- and binuclear-stage pollen (Figure 7G,H), while blue coloration was observed in mature trinucleate pollen (Figure 7I), indicating that the GUS gene is specifically expressed in trinucleate pollen. In summary, proRcDIR12 can drive the specific expression of GUS in mature trinucleate pollen.

Figure 5. Tow-type *Arabidopsis* GUS staining (Bar = 1 cm). (**A**) Wild-type plants; (**B**) 45-day-old proRcDIR12::*GUS* transgenic *Arabidopsis*; (**C**) 25-day-old proRcDIR12::GUS transgenic *Arabidopsis*; (**D**) 15-day-old proRcDIR12::*GUS* transgenic *Arabidopsis*. Arrows in panels (**B**,**C**) indicate blue-stained floral buds.

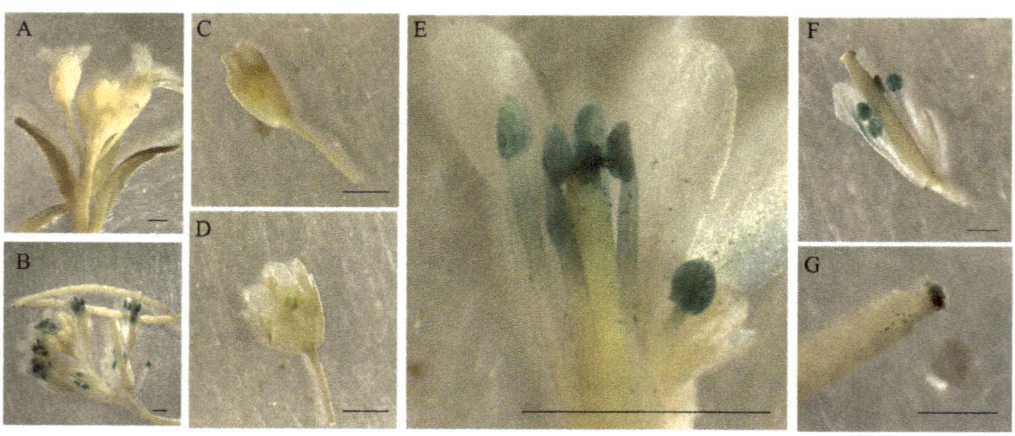

Figure 6. GUS staining of wild-type and proRcDIR12::*GUS* transgenic *Arabidopsis* (Bar = 1 cm). (**A**) GUS staining of wild-type *Arabidopsis* inflorescences; (**B**) proRcDIR12::*GUS* transgenic *Arabidopsis* inflorescence GUS staining; (**C**) proRcDIR12::*GUS* staining of transgenic *Arabidopsis* bracts; (**D**) proR-cDIR12::*GUS* staining of transgenic *Arabidopsis* primary-flowering flowers; (**E**) proRcDIR12::*GUS* transgenic *Arabidopsis* anther GUS staining; (**F**) proRcDIR12::*GUS* staining for the initiation of GUS transgenic *Arabidopsis* horn fruit development; (**G**) proRcDIR12::*GUS* staining with GUS staining of young *Arabidopsis* horn fruit with stigma pollen.

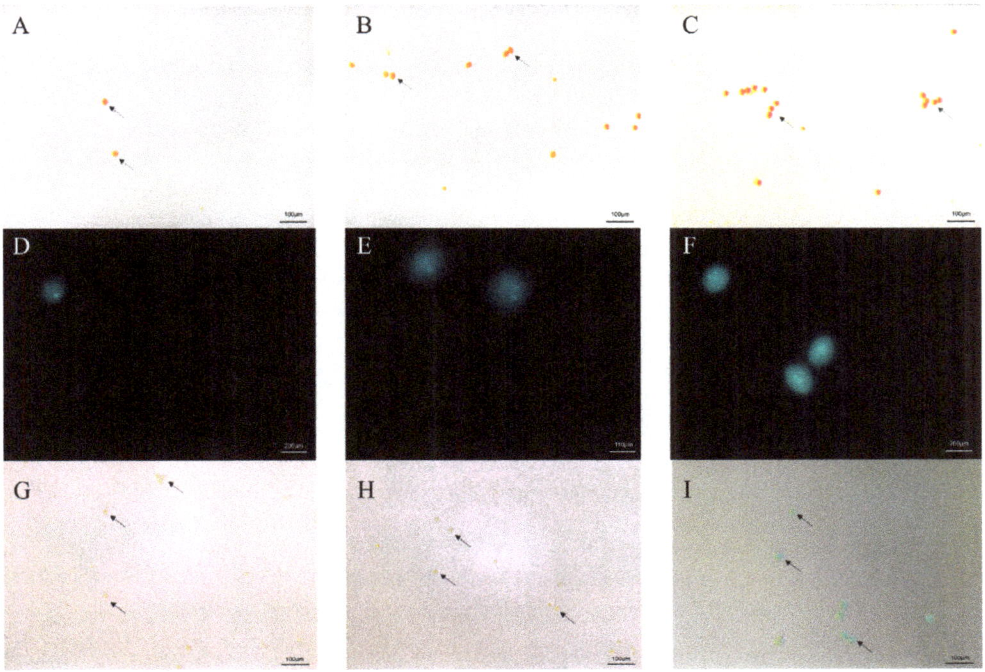

Figure 7. ProRcDIR12::*GUS* transgenic *Arabidopsis* pollen staining at three periods. (**A–C**) Pollen viability staining at three nuclear stages of proRcDIR12::*GUS* transgenic *Arabidopsis*. (**A**) Primary nuclear pollen; (**B**) dinuclear pollen; (**C**) trinuclear pollen. (**D–F**) DAPI staining of pollen at three nuclear stages of proRcDIR12::*GUS* transgenic *Arabidopsis*. (**D**) Primary nuclear pollen; (**E**) dinuclear pollen; (**F**) trinuclear pollen. (**G–I**) GUS staining of pollen at three nuclear stages of proRcDIR12::*GUS* transgenic *Arabidopsis*. (**G**) Primary nuclear pollen, (**H**) dinuclear pollen, (**I**) trinuclear pollen. Arrows in panels (**A–C**,**G–I**) indicate pollen grains.

4. Discussion

The rose serves as the type plant for the Rosaceae family and holds significant importance for studying post-pollination floral performance and breeding [40]. During rose breeding, pollen viability, germination rate, and pollen tube growth are critical traits [41], with post-pollination pollen viability directly determining hybridization success rates [42]. Notably, pollen morphological characteristics not only influence the dispersal and fertilization efficiency of plants such as *Bougainvillea* but also provide key evidence for taxonomic research [43]. In *A. thaliana*, normal pollen morphological development, efficient germination capacity, and rapid directed pollen tube growth are considered key factors influencing double fertilization success [44–46].

DIRs are widely present in various plants, including ferns, gymnosperms, and angiosperms [47]. These proteins play multifaceted roles in plant growth and development, disease resistance, and stress tolerance. By regulating the synthesis of lignin and lignans, they enhance the structural stability of plant cell walls, thereby improving plant tolerance to biotic and abiotic stresses. For instance, overexpression of the *GhDIR1* gene in *G. hirsutum* not only promotes the biosynthesis of lignans but also enhances resistance to *Verticillium dahliae*, thereby preventing the spread of the pathogen [48]. In *A. thaliana*, the loss of function of *AtDIR10/ESB1* results in the impaired formation of the Casparian strip in the roots and the abnormal deposition of lignin and suberin [49]. The diversity and importance of these functions make DIRs significant targets in research on plant stress tolerance and growth and development. However, their specific functional networks in roses still need to be further analyzed.

This study comprehensively presents information on the amino acid length, molecular weight, isoelectric point, instability index, and hydrophobicity of the 33 RcDIRs. More than 50% of *RcDIRs* are predicted to be localized in the chloroplasts, suggesting that these genes may play important roles in photosynthesis or chloroplast-related metabolic processes. There is extensive collinearity between the DIR gene families of *R. chinensis* and *A. thaliana*, with highly conserved chromosomal positions and arrangement orders for some genes, reflecting the core functional constraints of these genes in plant development or environmental adaptation. The cis-acting element results indicate that each *RcDIR* gene contains at least three types of functional elements, with the highest frequency of light response and abiotic stress response elements. The frequency of biotic stress elements is relatively lower. Three genes (*RcDIR15, RcDIR24, RcDIR31*) lack elements related to growth and development. Notably, *RcDIR30* contains the highest number of cis-acting elements, which may indicate its significant role in regulating various biological processes in plants.

The formation of gene clusters is usually associated with tandem duplication events, which are considered an important driving force for the expansion of gene families [50,51]. Studies have shown that tandemly duplicated gene clusters tend to form chromatin topological associating domains, which may facilitate the occurrence of tandem duplication events by promoting local chromatin interactions [52]. In the *RcDIR* gene family, the close arrangement patterns of different gene clusters suggest the key role of tandem duplication events in family expansion. Similar phenomena have been reported in the *DIR* gene families of *Nicotiana tabacum* and *Solanum lycopersicum* [53,54], and this mechanism may be a conserved evolutionary strategy for the functional divergence and enhanced environmental adaptability of plant DIR genes. The DIR-e subgroup has evolved stably across different species [55], and this subgroup includes *ESB1*, which plays a role in the construction of the Casparian strip by participating in lignin deposition, although its specific biochemical function has not been fully elucidated [49]. This study found that the DIR-e subgroup in *R. chinensis* contains an independent branch composed of RcDIR7-RcDIR12, and this branch is the largest tandem gene cluster in *R. chinensis*, possibly indicating that this branch has undergone a unique evolutionary process in *R. chinensis*, thereby acquiring species-specific functions. This unique evolution may enable it to play a key role in the specific physiological processes of *R. chinensis*, but its specific mechanism in pollen development still needs to be verified by experiments.

Hybrid breeding is the main breeding method for roses, and pollen sterility directly affects breeding efficiency and limits the cultivation of new hybrid rose varieties. Ensuring the maturity and fertility of pollen is one of the foundations for carrying out hybrid breeding work in roses, and research on the mechanism of pollen maturity in roses lays a theoretical foundation for this work. The development of anthers and pollen plays an important role in the biological processes of plant pollination and fruiting. Some genes that affect the formation of anthers and pollen are not only expressed in the anthers but also in other parts of the plant, such as the promoter of *G. hirsutum* casein kinase I (GhCKI), which is expressed in mature anthers but also in petals and sepals [56]. Some genes that show anther-specific expression, such as the auxin response factor 17 (ARF17), are expressed in the inner wall of the anther but not in the pollen, and the *A. thaliana arf17* mutant exhibits defects in the lignification of the inner wall of the anther, leading to anther dehiscence defects [57]. Researchers have also found some genes, such as the *G. hirsutum* GhWRKY22 transcription factor, which are expressed in both mature and young pollen, but the expression is weaker in young pollen and stronger in mature pollen [58]. This study analyzed the promoter regions of *RcDIRs* and found that they contain multiple pollen-specific expression elements, including AAATGA, POLLEN1LELAT52, and GTGANTG10. Focusing on *RcDIR12*, a member of the tandem gene cluster RcDIR7-RcDIR12, this study found that proRcDIR12 is specifically expressed in mature trinucleate pollen and not expressed in other locations, suggesting that *RcDIR12* may play an important role in the maturation process of plant pollen, providing a new perspective for the study of the mechanism of pollen maturity in roses. This study primarily focused on 'Old Blush' as the research subject. The conservation of proRcDIR12 activity across different rose varieties remains to be further validated. Additionally, although the proRcDIR12 sequence contains stress response elements, suggesting its potential involvement in stress response regulation, the systematic detection of *RcDIR12* expression characteristics and promoter activity under various stress conditions has not yet been conducted. Further research is needed to explore its functional conservation across different varieties and stress conditions.

Supplementary Materials: The following supporting information can be downloaded at: https://www.mdpi.com/article/10.3390/horticulturae11070717/s1, Figure S1: Agarose gel electrophoresis results of the constructed recombinant expression vector pBI121-proRcDIR12::GUS. Figure S2: Schematic representation of the recombinant expression vector pBI121-proRcDIR12::GUS. Figure S3: PCR identification of the proRcDIR12 promoter in transgenic Arabidopsis thaliana. Table S1. CDS sequences of the 33 RcDIRs and the RcDIR12 promoter sequence. Table S2: Pollen promoter elements of RcDIR gene family members.

Author Contributions: Conceptualization, Q.D. and Q.Y.; methodology, Z.W.; software, Y.Z.; validation, S.G., Y.P. and Q.L.; formal analysis, Y.H.; investigation, Q.D.; resources, Q.Y.; data curation, Q.D.; writing—original draft preparation, Q.D.; writing—review and editing, Q.L. and Y.H.; visualization, Y.Z.; supervision, Y.H.; project administration, Y.H.; funding acquisition, Y.H. All authors have read and agreed to the published version of the manuscript.

Funding: This research was funded by the National Natural Science Foundation of China (32471938, 31800594), China; the National Key Research and Development Project of China (2023YFD1200105), China; and the Beijing Natural Science Foundation (6174045), China.

Data Availability Statement: Data are contained within the article and Supplementary Materials.

Conflicts of Interest: The authors declare no conflicts of interest.

References

1. Hibrand, S.L.; Ruttink, T.; Hamama, L.; Kirov, I.; Lakhwani, D.; Zhou, N.N.; Bourke, P.M.; Daccord, N.; Leus, L.; Schulz, D.; et al. A high-quality genome sequence of *Rosa chinensis* to elucidate ornamental traits. *Nat. Plants* **2018**, *4*, 473–484. [CrossRef] [PubMed]
2. Zhang, Z.; Yang, T.; Liu, Y.; Wu, S.; Sun, H.; Wu, J.; Li, Y.; Zheng, Y.; Ren, H.; Yang, Y.; et al. Haplotype-resolved genome assembly and resequencing provide insights into the origin and breeding of modern rose. *Nat. Plants* **2024**, *10*, 1659–1671. [CrossRef] [PubMed]
3. Wei, S.; Ma, L. Comprehensive insight into tapetum-mediated pollen development in *Arabidopsis thaliana*. *Cells* **2023**, *12*, 247. [CrossRef]
4. Kilic, T. Identifying successful combinations by fertility index in old garden roses and hybrid tea roses crosses. *PeerJ* **2023**, *11*, e15526. [CrossRef]
5. NaRa, J.; Ki Young, P. Rose pollen management methods to improve productivity. *Agronomy* **2022**, *12*, 1285. [CrossRef]
6. Haberle, V.; Stark, A. Eukaryotic core promoters and the functional basis of transcription initiation. *Nat. Rev. Mol. Cell Biol.* **2018**, *19*, 621–637. [CrossRef]
7. Chen, L.; Miao, Y.; Wang, C.; Su, P.; Li, T.; Wang, R.; Hao, X.; Yang, G.; He, G.; Gao, C. Characterization of a novel pollen-specific promoter from wheat (*Triticum aestivum* L.). *Plant Mol. Biol. Rep.* **2012**, *30*, 1426–1432. [CrossRef]
8. Yan, S.; Wang, Z.; Liu, Y.; Li, W.; Wu, F.; Lin, X.; Meng, Z. Functional architecture of two exclusively late stage pollen-specific promoters in rice (*Oryza sativa* L.). *Plant Mol. Biol.* **2015**, *88*, 415–428. [CrossRef]
9. Davin, L.B.; Wang, H.B.; Crowell, A.L.; Bedgar, D.L.; Martin, D.M.; Sarkanen, S.; Lewis, N.G. Stereoselective bimolecular phenoxy radical coupling by an auxiliary (dirigent) protein without an active center. *Science* **1997**, *275*, 362–366. [CrossRef]

10. Halls, S.C.; Lewis, N.G. Secondary and quaternary structures of the (+)-pinoresinol-forming dirigent protein. *Biochemistry* **2002**, *41*, 9455–9461. [CrossRef]
11. Zhao, Z.; Guan, Y.; Qin, T.; Zheng, H.; Wang, H.; Xu, W.; Gu, W.; Yu, D.; Wei, J.; Hu, Y. A dirigent gene, *ZmDIR11*, positively regulates drought tolerance in maize. *Agronomy* **2025**, *15*, 604. [CrossRef]
12. Zhang, Q.; Luo, N.; Dai, X.; Lin, J.; Ahmad, B.; Chen, Q.; Lei, Y.; Wen, Z. Ectopic and transient expression of *VvDIR4* gene in *Arabidopsis* and grapes enhances resistance to anthracnose via affecting hormone signaling pathways and lignin production. *BMC Genom.* **2024**, *25*, 895. [CrossRef] [PubMed]
13. Gong, L.; Li, B.; Zhu, T.; Xue, B. Genome-wide identification and expression profiling analysis of *DIR* gene family in *Setaria italica*. *Front. Plant Sci.* **2023**, *14*, 1243806. [CrossRef] [PubMed]
14. Duan, W.; Xue, B.; He, Y.; Liao, S.; Li, X.; Li, X.; Liang, Y. Genome-wide identification and expression pattern analysis of dirigent members in the genus *Oryza*. *Int. J. Mol. Sci.* **2023**, *24*, 7189. [CrossRef]
15. Paniagua, C.; Bilkova, A.; Jackson, P.; Dabravolski, S.; Riber, W.; Didi, V.; Houser, J.; Gigli-Bisceglia, N.; Wimmerova, M.; Budínská, E.; et al. Dirigent proteins in plants: Modulating cell wall metabolism during abiotic and biotic stress exposure. *J. Exp. Bot.* **2017**, *68*, 3287–3301. [CrossRef]
16. Yang, W.; Duan, H.; Yu, K.; Hou, S.; Kang, Y.; Wang, X.; Hao, J.; Liu, L.; Zhang, Y.; Luo, L.; et al. Integrative dissection of lignin composition in Tartary buckwheat seed hulls for enhanced dehulling efficiency. *Adv. Sci.* **2024**, *11*, e2400916. [CrossRef]
17. Yuan, G.; Zou, T.; He, Z.; Xiao, Q.; Li, G.; Liu, S.; Xiong, P.; Chen, H.; Peng, K.; Zhang, X.; et al. *SWOLLEN TAPETUM AND STERILITY* 1 is required for tapetum degeneration and pollen wall formation in rice. *Plant Physiol.* **2022**, *190*, 352–370. [CrossRef]
18. Khan, A.; Li, R.; Sun, J.; Ma, F.; Zhang, H.; Jin, J.; Ali, M.; Haq, S.; Wang, J.; Gong, Z.; et al. Genome-wide analysis of dirigent gene family in pepper (*Capsicum annuum* L.) and characterization of *CaDIR7* in biotic and abiotic stresses. *Sci. Rep.* **2018**, *8*, 5500. [CrossRef]
19. Horton, P.; Park, K.J.; Obayashi, T.; Fujita, N.; Harada, H.; Adams-Collier, C.J.; Nakai, K. WoLF PSORT: Protein localization predictor. *Nucleic Acids Res.* **2007**, *35*, W585–W587. [CrossRef]
20. Chen, C.; Wu, Y.; Li, J.; Wang, X.; Zeng, Z.; Xu, J.; Liu, Y.; Feng, J.; Chen, H.; He, Y.; et al. TBtools-II: A "one for all, all for one" bioinformatics platform for biological big-data mining. *Mol. Plant* **2023**, *16*, 1733–1742. [CrossRef]
21. Tamura, K.; Stecher, G.; Kumar, S. MEGA11: Molecular evolutionary genetics analysis version 11. *Mol. Biol. Evol.* **2021**, *38*, 3022–3027. [CrossRef] [PubMed]
22. Letunic, I.; Bork, P. Interactive Tree of Life (iTOL) v6: Recent updates to the phylogenetic tree display and annotation tool. *Nucleic Acids Res.* **2024**, *52*, W78–W82. [CrossRef] [PubMed]
23. Waterhouse, A.M.; Procter, J.B.; Martin, D.M.A.; Clamp, M.; Barton, G.J. Jalview Version 2—A multiple sequence alignment editor and analysis workbench. *Bioinformatics* **2009**, *25*, 1189–1191. [CrossRef]
24. Bailey, T.L.; Boden, M.; Buske, F.A.; Frith, M.; Grant, C.E.; Clementi, L.; Ren, J.; Li, W.W.; Noble, W.S. MEME SUITE: Tools for motif discovery and searching. *Nucleic Acids Res.* **2009**, *37*, W202–W208. [CrossRef]
25. Hu, B.; Jin, J.; Guo, A.; Zhang, H.; Luo, J.; Gao, G. GSDS 2.0: An upgraded gene feature visualization server. *Bioinformatics* **2015**, *31*, 1296–1297. [CrossRef]
26. Abramson, J.; Adler, J.; Dunger, J.; Evans, R.; Green, T.; Pritzel, A.; Ronneberger, O.; Willmore, L.; Ballard, A.J.; Bambrick, J.; et al. Accurate structure prediction of biomolecular interactions with AlphaFold 3. *Nature* **2024**, *630*, 493–500. [CrossRef]
27. DeLano, W.L. Pymol: An open-source molecular graphics tool. *CCP4 Newsl. Protein Crystallogr.* **2002**, *40*, 82–92.
28. Lescot, M.; Déhais, P.; Thijs, G.; Marchal, K.; Moreau, Y.; de Peer, Y.V.; Rouzé, P.; Rombauts, S. PlantCARE, a database of plant cis-acting regulatory elements and a portal to tools for in silico analysis of promoter sequences. *Nucleic Acids Res.* **2002**, *30*, 325–327. [CrossRef]
29. Clough, S.J.; Bent, A.F. Floral dip: A simplified method for *Agrobacterium*-mediated transformation of *Arabidopsis thaliana*. *Plant J.* **1998**, *16*, 735–743. [CrossRef]
30. Alvarez-Buylla, E.R.; Benítez, M.; Corvera-Poiré, A.; Chaos Cador, Á.; de Foltel, S.; Gamboa de Buen, A.; Garay-Arroyo, A.; García-Ponce, B.; Jaimes-Miranda, F.; Pérez-Ruiz, R.V.; et al. Flower development. *Arab. Book* **2010**, *8*, e127. [CrossRef]
31. Ross, K.J.; Fransz, P.; Jones, G.H. A light microscopic atlas of meiosis in *Arabidopsis thaliana*. *Chromosome Res.* **1996**, *4*, 507–516. [CrossRef] [PubMed]
32. Kim, K.-W.; Moinuddin, S.G.A.; Atwell, K.M.; Costa, M.A.; Davin, L.B.; Lewis, N.G. Opposite stereoselectivities of dirigent proteins in *Arabidopsis* and *Schizandra* species. *J. Biol. Chem.* **2012**, *287*, 33957–33972. [CrossRef] [PubMed]
33. Effenberger, I.; Zhang, B.; Li, L.; Wang, Q.; Liu, Y.; Klaiber, I.; Pfannstiel, J.; Wang, Q.; Schaller, A. Dirigent proteins from cotton (*Gossypium* sp.) for the atropselective synthesis of gossypol. *Angew. Chem. Int. Ed.* **2015**, *54*, 14660–14663. [CrossRef] [PubMed]
34. Lin, J.; Fang, X.; Li, J.; Chen, Z.; Wu, W.; Guo, X.; Liu, N.; Huang, J.; Chen, F.; Wang, L.; et al. Dirigent gene editing of gossypol enantiomers for toxicity-depleted cotton seeds. *Nat. Plants* **2023**, *9*, 605–615. [CrossRef]

35. Corbin, C.; Drouet, S.; Markulin, L.; Auguin, D.; Lainé, É.; Davin, L.B.; Cort, J.R.; Lewis, N.G.; Hano, C. A genome-wide analysis of the flax (*Linum usitatissimum* L.) dirigent protein family: From gene identification and evolution to differential regulation. *Plant Mol. Biol.* **2018**, *97*, 73–101. [CrossRef]
36. Wang, Y.; Tang, H.; DeBarry, J.D.; Tan, X.; Li, J.; Wang, X.; Lee, T.-H.; Jin, H.; Marler, B.; Guo, H.; et al. MCScanX: A toolkit for detection and evolutionary analysis of gene synteny and collinearity. *Nucleic Acids Res.* **2012**, *40*, e49. [CrossRef]
37. Weterings, K.; Schrauwen, J.; Wullems, G.; Twell, D. Functional dissection of the promoter of the pollen-specific gene *NTP303* reveals a novel pollen-specific, and conserved *cis*-regulatory element. *Plant J.* **1995**, *8*, 55–63. [CrossRef]
38. Manimaran, P.; Reddy, M.R.; Rao, T.B.; Mangrauthia, S.K.; Sundararam, R.M.; Balachandran, S.M. Identification of *cis*-elements and evaluation of upstream regulatory region of a rice anther-specific gene, *OSIPP3*, conferring pollen-specific expression in *Oryza sativa* (L.) ssp. indica. *Plant Reprod.* **2015**, *28*, 133–142. [CrossRef]
39. Twell, D.; Yamaguchi, J.; Wing, R.A.; Ushiba, J.; McCormick, S. Promoter analysis of genes that are coordinately expressed during pollen development reveals pollen-specific enhancer sequences and shared regulatory elements. *Genes Dev.* **1991**, *5*, 496–507. [CrossRef]
40. Caser, M.; Dente, F.; Ghione, G.; Mansuino, A.; Giovannini, A.; Scariot, V. Shortening of selection time of *Rosa hybrida* by in vitro culture of isolated embryos and immature seeds. *Propag. Ornam. Plants.* **2014**, *14*, 139–144.
41. Nadeem, M.; Akond, M.; Riaz, A.; Qasim, M.; Younis, A.; Farooq, A. Pollen morphology and viability relates to seed production in hybrid roses. *Plant Breed. Seed Sci.* **2014**, *68*, 25–38. [CrossRef]
42. Macovei, A.; Caser, M.; Donà, M.; Valassi, A.; Giovannini, A.; Carbonera, D.; Scariot, V.; Balestrazzi, A. Prolonged cold storage affects pollen viability and germination along with hydrogen peroxide and nitric oxide content in *Rosa hybrida*. *Not. Bot. Hort. Agrobot. Cluj-Na* **2016**, *44*, 6–10. [CrossRef]
43. Tripathi, S.; Singh, S.; Roy, K.R. Pollen morphology of *Bougainvillea* (*Nyctaginaceae*): A popular ornamental plant of tropical and sub-tropical gardens of the world. *Rev. Palaeobot. Palynol.* **2017**, *239*, 31–46. [CrossRef]
44. Honys, D.; Twell, D. Comparative analysis of the *Arabidopsis* pollen transcriptome. *Plant Physiol.* **2003**, *132*, 640–652. [CrossRef]
45. Gibalová, A.; Reňák, D.; Matczuk, K.; Dupláková, N.; Cháb, D.; Twell, D.; Honys, D. *AtbZIP34* is required for *Arabidopsis* pollen wall patterning and the control of several metabolic pathways in developing pollen. *Plant Mol. Biol.* **2009**, *70*, 581–601. [CrossRef]
46. Vogler, F.; Schmalzl, C.; Englhart, M.; Bircheneder, M.; Sprunck, S. Brassinosteroids promote *Arabidopsis* pollen germination and growth. *Plant Reprod.* **2014**, *27*, 153–167. [CrossRef]
47. Laurence, B.; Davin, M.; Jourden, A.; Patten, A. Dissection of lignin macromolecular configuration and assembly: Comparison to related biochemical processes in allyl/propenyl phenol and lignan biosynthesis. *Nat. Prod. Rep.* **2008**, *25*, 1015–1090.
48. Shi, H.; Liu, Z.; Zhu, L.; Zhang, C.; Chen, Y.; Zhou, Y.; Li, F.; Li, X. Overexpression of cotton (*Gossypium hirsutum*) *dirigent 1* gene enhances lignification that blocks the spread of *Verticillium dahliae*. *Acta Biochim. Biophys. Sin.* **2012**, *44*, 555–564. [CrossRef]
49. Hosmani, P.S.; Kamiya, T.; Danku, J.; Salt, D.E. Dirigent domain-containing protein is part of the machinery required for formation of the lignin-based Casparian strip in the root. *Proc. Natl. Acad. Sci. USA* **2013**, *110*, 14498–14503. [CrossRef]
50. Jiang, M.; Chu, Z. Comparative analysis of plant *MKK* gene family reveals novel expansion mechanism of the members and sheds new light on functional conservation. *BMC Genom.* **2018**, *19*, 407. [CrossRef]
51. Qu, J.; Liu, L.; Guo, Z.; Li, X.; Pan, F.; Sun, D.; Yin, L. The ubiquitous position effect, synergistic effect of recent generated tandem duplicated genes in grapevine, and their co-response and overactivity to biotic stress. *Fruit Res.* **2023**, *3*, 16. [CrossRef]
52. Ma, N.; Li, X.; Ci, D.; Zeng, H.; Zhang, C.; Xie, X.; Zhong, C.; Deng, X.W.; Li, D.; He, H. Chromatin topological domains associate with the rapid formation of tandem duplicates in plants. *Nat. Commun.* **2019**, *10*, 484. [CrossRef] [PubMed]
53. Li, T.; Luo, W.; Du, C.; Lin, X.; Lin, G.; Chen, R.; He, H.; Wang, R.; Lu, L.; Xie, X. Functional and evolutionary comparative analysis of the *DIR* gene family in *Nicotiana tabacum* L. and *Solanum tuberosum* L. *BMC Genom.* **2024**, *25*, 671.
54. Saddique, M.A.B.; Guan, G.; Hu, B.; Khan, M.; Amjad, M.D.; Abbas, S.; Hussain, Z.; Maqsood, M.F.K.; Luo, X.; Ren, M. Genome-wide computational analysis of the dirigent gene family in *Solanum lycopersicum*. *Proteome Sci.* **2024**, *22*, 10. [CrossRef]
55. Liu, Z.; Wang, X.; Sun, Z.; Zhang, Y.; Meng, C.; Chen, B.; Wang, G.; Ke, H.; Wu, J.; Yan, Y.; et al. Evolution, expression and functional analysis of cultivated allotetraploid cotton *DIR* genes. *BMC Plant Biol.* **2021**, *21*, 89. [CrossRef]
56. Li, Y.; Li, Y.; Su, Q.; Wu, Y.; Zhang, R.; Li, Y.; Ma, Y.; Ma, H.; Guo, X.; Zhu, L.; et al. High temperature induces male sterility via *MYB66-MYB4*-Casein kinase I signaling in cotton. *Plant Physiol.* **2022**, *189*, 2091–2109. [CrossRef]
57. Xu, X.; Wang, B.; Feng, Y.; Xue, J.; Qian, X.; Liu, S.; Zhou, J.; Yu, Y.; Yang, N.; Xu, P.; et al. *AUXIN RESPONSE FACTOR17* directly regulates *MYB108* for anther dehiscence. *Plant Physiol.* **2019**, *181*, 645–655. [CrossRef]
58. Wang, Y.; Li, Y.; He, S.; Gao, Y.; Wang, N.; Lu, R.; Li, X. A cotton (*Gossypium hirsutum*) WRKY transcription factor (*GhWRKY22*) participates in regulating anther/pollen development. *Plant Physiol. Biochem.* **2019**, *141*, 231–239. [CrossRef]

horticulturae

MDPI

Article

Terpene Synthase (TPS) Family Member Identification and Expression Pattern Analysis in Flowers of *Dendrobium chrysotoxum*

Yanni Yang [1], Jianying Gong [2], Rongrong Nong [1], Qiao Liu [1], Ke Xia [1], Shuo Qiu [1,*,†] and Zaihua Wang [3,*,†]

[1] Guangxi Key Laboratory of Plant Functional Phytochemicals and Sustainable Utilization, Guangxi Institute of Botany, Guangxi Zhuang Autonomous Region and Chinese Academy of Sciences, Guilin 541006, China; yangyanni219@126.com (Y.Y.); 18477691913@163.com (R.N.); liuqiao20200101@163.com (Q.L.); xiake4502@126.com (K.X.)

[2] Guangxi Zhuang Autonomous Region Forestry Research Institute, No. 23 Yongwu Road, Nanning 530022, China; gongjianying2020@163.com

[3] Environmental Horticulture Research Institute, Guangdong Academy of Agricultural Sciences/Guangdong Provincial Key Lab of Ornamental Plant Germplasm Innovation and Utilization, Guangzhou 510640, China

* Correspondence: qiushuo001@163.com (S.Q.); wangzaihua@163.com (Z.W.); Tel.: +86-773-3550103 (S.Q.), Tel.: +86-020-85161185 (Z.W.); Fax: +86-773-3550067 (S.Q.)

† These authors contributed equally to this work.

Abstract: Flower fragrance is a crucial ornamental and economic trait of *Dendrobium chrysotoxum*, and the most abundant and diverse aroma-active compounds are terpenes. Terpene synthase (TPS) is the ultimate enzyme for the biosynthesis of various types of terpenes, and TPS genes were identified as the key regulators governing the spatiotemporal release of volatile terpene compounds. Until recently, the TPS gene family in *D. chrysotoxum* has remained largely unexplored. Our study characterizes the TPS genes in *D. chrysotoxum* and identifies 37 *DcTPS* gene family members. It helped identify the *DcTPS* genes, gene characteristics, the phylogeny relationship, conserved motif location, gene exon/intron structure, *cis-elements* in the promoter regions, protein–protein interaction (PPI) network, tissue specific expression and verification of the expression across different flowering stages and floral organs. Three highly expressed *DcTPS* genes were cloned, and their functions were verified using a transient expressed in tobacco leaves. Further functional verification showed that the proteins encoded by these genes were enzymes involved in monoterpene synthesis, and they were all involved in the synthesis of linalool. This study comprehensively expatiates on the TPS gene family members in *D. chrysotoxum* for the first time. These data will help us gain a deeper understanding of both the molecular mechanisms and the effects of the TPS genes. Furthermore, the discovery that three TPS-b genes (*DcTPS* 02, 10, 32) specifically drive linalool-based scent in *D. chrysotoxum*, will provide new insights for expanding the TPS-b subfamily in orchids and identifying the linalool synthases contributing to orchid fragrance.

Keywords: karst plant; flower fragrance; terpenes; TPS genes family; gene expression

1. Introduction

Dendrobium chrysotoxum Lindl. belongs to the Orchidaceae family, and its floral fragrance is one of its most important ornamental characteristics. In addition, floral fragrances also have the function of attracting pollinators [1], defending against natural enemies and resisting abiotic stress in various ecological habitats [2]. The high volatility organic compounds (VOCs) promote the aroma in orchids, and the floral components include a large number of generally lipophilic plant products with a molecular mass of less than 300, which can be classified into three categories based on their independent origins: terpenes, benzenoid aromatics, and fatty acid derivatives [3]. Among them, terpenes are one of the most diverse VOCs, and their molecular weight is relatively low, e.g., 5-carbon isoprene, 10-carbon monoterpenes, 15-carbon sesquiterpenes, and 20-carbon diterpenes constitute the largest proportion of volatile components in flowers [3–5]. To date, more than 80,000 terpenes have been identified from *Dendrobium*. The volatile floral scent compounds in *Dendrobium* flowers are mainly 10-carbon monoterpenes like linalool, β-ocimene, pinene, ylangene and limonene [6–9].

In plants, the intricate metabolic pathways of volatile terpenes have been thoroughly elucidated and extensively explored. The formation of terpenes occurs through two primary pathways: the mevalonate pathway (MVA pathway) and the 2C-methyl-D-erythritol-4-phosphate pathway (MEP pathway) [10]. The MVA pathway occurs in the cytoplasm, endoplasmic reticulum and peroxisomes, giving rise to the synthesis of 15-carbon sesquiterpenoids. The MEP pathway occurs in the plastid and is mainly responsible for the synthesis of 10-carbon monoterpenes and 20-carbon diterpenoids [11,12]. The 5-carbon compounds isopentenyl diphosphate (IPP) and dimethylallyl diphosphate (DMAPP) serve as precursors, and a sequence of enzymes associated with volatile terpene catalyzed reactions have been identified in both pathways [10–13]. Terpene synthases (TPSs) serve as the final enzymes that transform the substrates farnesyl diphosphate (FPP), farnesyl diphosphate (GPP) and all-trans-geranylgeranyl diphosphate (GGPP) into kinds of sesquiterpene, monoterpenes and diterpene, which are regarded as pivotal in the two pathways [14,15].

Recent studies have verified that TPSs are highly differentiated gene families, and are present in the genomes of all angiosperms and gymnosperms [16]. The applied bioinformatics techniques are classified into seven TPS subfamilies based on phylogenic analysis: TPS-a (encodes sesquiterpenes synthase), TPS-b (encodes cyclic monoterpenes and hemiterpenes synthase), TPS-c (catalyzes copalyl diphospate synthases), TPS-d (gymnosperm-specific), TPS-e/f (encodes copalyl diphosphate/kaurene synthases), TPS-g (encodes acyclic monoterpenes) and TPS-h (lycopod-specific) [14]. Despite intriguing differences between the TPS subfamilies, most full-length TPSs have two identical conserved domains: PF19086 (C-terminal) and PF01397 (N-terminal), defined in the Pfam (http://pfam.xfam.org/, accessed on 18 May 2024) database [17,18]. In addition, TPS harbors three conserved motif structures such as an N-terminal domain containing R(R)X8W, a C-terminal domain containing a DDxxD and an NSE/DTE motif [15].

The research on various terrestrial plants confirmed that the TPSs are encoded by a specific gene family involved in the biosynthesis of terpenoids [14,19]. These TPS genes have been successfully extracted from nutrient-rich tissues, as well as from fruits and flowers. The genome-wide analyses of TPSs have been identified in *Arabidopsis thaliana* [20], *Solanum lycopersicum* [21], *Vitis vinifera* [22], *Gossypium hirsutum* [23], *Brachypodium distachyon* [24] and *Liriodendron chinense* [25] when exploring their defensive functions after herbivore damage in different organs except flowers. On the other hand, the researchers note that these genes might be equally important in the biosynthesis of flower fragrance, as they carry out essential functions in attracting insects and guiding pollinators to determine the reproductive syndromes of plants [26]. Research has demonstrated that pollinators often show different preferences for floral traits, such as those related to scent and color, which might lead to reproductive isolation in some species [27–29]. The most well-studied TPS enzyme of floral scent biosynthesis is linalool synthase (LIS), which converts geranyl diphosphate to linalool [30,31]. The initial linalool can be further modified into other terpene synthase products [32,33].

Orchids are one of the largest families of flowering plants in the plant kingdom, and various terpenes are predominant volatile compounds in orchids. Monoterpenes, including compounds like linalool, pinene, ylangene and limonene, serve as primary contributors to the floral scent in orchid flowers [31]. To date, only a few TPS genes have been identified; thus, genome-wide TPS identification is limited in orchids. In recent reports, TPS genes have been identified in *Dendrobium officinale* (34 TPSs) [34], *Cymbidium faberi* (32 TPSs) [35], and *Freesia* folwers (31 terpenes) [16]. Some TPS genes have also been functionally validated, for example, the *PbTPS5* and *PbTPS10* genes were integral to the biosynthesis of monoterpenes in *Phalaenopsis bellina* [36]. In *Freesia* hybrid flowers, *FhTPS1* is responsible for catalyzing the production of linalool. In contrast, *FhTPS4*, *FhTPS6*, and *FhTPS7* have the ability to interact with both GPP and FPP substrates [16].

Until recently, the TPS gene family in *D. chrysotoxum* has not been well-characterized. In the current study, all *DcTPS* family members were identified and characterized from *D. chrysotoxum*, using a previous *D. chrysotoxum* genomic database and a comprehensive analysis of members was performed. It facilitated the identification of *DcTPS* genes, characterization of their gene features, analysis of phylogenetic relationships, determination of conserved motif locations, examination of gene exon/intron structures, investigation of *cis-elements* in promoter regions, construction of protein–protein interaction (PPI) networks, evaluation of the tissue-specific expression, and verification of expression patterns under different flowering stages and in various floral organs. Three highly expressed *DcTPS* genes were cloned, and their functions were verified by transient expression in tobacco leaves. On this basis, the *DcTPS* genes reported in this study are helpful to further understand the molecular mechanism of monoterpene synthesis. In addition, our results will provide valuable candidate genes for scent modification in *D. chrysotoxum* and other orchids.

2. Methods

2.1. Identification of TPS Genes in D. chrysotoxum

To identify the TPS genes, the latest recently released whole-genome annotations of *D. chrysotoxum* were downloaded from NCBI (https://ftp.ncbi.nlm.nih.gov/genomes/all/GCA/019/514/585/GCA_019514585.1_ASM1951458v1/, accessed on 19 May 2024). The annotated protein databases were scanned using HMMER 3.0 (http://hmmer.org/, accessed on 19 May 2024) with the Hidden Markov model (HMM) of TPS N-terminal domain (PF01397) and TPS C-terminal domain (PF19086), which were downloaded from Pfam (http://pfam.xfam.org/, accessed on 19 May 2024). By obtaining proteins from the TPS HMM, a high-quality protein set (E-value $< 1 \times 10^{-20}$) was arranged and used to construct a *D. chrysotoxum*-specific TPS HMM by hmmbuild in HMMER 3.0. The *D. chrysotoxum*-specific TPS HMM was used to align all protein sequences with an E-value lower than 1×10^{-5}. In order to avoid any non-specific sequences outside the TPS cluster, all A. thaliana TPS proteins were used as queries to explore the *D. chrysotoxum* database with default parameters. Using the Pfam database and Conserved Domain Database (CDD, https://www.ncbi.nlm.nih.gov/Structure/cdd/wrpsb.cgi, accessed on 19 May 2024) for filtering redundant sequences. Only a total full length of TPS domain sequences were selected as DcTPS protein for subsequent analysis. As a result, a total of 37 highly similar TPS genes were identified in the *D. chrysotoxum* genome.

2.2. Characteristics of TPS Genes

To characterize and align the *DcTPS* sequences, the ProtParam tool (https://www.expasy.org/, accessed on 19 May 2024) was employed to predict the physicochemical properties of the *DcTPS* proteins, including protein length, molecular weight (MW), predicted theoretical isoelectric point (pI), instability index, aliphatic index, and grand average of hydropathicity of the encoded proteins.

2.3. Phylogenetic, Conserved Motifs and Gene Structure Analysis

The ClustalW algorithm was used to select the target protein sequences, Supplementary Table: Tables S1 and S3 list the gene IDs of the *DcTPS* (37) and *AtTPS* (32) members. In the MEGA-7.0 (https://www.megasoftware.net/, accessed on 20 May 2024) software, an unrooted phylogenetic tree was constructed using the maximum likelihood statistics method. The bootstrap-replicates number was set to 1000 repetitions using the Poisson model. The generated tree was redrawn and annotated by the FigTree v1.4.3 software. Based on the well-established division in *A. thaliana*, the *D. chrysotoxum* TPS members were further categorized into multiple subcategories.

The MEME program (http://meme-suite.org/tools/meme, accessed on 2 July 2024) was employed to detect the conserved motifs. All other default parameters were kept at their standard values, except that the maximum number was established at 20. The intricate exon–intron architecture of the *DcTPS* genes was visualized through the Gene Structure Display Server (GSDS) 2.0 (https://gsds.gao-lab.org/, accessed on 3 July 2024) program, offering a vivid glimpse into its complex genomic blueprint.

2.4. Cis-Elements in the Promoter Regions Analysis

The 2000 bp upstream regions of the candidate *DcTPS* genes were utilized for analyzing *cis-elements* regulatory elements in their promoters. Plant care software (http://bioinformatics.psb.ugent.be/webtools/plant-care/html/, accessed on 13 June 2024) was used for searching registry. The Gene Structure Display Server (GSDS) 2.0 (http://gsds.cbi.pku.edu.cn/, accessed on 8 July 2024) was utilized to generate the figure.

2.5. Protein–Protein Interaction (PPI) Network Prediction Analysis

To delve deeper into the relationships among the *DcTPS* genes, we utilized the interologues of *A. thaliana* to predict PPI networks. The PPI network diagram was constructed using the STRING software (Version 11.0), with a confidence score threshold of 0.4.

2.6. GO Classification and Enrichment Analysis

The transcriptome data of *D. chrysotoxum* were retrieved from the NCBI database (https://www.ncbi.nlm.nih.gov/datasets/taxonomy/161865/, accessed on 25 July 2024) and 37 *DcTPS* genes were screened. Transcript abundance was assessed using the read per million mapping (FPKM) values per kilobase transcript of the *DcTPS* genes. All the genes and transcripts obtained from the transcriptome assembly were compared with gene ontology (GO) databases to obtain the functional information of *DcTPS* genes comprehensively and made statistics on annotation in the databases. Six samples (leaves, pseudobulbs, petals, sepals, labellums and gynandrium) were completed, and a total of 49.77 Gb of Clean Data was obtained, and the Clean Data of each sample reached 7.15 Gb, and the percentage of Q30 base was more than 92.09%. Trinity is used to assemble clean data of all samples from scratch, and the assembly results are optimized and evaluated. The results show that the number of unigene assembled is 72,621, the number of Transcript is 122,375, and the average length of N50 is 1326 bp, and Salmon Version 0.14.1 was used to quantify gene expression, with the default settings. The software Goatools (v0.6.5) was used to conduct GO enrichment analysis, so as to obtain the main GO functions of the *DcTPS* genes. The Fisher exact test was applied, and if the adjusted p-value was less than 0.05, the GO function was deemed to be significantly enriched.

2.7. Cultivation of D. chrysotoxum

The plant material selected in this study was *D. chrysotoxum*, from which the flowers are both highly scented and beautiful. *D. chrysotoxum* was bought from Zhejiang Senhe Seed Co., Ltd. (Hangzhou, Zhejiang Province, China) and was one of the cultivars popularized for planting in China. The cultivar used in this experiment was obtained from the laboratory of Professor Shuo Qiu of the Guangxi Institute of Botany (Guilin, Guangxi Province, China), complying with Chinese legislation. The growth was observed under natural light and temperatures about 20–25 °C. Fresh flowers of *D. chrysotoxum* were sampled in different fluorescence stages (bud stage, first flowering stage, full bloom stage, and declining stage) and flower parts (leaves, pseudobulbs, petals, sepals, labellums and gynandrium). The flowering process of *Dendrobium* can be divided into four stages according to the changes of flower organ development: bud stage, petals not open; first flowering stage, petals begin to unfold; full bloom stage, petals fully open; declining stage, the petals become smaller and gradually fall. All samples with three replicates were immediately frozen in liquid nitrogen for storage at −80 °C until needed. The selection of different flowering stages and different flower parts are shown in Figures 1 and 2.

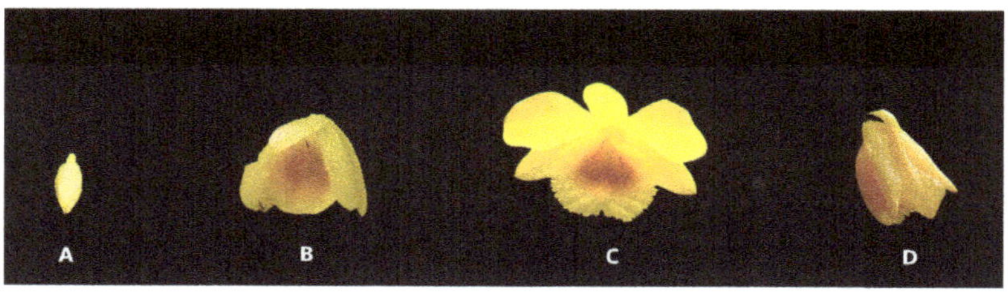

Figure 1. Different flowering stages of *D. chrysotoxum*: (**A**) bud stage; (**B**) first flowering stage; (**C**) full bloom stage; (**D**) declining stage.

Figure 2. Different flower parts of *D. chrysotoxum*: (**A**) plants of *D. chrysotoxum*; (**B**) leaves; (**C**) pseudobulbs; (**D**) petals, sepals, labellums and gynandrium.

2.8. RNA Extraction and qRT-PCR Expression Analysis

Total RNA was extracted from each sample using an RNA extraction kit (Huayueyang Biotechnology (Beijing) Co., Ltd., Beijing, China). Subsequently, 1 μg of total RNA was reverse-transcribed into cDNA using the Takara cDNA Synthesis Kit (Takara Biotechnology (Dalian) Co., Ltd., Dalian, China). Primers for qRT-PCR were designed using Primer 5.0 software (Supplementary Table: Table S10). The qRT-PCR was performed using ACT-1 housekeeping gene as endogenous reference. The PCR was mixed by Roche Lightcycler 480 Real-time PCR System in 10 μL with SYBR Green PCR reactions Master Mix kit (Vazyme, Nanjing, China), following the manufacturer's protocol. Each reaction was performed with three replicates. The relative gene expression level was calculated by the $2^{-\Delta\Delta CT}$ method, using the Thermal Cycler software (version 7500 Real-Time PCR Systems).

2.9. Transient Expression of DcTPS Genes in Nicotiana benthamiana

The CDS of *DcTPS02*, *DcTPS10* and *DcTPS32* were amplified using specific primers (listed in Supplementary Table S11) and cloned into the pBI121 binary vector. The resulting constructs were subsequently introduced into the Agrobacterium strain GV3101 for further transformation (Supplementary Figures S1 and S2). These *Agrobacterium* strains carrying *DcTPSs* and strains carrying the gene encoding the viral suppressor p19 infiltrated four-week-old *Nicotiana benthamiana* youngest leaves [16,37,38]. When the freshly cultured *Agrobacterium* reached an optical density (OD600 nm) of 0.6 to 0.8, they were harvested by centrifugation and re-suspended in an infiltration buffer containing 2-(N-morpholino) ethanesulfonic acid (10 mM) and magnesium chloride (10 mM). When freshly grown *Agrobacterium* cultures reached an OD600 nm of 0.6–0.8, they were centrifuged and resuspended in 2-(N-morpholino) ethanesulfonic acid (10 mM) and $MgCl_2$ (10 mM) infiltration media. The above substances were incubated for 2 h at 28 °C in a non-shaking environment. The cultures containing target gene and p19 were mixed in a 1:1 ratio before infiltration into *N. benthamiana* leaves. The plants were kept in a growing chamber at 22 °C with 16 h of light/8 h of darkness for 5 days. After transformation, the infected leaves were collected and placed in 20 mL SPME vial for GC–MS analysis of volatile compounds. Leaves infiltrated with empty binary vectors were used as a control. New product peaks were observed in contrast to controls and were identified by comparing mass spectra with the NIST 2005 Mass Spectral Library.

2.10. GC–MS Analysis

Ten *D. chrysotoxum* fresh flowers during the bloom stage were enclosed and sampled in 40 mL brown headspace sampling bottles with three biological replicates. The manual solid-phase microextraction (SPME) injector, along with the 50/30 μm PDMS/CAR/DVB fiber (SUPELCO, Inc., Bellefonte, PA, USA), was introduced into the 6890N-5975B Gas Chromatography–Mass Spectrometry (GC–MS) system (Agilent Technologies, Santa Clara, California, USA). This process was carried out at a temperature of 250 °C for a duration of 30 min. Fiber heads were placed into brown headspace sampling bottles, and the headspace was extracted at 40 °C for 30 min to facilitate the extraction process. Following this, the fiber head was taken out and inserted into the GC–MS injection port. After a 5 min analysis, the samples were injected for additional examination [39].

The HP5-MS quartz capillary column (30 m × 0.25 mm × 0.25 μm) was operated at a flow rate of 0.8 mL·min^{-1}. High-purity helium (99.999%) was used as the carrier gas, and the injection mode was set to splitless. The temperature program was set as follows: 40 °C for 3 min, 3 °C min^{-1} up to 73 °C for 3 min, then 220 °C at 5 °C min^{-1} for 2 min. The electron ionization (EI) ion source temperature was 230 °C, and the electron energy was 70 eV. The GC–MS transmission line temperature was 250 °C, and the scanning range was 40–450 amu [40]. According to the total ionization chromatography of GC–MS, the retention time was compared with the National Institute of Standards and Technology (https://www.nist.gov/, accessed on 7 June 2024) to determine the volatile compounds detected during the experiment [41]. Xcalibur1.2 software was used to conduct quantitative analysis according to peak area normalization method, and the phase pair content of each chemical component was obtained, respectively.

3. Results

3.1. Identification and Characterization of TPS Gene Family in D. chrysotoxum Genome Resources

A total of thirty-seven non-redundant terpene synthase genes (*DcTPSs*) were identified from the *D. chrysotoxum* genome and subjected to subsequent analyses. As shown in Table 1, the physical and chemical properties of the *DcTPS* genes were systematically analyzed. The protein sequence of the *DcTPS* genes varied in length from 360 (*DcTPS31*) to 1305 (*DcTPS21*) amino acid (aa) residues, with a corresponding molecular weight (MW) ranging from 41.82 to 152.19 kDa. The theoretical isoelectric point (pI) of *DcTPSs* ranged from 5.19 to 9.79, with an average pI of 6.12, suggesting that most *DcTPS* proteins were slightly acidic. With the exception of six *DcTPS* proteins (*DcTPS18*, 36, 37, 11, 8 and 25), all the other thirty-one *DcTPS* proteins were considered unstable (Instability index > 40). The content of aliphatic acids in *DcTPS* genes was high, and the aliphatic index ranged from 83.17 to 99.83. Because of the low average hydropathy value (<−0.084), most *DcTPS* were predicted to be hydrophilic.

Table 1. Physical and chemical characteristics of *DcTPS* genes.

DcTPS s Name	ProteinLength (a.a)	Molecular Weight Mw/kDa	Theoretical pI	Instability Index	Aliphatic Index	Grand Average of Hydropathicity
DcTPS 1	510	58.84	8.16	40.98	88.14	−0.174
DcTPS 2	427	48.79	5.71	43.00	93.70	−0.085
DcTPS 3	1286	150.38	5.99	45.44	92.86	−0.242
DcTPS 4	670	77.97	6.56	45.94	96.06	−0.195
DcTPS 5	530	62.76	6.94	47.18	94.38	−0.206
DcTPS 6	512	60.52	5.58	46.53	96.93	−0.215
DcTPS 7	553	64.91	5.55	40.07	97.32	−0.263
DcTPS 8	510	58.45	6.00	39.76	97.18	−0.105
DcTPS 9	438	51.35	5.30	45.07	97.72	−0.205
DcTPS 10	598	69.48	5.21	47.26	92.81	−0.236
DcTPS 11	764	88.92	7.07	39.42	93.69	−0.280
DcTPS 12	458	53.88	6.07	44.15	93.89	−0.191
DcTPS 13	606	71.03	6.08	45.46	88.18	−0.314
DcTPS 14	648	75.92	5.48	41.55	93.49	−0.249
DcTPS 15	940	109.39	5.69	40.37	93.46	−0.160
DcTPS 16	627	73.15	5.61	46.33	97.99	−0.134
DcTPS 17	657	76.62	5.23	42.66	97.20	−0.157
DcTPS 18	395	46.25	5.49	36.83	97.44	−0.137
DcTPS 19	619	72.61	6.41	41.03	93.47	−0.262
DcTPS 20	834	97.32	7.12	47.90	86.24	−0.359
DcTPS 21	1305	152.19	8.46	43.40	86.08	−0.349
DcTPS 22	608	71.10	7.88	41.18	85.84	−0.375
DcTPS 23	563	65.79	5.68	49.30	84.01	−0.334
DcTPS 24	842	98.76	5.37	40.22	94.87	−0.113
DcTPS 25	504	58.97	5.29	39.83	93.27	−0.221
DcTPS 26	486	56.69	5.19	42.22	97.08	−0.163
DcTPS 27	550	64.78	5.65	49.26	95.20	−0.216
DcTPS 28	446	52.78	5.54	46.71	94.01	−0.238
DcTPS 29	499	58.93	5.70	44.76	98.48	−0.156
DcTPS 30	923	105.17	6.13	45.04	93.69	−0.084
DcTPS 31	360	41.82	9.79	57.04	83.17	−0.480
DcTPS 32	778	91.89	5.98	44.67	89.29	−0.355
DcTPS 33	598	70.24	5.91	42.60	91.84	−0.270
DcTPS 34	617	72.14	5.86	51.28	93.42	−0.243
DcTPS 35	539	63.33	5.51	46.90	99.83	−0.194
DcTPS 36	549	64.36	5.69	38.12	92.66	−0.397
DcTPS 37	509	59.57	5.54	39.41	86.07	−0.300

The gene IDs for the *DcTPS* members are provided in Supplementary Table: Table S1.

All *DcTPS* proteins possessed two conserved domains characteristic of terpene synthases:

PF19086 (C-terminal) and PF01397 (N-terminal). A summary of the *DcTPS* genes, along with their corresponding HMM profiles, is presented in Supplementary Table: Table S2.

3.2. Phylogenetic Relationship of DcTPS Family

To further explore the evolutionary relationships within the TPS gene families, a total of 69 TPS proteins were analyzed, including 37 from *D. chrysotoxum* and 32 from *A. thaliana* (Supplementary Table: Table S3). An unrooted phylogenetic tree was constructed using the MEGA-7.0 software.

All the 69 TPS full-length protein sequences were clearly clustered into five categories with a bootstrap value of 100%. These categories could be further classified into TPS-a, TPS-b, TPS-c, TPS-e/f and TPS-g subgroups (Figure 3). Among these, the TPS-c subgroup contains the least number of *DcTPS*s (6), while TPS-b has the most numbers of *DcTPS*s (17), and the TPS-a subgroup contains *DcTPS*s (14).

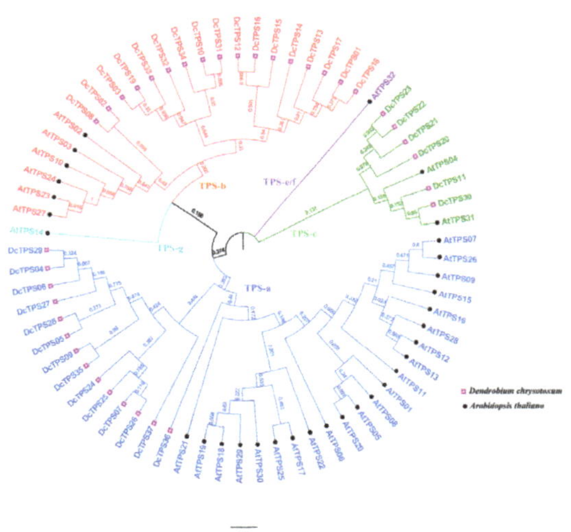

Figure 3. Phylogenetic analysis of *AtTPS* and *DcTPS* proteins. Subgroups are indicated by different colors. The phylogenetic tree was constructed using the Maximum Likelihood statistical method, with 1000 bootstrap replications under the Poisson model. The resulting tree was subsequently visualized and annotated using FigTree v1.4.3 software.

3.3. Conserved Motifs and Gene Structure Analysis

The diversity of *DcTPS* is crucial for a deeper understanding of the characteristics and functions of this family in *D. chrysotoxum*. Evolutionary mechanisms, conserved motifs, and gene intron/exon distribution were identified within the 37 *DcTPS* protein sequences. A total of 10 conserved motifs (referred to as motifs 1–10) were finally identified by MEME (http://meme-suite.org/tools/meme, accessed on 20 June 2024). The lengths of the conserved motifs varied between 21 and 41 amino acids, and the detailed features of these motifs are summarized in Supplementary Table: Table S4. As shown in Figure 4A, all the *DcTPS* genes are divided into three subfamilies, TPS-a, TPS-b and TPS-c. In Figure 4B, motifs from the same subfamily show a similar pattern, further verifying their evolutionary relationship. The numbers of *DcTPS*s motifs ranged in length from 5 to 10. The majority of the members of the TPS-a gene family contained the largest number of motifs, with 10, while *DcTPS*31 had only 5 motifs. Interestingly, we found that most of the genes with variation in the number of these motifs were in the TPS-b subfamily, suggesting the diversity and extension of this subgroup. Thirty-six *DcTPS*s (except for *DcTPS*31) contained the DDxxD motif (motif 1), and thirty-three *DcTPS*s (except for *DcTPS*9, 11, 20, 30) contained the RRX8W motif (motif 9). These findings provided additional evidence to support specific evolutionary characteristics of the *DcTPS* gene family. Moreover, the motifs and arrangement sequences that are distinctive to each *DcTPS* group may be considered to have a special function for *DcTPS* proteins.

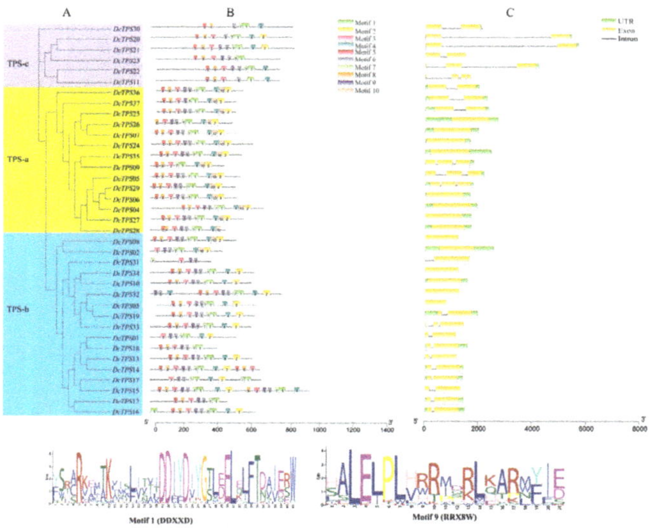

Figure 4. Arrangement of conserved motifs and structural features in *DcTPS* genes: (**A**) a phylogenetic tree was generated using the neighbor-joining method with the MEGA-7.0 software; (**B**) distribution of conserved motifs within *DcTPS* genes. Various motifs (1–10) are indicated by distinct colors; (**C**) structures of the 37 *DcTPS* genes. Exons are depicted as yellow boxes, while introns are shown as black lines. The green boxes indicate the untranslated regions (UTRs) within gene sequences. The logo of motif 1 and motif 9 are at the bottom of the image.

The evolutionary investigation through the structural diversity of polygenic families constituted a crucial component of the study.

For this purpose, a detailed structural analysis of the *DcTPS* genes was conducted by creating an exon–intron map (Figure 4C). Variations in the number of introns was observed among the 37 *DcTPS* members, ranging from 0 to 3. Of the 37 *DcTPS* genes, 14 lacked introns, 15 possessed a single intron, and 8 had between 2 and 3 introns. Additionally, members within the same subfamily of the phylogenetic tree exhibited similar protein-conserved motifs and intron counts. For example, the TPS-a subfamily had no introns, the TPS-b had one–two introns, and the TPS-c gene had two–three introns, respectively. Notably, conserved motif arrangement of intron–exon was found among most of the subfamilies but some differences were observed as well. Taken as an example, *DcTPS* 20/21 exhibits identical exon and intron phase counts but demonstrates variation in intron length. Such elements could possibly account for the variations in the size and structure of the *DcTPS* genes.

3.4. Promoters Cis-Elements Analysis

To gain a deeper understanding of the potential regulatory mechanisms of *DcTPS* genes, an investigation of *cis-elements* was conducted in the promoter region (2000 bp upstream of the *DcTPS*s translation initiation site, ATG). The promoter sequences were analyzed using the online software Plantcare database [42]. In total, we identified 824 cis-acting regulatory elements and classified them into 16 functions, which are listed in Supplementary Table: Table S5. Using TBtools software (Version 2.1), these cis-acting elements and their locations were generated in Figure 5A. These *cis-elements* could be divided into three groups: elements related to plant growth and development (Figure 5B), elements related to plant phytohormone responsiveness (Figure 5C), and stress responsiveness (Figure 5D). The plant phytohormone responsiveness category (258/824) contained ABRE, AuxRR-core and CGTCA-motif, etc., and the ABRE motif (79/258) was an essential element in the promoter that was related to abscisic acid responsiveness. The stress responsiveness category (135/824) contained nine cis-acting regulatory elements and just consisted of six motifs. The MBS motif (35/135) was an essential element in the promoter that was related to drought-inducibility. Most of the *cis-elements* were in the plant growth and development category (420/824), which contained twenty-five cis-acting regulatory elements. Among them, the Box 4 motifs, G-box motifs and TCT motifs accounted for the largest proportion (50.48%), which are associated with light response. In addition, the motifs related to light response accounted for 44.17% (364/824) of the total motifs. These findings indicated that the expression pattern of the *DcTPS* genes may be regulated by the light period.

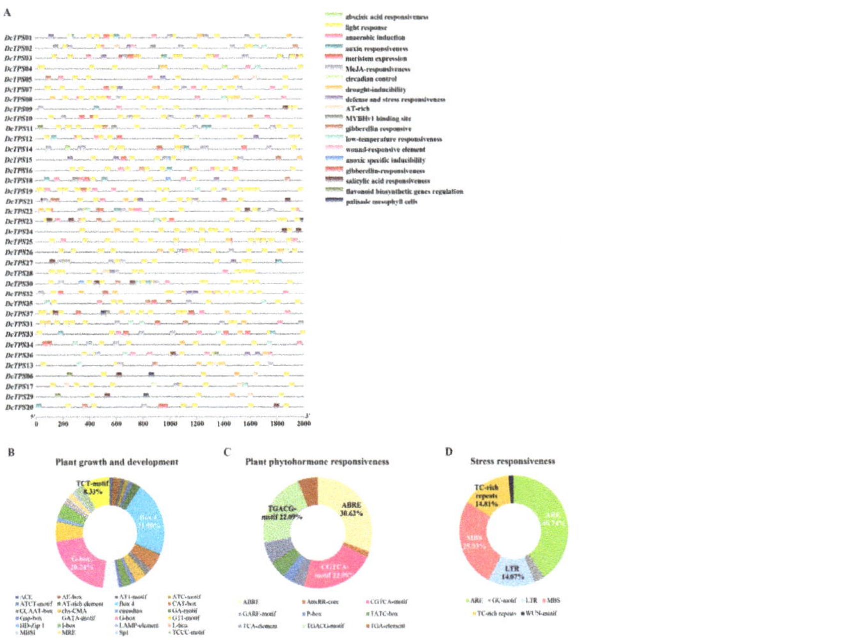

Figure 5. Promoter analysis of *DcTPS* genes: (**A**) prediction of *cis-elements* in the promoter regions. The number at the bottom represents the distance to the translation start codon, ATG; (**B**) the proportion of different *cis-elements* elements in plant growth and development; (**C**) the proportion of different *cis-elements* elements in plant phytohormone responsiveness; (**D**) the proportion of different *cis-elements* elements in and stress responsiveness.

3.5. Protein–Protein Interaction (PPI) Network Analysis

We used *A. thaliana* orthologue proteins to construct a PPI network using STRING software to further understand the association of the *DcTPS* proteins and predict their relationships. For reliability, only nine *DcTPS* proteins with the highest combined score were selected (Table S6). As shown in Figure 6, the *DcTPS*02 and *DcTPS*10 proteins are more closely related to other family members, both of which are members of the TPS-b subfamily. These interaction networks may provide important clues for understanding the function of unknown proteins, indicating the greater importance of these genes in the PPI network.

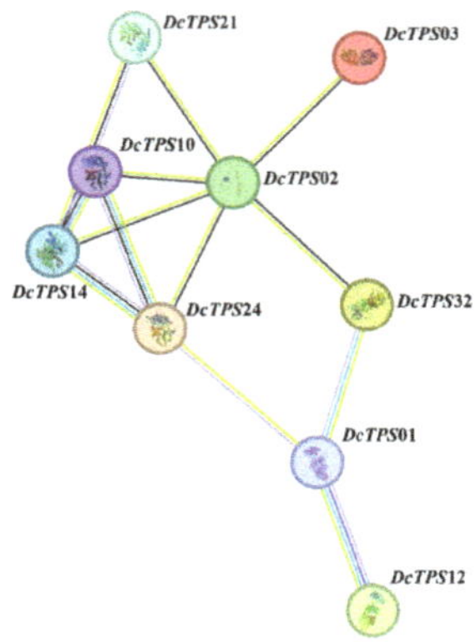

Figure 6. Protein–protein interaction (PPI) network analysis of the *DcTPS*.

3.6. GO Classification and Enrichment of DcTPS Genes

To further reveal the functions of the *DcTPSs*, gene ontology functional classification analysis was performed (Table S7). According to the classification results (Figure 7A), the molecular function annotation terms contained the majority of *DcTPS* genes (35/37), which were involved in catalytic activity and binding. GO enrichment analysis was carried out to better understand the preferred functional characteristics of the *DcTPS* genes, and these genes were mostly enriched in the terpenoid metabolic process, diterpenoid biosynthetic process, and diterpenoid metabolic process (Table S8, Figure 7B).

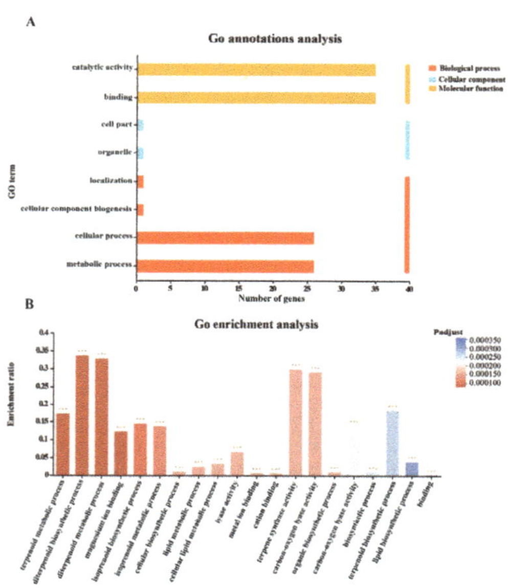

Figure 7. GO ontology annotations and enrichment analysis of *DcTPSs*. (**A**) GO ontology annotations analysis; (**B**) GO enrichment analysis. The *** in the figure indicate the statistical significance (adjusted *p*-value < 0.001).

3.7. Different Tissues Expression Analysis of DcTPS Genes

Abundant evidence has confirmed that TPS genes play an important role in plant growth and development. To understand the physiological function of *DcTPS* genes better, an RNA sequencing transcriptome database of different tissues including leaves, pseudobulbs, petals, sepals, labellums and gynandrium was established to study the expression patterns. A heat map of the hierarchical clustering was generated to show the expression profiles of the *DcTPS* genes in different tissues (Figure 8). The expression of four *DcTPS* genes (*DcTPS*05, 09, 30, 36) was not detected in any of the tissues analyzed, which may be due to differences in spatio-temporal expression patterns. The expression of *DcTPS*s revealed a tissue and organ-specific pattern, which showed a relatively high expression level in floral organs (petals, sepals, labellums and gynandrium), but they had a lower expression in leaves and pseudobulbs. Specifically, *DcTPS*02, *DcTPS*01, *DcTPS*10, *DcTPS*32, *DcTPS*03 and *DcTPS*18 were highly detected in floral organs, while *DcTPS*27, *DcTPS*20, *DcTPS*31 and *DcTPS*12 exhibited a high level of expression in leaves or pseudobulbs. Overall, the expression of TPS genes in flower organs was higher than in other tissues, and the differential expression of these *DcTPS* genes that were highly articulated in flower organs may have an important function in the flower fragrance of *D. chrysotoxum*, which needs to be further verified.

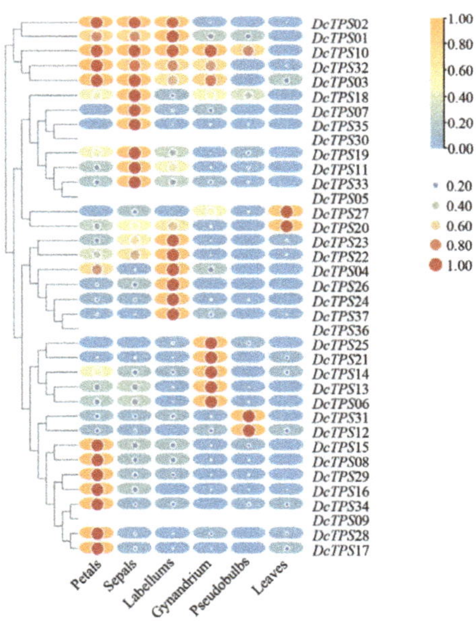

Figure 8. Differential tissue expression patterns of *DcTPS* genes. On the left is the tree diagram of sample clustering. The color in the figure represents the expression value of this gene after standardized treatment in each sample. The change from orange to blue within the round-rectangle shape indicated high to low expression levels. The change from red to blue within the circle shape indicated the expression quantity size by area. The numbers in the upper right color represent trends in gene expression levels. The FPKM expression values are listed in supplementary Table S9 and the heatmap was generated in TBtools [43].

3.8. Different Fluorescence Stages' Expression Analysis of DcTPS Genes

To further investigate the potential roles of *DcTPS* in flowers, the real-time reverse transcription quantitative PCR expression was confirmed to investigate *DcTPS* gene expression patterns at four floral developmental stages, and the sequences of primers used in qRT-PCR were listed in Table S10. Based on the FPKM values of genes highly expressed in flower organs, we selected six candidate genes (*DcTPS*02, *DcTPS*01, *DcTPS*10, *DcTPS*32, *DcTPS*03 and *DcTPS*18). The qRT-PCR results in Figure 9 show that the expression profiles of the six genes were expressed differently among the four fluorescence stages but that all these genes were mainly expressed in the full bloom stage. In particular, *DcTPS*02 had the highest expression levels during the full bloom stage. Moreover, *DcTPS*10 and *DcTPS*32 also had higher gene expression levels, which were similar to the tissue expression patterns displayed in Figure 8. Combined with the results of PPI (Figure 6), *DcTPS*02, *DcTPS*10 and *DcTPS*32 may play a crucial role in floral formation and synthesis.

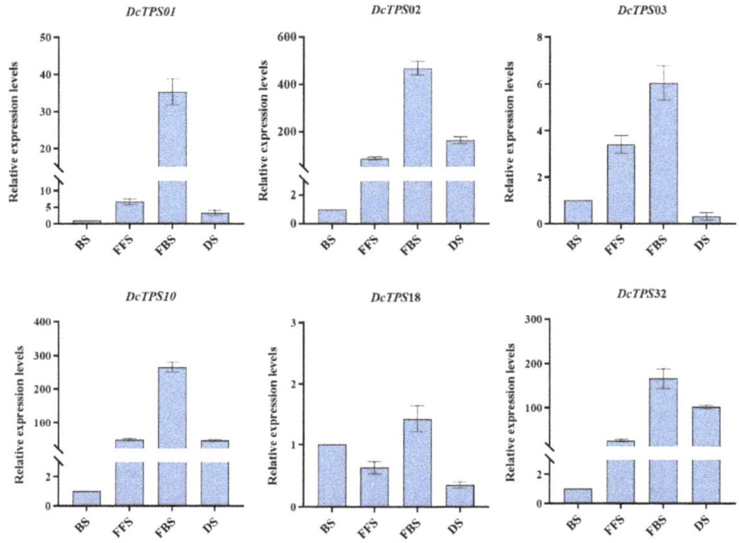

Figure 9. Analysis of *DcTPS* genes relative expression in different fluorescence stages. BS, Bud stage; FFS, First flowering stage; FBS, Full bloom stage; DS, Declining stage. The error bars indicate three biological replicates. Data are presented as means ± SD.

3.9. Functional Identification of DcTPS Genes in Plants

To explore the dominant roles of *DcTPS* genes in the terpene biological processes, three full-length CDS sequences of *DcTPS*02, *DcTPS*10 and *DcTPS*32 were cloned into vector pBI121 (primers used in clone were listed in Table S11) and transiently expressed in tobacco leaves (Supplementary Figures S1 and S2). The samples of transgenic tobacco leaves were taken five days after inoculation and the major emitted terpenes were analyzed by GC–MS and verified using the NIST Mass Spectral Library. As shown in Figure 10, the release of linalool in *DcTPS*10 and *DcTPS*02 was relatively high, reaching 100% and 93.03%, respectively. *DcTPS*32 was expressed in *D. chrysotoxum* at a lower level, and its expression was found to be correlated with the release of three products: α-pinene (15.27%), β-pinene (42.08%) and linalool (42.65%). Therefore, *DcTPS*02, *DcTPS*10 and *DcTPS*32 have been proven to be enzymes involved in monoterpene synthesis, and all are closely related to the synthesis of linalool.

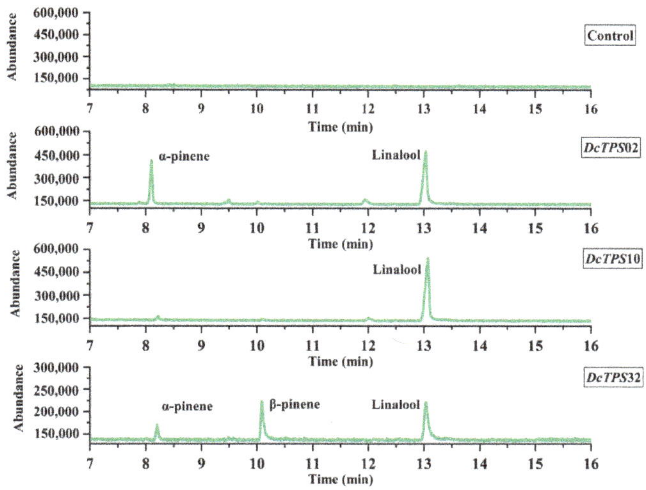

Figure 10. Functional characterization of *DcTPSs* in tobacco leaves by *Agrobacterium*-mediated infiltration. The X-axis represents the retention time of the peak outflow, and the Y-axis represents integrated area of the chromatographic peak. Reactions only performed with empty binary vectors were used as the blank control (these data were listed in Supplementary Table S12). New product peaks were observed in contrast to controls and were identified by comparing with the NIST 2005 Mass Spectral Library.

4. Discussion

4.1. Comprehensively Investigated TPS Gene Family in D. chrysotoxum

Dendrobium chrysotoxum is one of the rare orchids in karst areas, with a variety of epiphytic and terrestrial growth forms [44]. It has a long history of cultivation because of its unique floral fragrance and beautiful flower shape, which has high ornamental value. Terpene Synthase (TPS) is a key enzyme and is responsible for producing the immense diversity of terpene derivatives as well as playing vital roles in floral fragrance and pollinators attraction. In recent years, in an attempt to understand TPS genes, some reports by different research groups have independently carried out genome-wide identification of the TPS gene family in *A. thaliana* (32 members) [20,45], *Selaginella moellendorfii* (14 members) [46], *Glycine max* (14 members), *Dendrobium officinale* (34 members) [34], *Cymbidium faberi* (32 members) [35] and *Freesia* (8 members) [16]. Nevertheless, the diversity (across various dimensions) and characterization of the TPS gene family in *D. chrysotoxum* remain inadequately explored in a comprehensive manner. On the basis of a previous study, we investigated the TPS genes in the *D. chrysotoxum* genome and a total of 37 TPS genes (*DcTPS*1-37) were identified. A comprehensive analysis was conducted on their physicochemical properties, covering protein length, molecular weight, theoretical isoelectric point, instability index, aliphatic index, and grand average of hydropathicity. In order to gain a deeper understanding of the *DcTPS* gene family, we carried out analyses focusing on their phylogenetic classification, conserved motifs, exon–intron organization, *cis-elements*, protein–protein interaction (PPI) network, and expression patterns in different tissues and fluorescence stages. Finally, we screened out the key floral expression *DcTPS* genes and identified their functions. This study represents the first thorough analysis of the evolution of the TPS family in *D. chrysotoxum*, and the resulting information is undoubtedly valuable for providing additional evidence to infer the potential functions of key TPS genes in terpene synthesis.

4.2. TPS-b Was a Dominant Subfamily in D. chrysotoxum

Phylogenetic analysis of the 69 TPS members (comprising 37 from *D. chrysotoxum*, 32 *A. thaliana*) were divided into five subgroups according to their sequence homology and classification from *A. thaliana*. In our study, *AtTPS*s were classified into five subclasses, while all *DcTPS*s belong to only three subclasses (TPS-a, TPS-b and TPS-c). Among them, the TPS-b subclass has 17 *DcTPS* genes, which was the largest subfamily among the *DcTPS*s. Similar findings were also found in *D. officinale* (16 of the 34 *DoTPS* genes were TPS-b genes) [34], *Freesia* (4/8) [16] and *C. faberi* (15/32) [35]. However, this was not consistent with *A. thaliana* [45], *Oryza sativa* [7], *Sorghum bicolor* [14], *Solanum lycopersicum* [21] and *Liriodendron chinense* [47], which have a dominant subfamily TPS-a. These results further confirmed that orchids contained more TPS genes in TPS-b, and TPS-b was also the most expanded classification in orchids, with a dominant subfamily [34,48].

The genes in the TPS-b subgroup are mainly responsible for the synthesis of monoterpenes. Monoterpene volatiles are the main terpenoids in orchids, which are dominant in the flower scent of orchids. Many studies have confirmed that TPS-b protein with GPP as substrate in orchids can produce large amounts of monoterpenes. For instance, in *Freesia*, *FhTPS*1 and *FhTPS*2 had the ability to convert GPP into linalol and α-terpinol [16]. In *Cymbidium faberi*, *CfTPS*18 could convert GPP to β-myrcene, geraniol, and α-pinene in *Escherichia coli* [35]. The *DoTPS*10 of *D. officinale*, uniquely converted GPP to linalool in *E. coli* BL21 [34].

4.3. TPS-b Subfamily Genes Have Conserved Motifs Encoding Monoterpenes

Thirty-six DcTPSs (except for DcTPS31) contained the DDxxD motif, and thirty-three DcTPSs (except for DcTPS9, 11, 20, and 30) contained the RRX8W motif. In plants, DDxxD was an aspartate-rich motif that could interact with divalent metal ions and participate in positioning the substrate for catalysis [49]. The RRX8W motif was involved in the production of cyclic monoterpenes and was absent in producing acyclic products [14]. In our present study, all the DcTPSs in the TPS-b contained the conserved RRX8W motif, which proved to be critical for floral fragrance release due to its association with encoding monoterpenes.

In general, genes grouped into the same subclasses tend to display comparable evolutionary features and share a consistent motif arrangement structure. Seventeen *DcTPS* genes were clustered into the TPS-b subclass, and analysis of the conserved motifs of the *D. chrysotoxum* proteins further corroborates the categorization of the *DcTPS* family. Except for *DcTPS*31, at least seven identical conserved motifs have been identified within the TPS-b subclass. These motifs are likely to be crucial for its functional role. It is important to highlight that *DcTPS*10 and *DcTPS*02 within the TPS-b subclass exhibited a closer relationship with other family members in the PPI network. These findings indicate that the genes within the TPS-b subfamily warrant additional exploration into their functional roles.

Different elements in promoter areas were found in all 37 *DcTPS* genes of *D. chrysotoxum*, and they belonged to three clusters. Most of the *cis-elements* were in the plant growth and development clusters, among which the number of Box 4, G-box and TCT motifs associated with light-response contained most of this cluster. The results indicated that the expression patterns of *DcTPS* may be regulated by light treatment and may respond to multiple environmental stresses.

GO annotation analysis indicated that the biological process contained most *DcTPS* genes, and these genes were mostly enriched in the terpenoid metabolic process. The function of the TPS genes in controlling terpenoid biosynthesis has been well-documented in many plant species. The formation of floral scent is a complex process that is associated with the *DcTPS* genes. In our study, most of the TPS-b subclass like *DcTPS*01, *DcTPS*02,

DcTPS03, *DcTPS10*, *DcTPS18* and *DcTPS32* had high expression levels in the floral organs, indicating that these genes are closely related to the formation of flower fragrance. The expression through qRT-PCR analysis of the above genes showed that *DcTPS02*, *DcTPS10* and *DcTPS03* had the highest expression levels during the full bloom stage. Functional identification of *DcTPS* genes showed that *DcTPS10* was associated with the release of linalool, *DcTPS02* was correlated with linalool and α-pinene, and *DcTPS32* had a very close link with α-pinene, β-pinene and linalool. According to our categorized information, these three genes were all in the TPS-b subclass, which is annotated as monoterpene synthases. Therefore, by selecting 37 *DcTPS* genes, we found that *DcTPS02*, *DcTPS10* and *DcTPS32* were closely related to the synthesis of linalool, which may play crucial roles in floral scent and attracting pollinators in *D. chrysotoxum*. The study of TPS family genes showed that these *DcTPS*s have common features, such as structural similarity, sequence conservation, etc., which makes the transferred *DcTPS* exhibit similar functions, and these transgenic plants showed improved biosynthesis of terpenes. However, the expression differences among different family members and the internal regulatory mechanisms of different terpenoids between genes still need to be further explored and analyzed.

This comprehensive study provides a molecular basis for further investigation of volatile terpenes in *D. chrysotoxum* flowers. The characterization of key *DcTPS* genes could also have genetic potential value for the formation of major terpenes in *D. chrysotoxum* and other related species.

5. Conclusions

Monoterpene volatiles are the main components of floral fragrance in orchidaceae plants, and TPS is the ultimate enzyme in their biosynthesis. The diversity in the TPS gene family structure specifies their vast range of functions. Until recently, the *DcTPS* genes' regulatory system during plant growth and development was poorly understood. In the current study, *DcTPS* family members were identified through gene family analysis and a comprehensive analysis of members was performed, including the location of conserved motifs, gene structure and promoter cis-acting elements, etc. Gene evolution and selection analysis elucidated the diversity and extension of the family members. In addition, we also investigated the expression of *DcTPS* genes in different tissues and fluorescence stages, and further selected highly expressed *DcTPS*s for functional identification in plants. The discovery of three TPS-b genes specifically driving linalool-based scent in *D. chrysotoxum* provides fundamental insights into the family, which can be utilized in future studies of the TPS-mediated molecular mechanisms underlying plant growth and development. This study may serve as a valuable reference for the functional exploration and molecular breeding of novel aromatic species, paving the way for groundbreaking advancements in this captivating field.

Supplementary Materials: The following supporting information can be downloaded at: https://www.mdpi.com/article/10.3390/horticulturae11060566/s1, Table S1. List of the TPSs sequences in *D. chrysotoxum*. Table S2. A catalog of *DcTPS* genes with their HMM profiles. Table S3. TPS genes used in *A. thaliana*. Table S4. Conserved motifs of DcTPSs. Table S5. Cis-elements in the promoter regions of DcTPS genes. Table S6. Protein–protein interaction (PPI) network analysis of DcTPS genes. Table S7. GO classification analysis of DcTPS genes. Table S8. GO enrichment analysis of DcTPS genes. Table S9. The FPKM values of DcTPS genes in differential tissues. Table S10. Sequences of primers used in qRT-PCR. Table S11. Primer sequence for DcTPS genes clone. Table S12. The release amounts of terpenes were analyzed by GC-MS; Figure S1: pBI121-EGFP plasmid map. Figure S2: Construction of pBI121-DcTPS vectors. Figure S3: Stability analysis of ACT-1 in cultivars of *D. chrysotoxum* in different fluorescence stages and different tissues.

Author Contributions: Conceptualization, Y.Y.; Methodology, J.G.; Software, Y.Y. and R.N.; Investigation, Q.L. and K.X.; Writing—original draft, Y.Y.; Writing—review and editing, S.Q. and Z.W.; Visualization, Y.Y. and R.N.; Project administration, S.Q.; Funding acquisition, S.Q. and Z.W. All authors have read and agreed to the published version of the manuscript.

Funding: This work was supported by the Natural Science Foundation of Guangxi Zhuang Autonomous Region (2025GXNSFAA069846), the National Natural Science Foundation of China (31560567), the Guangdong Province Modern Agriculture Industry Common Key Technology R&D Innovation Team Construction Project, Facilities and Equipment (Planting Direction) Common Key Technologies (2024CXTD01), the Start-up Fund of Innovation Team of Guangxi Academy of Sciences for Innovation and Utilization of Germplasm in Horticultural Crops (CQZ-E-1919), the Guangxi Light of the Future Visiting Research Program Project (Scientific and Technological Talents Specialized project, 2024), the Fundamental Research Fund of Guangxi Institute of Botany (23011), the Guilin Scientific Research and Technology Development Plan (20220135-1), the Guangxi Key Research and Development Program (Guike AB24010138) and the fund of the Guangxi Key Laboratory of Plant Functional Phytochemicals and Sustainable Utilization (ZRJJ2023-1 and ZRJJ2024-13).

Data Availability Statement: The original contributions presented in the study are included in the article/Supplementary Material, further inquiries can be directed to the corresponding authors.

Conflicts of Interest: The authors declare no conflicts of interest.

Abbreviations

aa amino acid

References

1. Sun, Z.; Song, Y.; Zeng, R. Advances in studies on intraspecific and interspecific relationships mediated by plant volatiles. *J. S. China Agric. Univ.* **2019**, *40*, 166–174. [CrossRef]
2. Wu, J.N.; Liu, Z.Y.; Wu, M.Q. Analysis of Volatile Components in the Flowers of Seven Species of Dendrobium. *Mol. Plant Breed.* **2023**, *3*, 1–16. Available online: http://kns.cnki.net/kcms/detail/46.1068.S.20230314.0905.004.html (accessed on 8 January 2022)
3. Dudareva, N.; Klempien, A.; Muhlemann, J.K.; Kaplan, I. Biosynthesis, function and metabolic engineering of plant3volatile organic compounds. *New Phytol.* **2013**, *198*, 16–32. [CrossRef]
4. Abbas, F.; Yu, Y.; Yue, R.; Yue, Y.; Amanullah, S.; Jahangir, M.M.; Fan, Y. Volatile terpenoids: Multiple functions, biosynthesis, modulation and manipulation by genetic engineering. *Planta Int. J. Plant Biol.* **2017**, *246*, 115–129. [CrossRef]
5. Yu, F.; Utsumi, R. Diversity, regulation, and genetic manipulation of plant mono- and sesquiterpenoid biosynthesis. *Cell. Mol. Life Sci.* **2009**, *66*, 3043–3052. [CrossRef] [PubMed]
6. Baek, Y.S.; Ramya, M.; An, H.R.; Park, P.M.; Lee, S.Y.; Baek, N.I.; Park, P.H. Volatiles Profile of the Floral Organs of a New Hybrid Cymbidium, 'Sunny Bell' Using Headspace Solid-Phase Microextraction Gas Chromatography-Mass Spectrometry Analysis. *Plants* **2019**, *8*, 251. [CrossRef]
7. Bohlmann, J.; Meyer-Gauen, G.; Croteau, R. Plant terpenoid synthases: Molecular biology and phylogenetic analysis. *Proc. Natl. Acad. Sci. USA* **1998**, *95*, 4126–4133. [CrossRef]
8. Huang, L.-M.; Huang, H.; Chuang, Y.-C.; Chen, W.-H.; Wang, C.-N.; Chen, H.-H. Evolution of Terpene Synthases in Orchidaceae. *Int. J. Mol. Sci.* **2021**, *22*, 6947. [CrossRef]
9. Flath, R.A.; Ohinata, K. Volatile Components of the Orchid Dendrobium superbum Rchb. f. *J. Agric. Food Chem.* **1982**, *30*, 841–842. [CrossRef]
10. Vranová, E.; Coman, D.; Gruissem, W. Network Analysis of the MVA and MEP Pathways for Isoprenoid Synthesis. *Annu. Rev. Plant Biol.* **2013**, *64*, 665–700. [CrossRef]
11. Pulido, P.; Perello, C.; Rodriguez-Concepcion, M. New Insights into Plant Isoprenoid Metabolism. *Mol. Plant* **2012**, *5*, 964–967. [CrossRef] [PubMed]
12. Simkin, A.J.; Guirimand, G.; Papon, N.; Courdavault, V.; Thabet, I.; Ginis, O.; Bouzid, S.; Giglioli-Guivarc'h, N.; Clastre, M. Peroxisomal localisation of the final steps of the mevalonic acid pathway in planta. *Plant Signal. Behav.* **2011**, *234*, 903–914. [CrossRef] [PubMed]
13. Sandra, I.; Krause, S.T.; Grit, K.; Jonathan, G.; Jörg, D.; Köllner, T.G. The organ-specific expression of terpene synthase genes contributes to the terpene hydrocarbon composition of chamomile essential oils. *Bmc Plant Biol.* **2012**, *12*, 84–97. [CrossRef]

14. Chen, F.; Tholl, D.; Bohlmann, J.; Pichersky, E. The family of terpene synthases in plants: A mid-size family of genes for specialized metabolism that is highly diversified throughout the kingdom. *Plant J.* **2011**, *66*, 212–229. [CrossRef]
15. Degenhardt, J.; Kllner, T.G.; Gershenzon, J. Monoterpene and sesquiterpene synthases and the origin of terpene skeletal diversity in plants. *Phytochemistry* **2009**, *69*, 1621–1637. [CrossRef]
16. Gao, F.; Liu, B.; Li, M.; Gao, X.; Fang, Q.; Liu, C.; Ding, H.; Wang, L.; Gao, X. Identification and Characterization of Terpene Synthase Genes Accounting for the Volatile Terpene Emissions in Flowers of Freesia × hybrida. *J. Exp. Bot.* **2018**, *69*, 4249–4265. [CrossRef] [PubMed]
17. Bateman, A.; Coin, L.; Durbin, R.; Finn, R.D.; Hollich, V.; Griffiths-Jones, S.; Khanna, A.; Marshall, M.; Moxon, S.; Sonnhammer, E.L.L.; et al. The Pfam Protein Families Database. *Nucleic Acids Res.* **2004**, *32*, D138–D141. [CrossRef]
18. Jiang, S.Y.; Jin, J.J.; Sarojam, R.; Ramachandran, S. A Comprehensive Survey on the Terpene Synthase Gene Family Provides New Insight into Its Evolutionary Patterns. *Genome Biol. Evol.* **2019**, *11*, 2078–2098. [CrossRef]
19. Trapp, S.C.; Croteau, R.B. Genomic Organization of Plant Terpene Synthases and Molecular Evolutionary Implications. *Genetics* **2001**, *158*, 811–832. [CrossRef]
20. Tholl, D.; Lee, S. Terpene Specialized Metabolism in Arabidopsis thaliana. *Arab. Book* **2011**, *9*, e0143. [CrossRef]
21. Falara, V.; Akhtar, T.A.; Nguyen, T.T.H.; Spyropoulou, E.A.; Bleeker, P.M. The tomato (*Solanum lycopersicum*) terpene synthase gene family. *Plant Physiol.* **2011**, *157*, 770–789. [CrossRef] [PubMed]
22. Martin, D.M.; Aubourg, S.; Schouwey, M.B.; Daviet, L.; Schalk, M.; Toub, O.; Lund, S.T.; Bohlmann, J. Functional Annotation, Genome Organization and Phylogeny of the Grapevine (Vitis vinifera) Terpene Synthase Gene Family Based on Genome Assembly, FLcDNA Cloning, and Enzyme Assays. *BMC Plant Biol.* **2010**, *10*, 226. [CrossRef] [PubMed]
23. Zhang, G.Q.; Liu, K.W.; Li, Z.; Lohaus, R.; Hsiao, Y.Y.; Niu, S.C.; Wang, J.Y.; Lin, Y.C.; Xu, Q.; Chen, L.J.; et al. The Apostasia genome and the evolution of orchids. *Nature* **2017**, *549*, 379–383. [CrossRef]
24. Kaundal, R.; Zhao, S.P.X. Combining Machine Learning and Homology-Based Approaches to Accurately Predict Subcellular Localization in *Arabidopsis*. *Plant Physiol.* **2010**, *154*, 36–54. [CrossRef]
25. Chou, K.C.; Shen, H.B. Plant-mPLoc: A Top-Down Strategy to Augment the Power for Predicting Plant Protein Subcellular Localization. *PLoS ONE* **2010**, *5*, e11335. [CrossRef]
26. Ojeda, D.I.; Santosguerra, A.; Olivatejera, F.; Valido, A.; Xue, X.; Marrero, A.; Caujapé-Castells, J.; Cronk, Q. Bird-pollinated Macaronesian Lotus (Leguminosae) evolved within a group of entomophilous ancestors with post-anthesis flower color change. *Perspect. Plant Ecol. Evol. Syst.* **2013**, *15*, 193–204. [CrossRef]
27. Byers, K.J.R.P.; Vela, J.P.; Peng, F.; Riffell, J.A.; Bradshaw, H.D. Floral volatile alleles can contribute to pollinator-mediated reproductive isolation in monkeyflowers (Mimulus). *Plant J. Cell Mol. Biol.* **2015**, *80*, 1031–1042. [CrossRef]
28. Knudsen, J.T.; Eriksson, R.; Gershenzon, J.; Ståhl, B. Diversity and distribution of floral scent. *Bot. Rev.* **2006**, *72*, 1. [CrossRef]
29. Ramsey, J.; Bradshaw, H.D.; Schemske, D.W. Components of reproductive isolation between the monkeyflowers mimulus lewisii and m. cardinalis (phrymaceae). *Evolution* **2003**, *57*, 1520–1534. [CrossRef]
30. Dudareva, N.; Murfitt, L.M.; Mann, C.J.; Gorenstein, N.; Kolosova, N.; Kish, C.M.; Bonham, C.; Wood, K. Developmental Regulation of Methyl Benzoate Biosynthesis and Emission in Snapdragon Flowers. *Plant Cell* **2000**, *12*, 949–961. [CrossRef]
31. Yang, Y.; Xia, K.; Wu, Q.; Lu, X.; Lu, S.; Zhao, Z.; Qiu, S. Combined Analysis of Volatile Compounds and Extraction of Floral Fragrance Genes in Two *Dendrobium* Species. *Horticulturae* **2023**, *9*, 745. [CrossRef]
32. Burkhardt, D.; Mosandl, A. Biogenetic Studies in Syringa vulgaris L.: Bioconversion of $^{18}O(^{2}H)$-Labeled Precursors into Lilac Aldehydes and Lilac Alcohols. *J. Agric. Food Chem.* **2003**, *51*, 7391–7395. [CrossRef] [PubMed]
33. Shalit, M.; Guterman, I.; Volpin, H.; Bar, E.; Tamari, T.; Menda, N.; Adam, Z.; Zamir, D.; Vainstein, A.; Weiss, D.; et al. Ester Formation in Roses. Identification of an Acetyl-Coenzyme A. Geraniol/Citronellol Acetyltransferase in Developing Rose Petals. *Plant Physiol.* **2003**, *131*, 1868–1876. [CrossRef] [PubMed]
34. Yu, Z.M.; Zhao, C.H.; Zhang, G.H.; Silva, J.A.T.; Duan, J. Genome-Wide Identification and Expression Profile of TPS Gene Family in *Dendrobium officinale* and the Role of *DoTPS*10 in Linalool Biosynthesis. *Int. J. Mol. Sci.* **2020**, *21*, 5419. [CrossRef]
35. Wang, Q.Q.; Zhu, M.J.; Yu, X.; Bi, Y.Y.; Zhou, Z.; Chen, M.K.; Chen, J.; Zhang, D.; Ai, Y.; Liu, Z.J.; et al. Genome-Wide Identification and Expression Analysis of Terpene Synthase Genes in *Cymbidium faberi*. *Front. Plant Sci.* **2021**, *12*, 751853. [CrossRef]
36. Chuang, Y.C.; Hung, Y.C.; Tsai, W.C.; Chen, H.H. PbbHLH4 regulates floral monoterpene biosynthesis in *Phalaenopsis* orchids. *J. Exp. Bot.* **2018**, *69*, 4363–4377. [CrossRef]
37. Green, S.A.; Chen, X.; Nieuwenhuizen, N.J.; Matich, A.J.; Wang, M.Y.; Bunn, B.J.; Yauk, Y.K.; Atkinson, R.G. Identification, functional characterization, and regulation of the enzyme responsible for floral (E)-nerolidol biosynthesis in kiwifruit (*Actinidia chinensis*). *J. Exp. Bot.* **2011**, *63*, 1951–1967. [CrossRef]
38. Nieuwenhuizen, N.J.; Wang, M.Y.; Matich, A.J.; Green, S.A.; Chen, X.; Yauk, Y.K.; Beuning, L.L.; Nagegowda, D.A.; Dudareva, N.; Atkinson, R.G. Two terpene synthases are responsible for the major sesquiterpenes emitted from the flowers of kiwifruit (*Actinidia deliciosa*). *J. Exp. Bot.* **2009**, *60*, 3203–3219. [CrossRef]

39. Feng, L.G.; Chen, C.; Sheng, L.X.; Liu, P.; Tao, J.; Su, J.L.; Zhao, L.Y. Comparative analysis of headspace volatiles of *Chinese Rosa rugosa*. *Molecules* **2010**, *15*, 8390–8399. [CrossRef]
40. Heller, S.R.; Milne, G.W. *EPA/NIH Mass Spectral Data Base*; U.S. Governwent Printing Office: Washington, DC, USA, 1980.
41. Zhang, H.; Chen, M.; Wang, X.; Dai, J.; Zhang, X.; Zhang, Z.; Zhang, X.; Tang, M.; Tang, J.; Gong, J.; et al. Transcriptome Analysis of *Rhododendron liliiflorum* H. Lév. Flower Colour Differences. *Horticulturae* **2023**, *9*, 82. [CrossRef]
42. Lescot, M.; Déhais, P.; Thijs, G.; Marchal, K.; Moreau, Y.; Van de Peer, Y.; Rouzé, P.; Rombauts, S. PlantCARE, a database of plant cis-acting regulatory elements and a portal to tools for in silico analysis of promoter sequences. *Nucleic Acids Res.* **2002**, *30*, 325–327. [CrossRef]
43. Chen, C.; Chen, H.; Zhang, Y.; Thomas, H.R.; Frank, M.H.; He, Y.; Xia, R. TBtools, a Toolkit for Biologists integrating various HTS-data handling tools with a user-friendly interface. *Cold Spring Harb. Lab.* **2020**, *13*, 1194–1202. [CrossRef]
44. Hsiao, Y.Y.; Pan, Z.J.; Hsu, C.C.; Yang, Y.P.; Hsu, Y.C.; Chuang, Y.C.; Shih, H.H.; Chen, W.H.; Tsai, W.C.; Chen, H.H. Research on Orchid Biology and Biotechnology. *Plant Cell Physiol.* **2011**, *52*, 1467–1486. [CrossRef] [PubMed]
45. Aubourg, S.; Lecharny, A.; Bohlmann, J. Genomic analysis of the terpenoid synthase (*AtTPS*) gene family of *Arabidopsis thaliana* *Mol. Genet. Genom.* **2002**, *267*, 730–745. [CrossRef] [PubMed]
46. Li, G.L.; Köllner, T.G.; Yin, Y.B.; Jiang, Y.F.; Chen, H.; Xu, Y.; Gershenzon, J.; Pichersky, E.; Chen, F. Nonseed plant Selaginella moellendorfii has both seed plant and microbial types of terpene synthases. *Proc. Natl. Acad. Sci. USA* **2012**, *109*, 14711–14715 [CrossRef]
47. Cao, Z.; Ma, Q.; Weng, Y.; Shi, J.; Chen, J.; Hao, Z. Genome-Wide Identification and Expression Analysis of TPS Gene Family in *Liriodendron chinense*. *Genes* **2023**, *14*, 770. [CrossRef]
48. Li, N.; Dong, Y.; Lv, M.; Qian, L.; Sun, X.; Liu, L.; Cai, Y.; Fan, H. Combined Analysis of Volatile Terpenoid Metabolism and Transcriptome Reveals Transcription Factors Related to Terpene Synthase in Two Cultivars of Dendrobium officinale Flowers *Front. Genet.* **2021**, *12*, 661296. [CrossRef]
49. Zeng, X.; Liu, C.; Zheng, R.; Cai, X.; Luo, J.; Zou, J.; Wang, C. Emission and Accumulation of Monoterpene and the Key Terpene Synthase (TPS) Associated with Monoterpene Biosynthesis in *Osmanthus fragrans Lour*. *Front. Plant Sci.* **2016**, *12*, 1232. [CrossRef]

horticulturae

Article

Bioinformatics Analysis and Expression Features of Terpene Synthase Family in *Cymbidium ensifolium*

Mengyao Wang [1,2,†], Baojun Liu [1,†], Jinjin Li [2], Ningzhen Huang [1], Yang Tian [2], Liting Guo [2], Caiyun Feng [2], Ye Ai [2,*] and Chuanming Fu [1,*]

1 Guangxi Institute of Botany, Guangxi Zhuang Autonomous Region and Chinese Academy of Sciences, Guilin 541006, China; wmyao0818@163.com (M.W.); bjliumail@126.com (B.L.); hnzhen2002@163.com (N.H.)
2 Key Laboratory of National Forestry and Grassland Administration for Orchid Conservation and Utilization, College of Landscape Architecture and Art, Fujian Agriculture and Forestry University, Fuzhou 350002, China; ljj1107@fafu.edu.cn (J.L.); tianyang_86@163.com (Y.T.); lt3548607031@163.com (L.G.); caiyun717@fafu.edu.cn (C.F.)
* Correspondence: aiye@fafu.edu.cn (Y.A.); fuchuanmingming@163.com (C.F.)
† These authors contributed equally to this work.

Abstract: Terpene synthases (TPSs) are crucial for the diversification of terpenes, catalyzing the formation of a wide variety of terpenoid compounds. However, genome-wide systematic characterization of *TPS* genes in *Cymbidium ensifolium* has not been reported. Within the genomic database of *C. ensifolium*, we found 30 *CeTPS* genes for this investigation. *CeTPS* genes were irregularly distributed throughout the seven chromosomes and primarily expanded through tandem duplications. The CeTPS proteins were classified into three TPS subfamilies, including 17 TPS-b members, 8 TPS-a members, and 5 TPS-c members. Conserved motif analysis showed that most CeTPSs contained DDxxD and RRX_8W motifs. *Cis*-element analysis of *CeTPS* gene promoters indicated regulation primarily by plant hormones and stress. Transcriptome analysis revealed that *CeTPS1* and *CeTPS18* had high expression in *C. ensifolium* flowers. qRT-PCR results showed that *CeTPS1* and *CeTPS18* were predominantly expressed during the flowering stage. Furthermore, CeTPS1 and CeTPS18 proteins were localized in the chloroplasts. These results lay the theoretical groundwork for future research on the functions of *CeTPSs* in terpenoid biosynthesis.

Keywords: terpene synthase; *Cymbidium ensifolium*; TPS gene family; expression patterns

1. Introduction

Terpenoids constitute the most diverse and structurally complex class of compounds in plants [1,2]. They are essential to the attraction of pollinators to plants, protection from pathogens and predators, and exchange of information [3,4]. Moreover, terpenoids are utilized extensively due to their distinct flavor and scent in various types of industries, including cosmetics, food, perfume, and pharmaceuticals [5,6]. Terpenoids, composed of isoprene (C5) units, include monoterpenes (C10), sesquiterpenes (C15), and diterpenes (C20) [7]. Plants produce C5 isoprenoid precursors through two distinct pathways: the methylerythritol phosphate (MEP) pathway in the plastid and the mevalonate (MVA) pathway in the cytoplasm. The isomers dimethylallyl diphosphate (DMAPP) and isopentenyl diphosphate (IPP) are produced via these two pathways [8]. Under the catalytic action of various isoprenyl diphosphate synthases (IPSs), different quantities of IPP and DMAPP synthesize the terpenoid precursors geranyl pyrophosphate (GPP), farnesyl pyrophosphate (FPP), and geranylgeranyl pyrophosphate (GGPP) (Figure 1) [9]. The structural characteristics of terpene synthases (TPSs) are fundamental to the generation of terpene diversity [10]. TPSs are classified as monoterpene synthases, sesquiterpene synthases, and diterpene synthases, which synthesize the skeletons of monoterpene, sesquiterpene, and diterpene compounds using GPP, FPP, and GGPP as precursor substrates [11]. The terpene skeleton forms more diverse compounds in the modification of cytochrome P450 monooxygenases, glycosyltransferase, and acyltransferase [9,12,13].

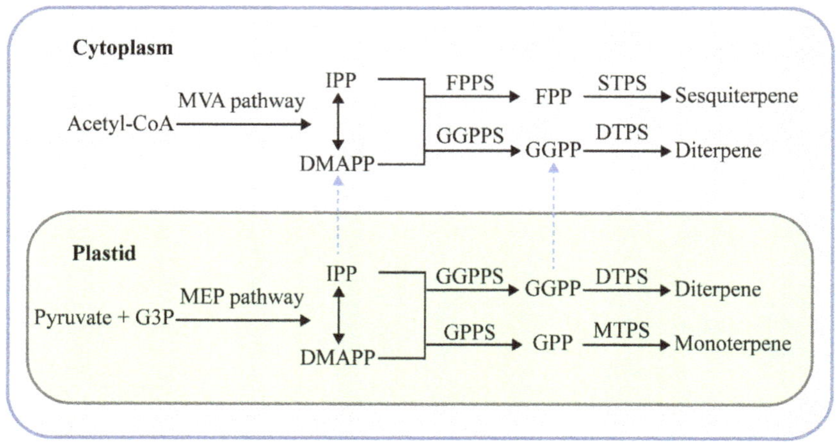

Figure 1. Pathways of terpenoid biosynthesis in plants. Two precursor substances of terpenoids are produced by the cytosolic mevalonate (MVA) and plastid methylerythritol phosphate (MEP) pathways. G3P: glyceraldehyde 3-phosphate; IPP: isopentenyl diphosphate; DMAPP: dimethylallyl diphosphate; FPPS: FPP synthase; GGPPS: GGPP synthase; GPPS: GPP synthase; FPP: farnesyl pyrophosphate; GGPP: geranylgeranyl pyrophosphate; GPP: geranyl pyrophosphate; STPS: sesquiterpene synthase; DTPS: diterpene synthase; MTPS: monoterpene synthase.

Most TPS proteins exhibit the RRX_8W motif near the N-terminus, as well as the DDxxD and NSE/DTE motifs near the C-terminus [14]. The TPS gene family can be divided into seven subfamilies, including angiosperm-specific subfamilies TPS-a, TPS-b, and TPS-g, the gymnosperm-specific subfamily TPS-d, and subfamilies TPS-c and TPS-e/f, which are shared by both angiosperms and gymnosperms [15]. The TPS-h subfamily is found exclusively in the *Selaginella moellendorffii* [16]. As the largest subfamily, TPS-a encodes sesquiterpene synthases in both monocot and dicot plant species. The TPS-b subfamily encodes monoterpene synthases with the RRX_8W motif, while the TPS-g subfamily, which lacks this motif, encodes both monoterpene and sesquiterpene synthases. The "DxDD" motif, rather than the "DDxxD" sequence, distinguishes the TPS-c subfamily, which encodes diterpene synthase enzymes [15]. There are several *TPS* genes found in the genomes of *Arabidopsis thaliana* [17], *Vitis vinifera* [18], Rosaceae [19], *Cinnamomum camphora* [20], and *Camellia sinensis* [21].

Orchidaceae is one of the most diverse families among the angiosperms [22]. Orchid evolution has been significantly influenced by the interactions between pollinators and orchids [23]. Terpenoids are one of the main components of floral scent in orchids. Floral scents enhance horticultural aesthetics and improve fertilization rates by attracting pollinators [24,25]. Monoterpenes and sesquiterpenes are the primary volatile compounds in the floral scents of *Phalaenopsis bellina* and *Cymbidium goeringii* [26,27]. Currently, the genomes of multiple Orchidaceae plants contain the TPS gene families, including *Apostasia shenzhenica*, *P. equestris* [28], *Dendrobium officinale* [29], *C. faberi* [30], *C. goeringii* [31], and *D. chrysotoxum* [32]. Several *TPS* genes in Orchidaceae are involved in terpenoid biosynthesis. For instance, the DoTPS10 protein in *D. officinale* converts GPP directly to linalool [29]. Transient overexpression of *DoGES1* in *Nicotiana benthamiana* leaves promotes the accumulation of geraniol [33]. The CfTPS18 protein in *C. faberi* can convert GPP into β-myrcene, geraniol, and α-pinene in *Escherichia coli* [30]. In *P. bellina*, TPS-b and TPS-e/f subfamily members are involved in monoterpene biosynthesis in flowers [34].

C. ensifolium is highly valued for its ornamental qualities, including fragrance, color, and flower form. Monoterpenes and sesquiterpenes are key components of the floral scent in *C. ensifolium* [35]. However, the molecular processes that drive terpenoid production in *C. ensifolium* remain unexplored. We identified 30 *CeTPS* gene family members and named them according to their chromosomal locations, then analyzed their physicochemical properties, phylogenetic relationships, gene structures, collinearity, and *cis*-regulatory elements. Expression patterns in various organs and flower developmental stages were examined using transcriptome sequencing data and quantitative real-time PCR (qRT-PCR) experiments, identifying genes potentially involved in terpenoid biosynthesis. Subcellular localization of key TPS proteins was also investigated. Our analysis provides a theoretical basis for future research on terpene metabolism and regulation in *C. ensifolium*.

2. Materials and Methods

2.1. Plant Materials

Plant materials for the study were sourced from the greenhouse at the Forest Orchid Garden, Fujian Agriculture and Forestry University. All the plant materials were cultivated under natural conditions, with temperatures ranging from 25 °C to 35 °C, humidity levels between 75% and 90%, and day length averaging 13 to 14 h. Samples of different organs (leaf, Le; root, Ro; pseudobulb, Ps; flower, Fl) of *C. ensifolium* were collected. Flowers from *C. ensifolium* were collected at three development stages (small bud, S1; large bud, S2; flowering, S3). Each sample was stored at −80 °C after cryopreserved in liquid nitrogen, ensuring their molecular integrity for accurate downstream analysis. Three replicates of each sample were obtained from diverse plant sources.

2.2. Identification and Physicochemical Properties of CeTPS

The NGDC database (https://ngdc.cncb.ac.cn/, accessed on 17 September 2023) provided the *C. ensifolium* genome sequencing and annotation data [35], while *A. thaliana*'s TPS gene family protein sequences were obtained from TAIR (https://www.arabidopsis.org/, accessed on 17 September 2023). Using the TBtools 1.132 BlastP program, the AtTPS protein sequences served as query sequences to extract homologous sequences from *C. ensifolium* [36]. We accessed the Pfam database (http://pfam.xfam.org/, accessed on 17 September 2023) to acquire Hidden Markov Models (HMMs) for TPS domains, PF01397 and PF03936, for accurate identification of TPS genes [37]. The Simple HMM Search tool in TBtools was used to identify protein sequences with two conserved structural domains, ensuring precise classification and analysis [36]. Duplicate protein sequences were eliminated after merging the results of the two searches. Candidate sequences were validated using SMART (http://smart.embl-heidelberg.de, accessed on 17 September 2023) and NCBI CDD (http://www.ncbi.nlm.nih.gov/cdd/, accessed on 17 September 2023) websites, and sequences lacking complete structural domains were removed to ensure accurate and reliable downstream analysis.

The molecular weight (MW), isoelectric point (pI), instability index (II), and grand average of hydropathicity (GRAVY) of CeTPS proteins were analyzed using the ExPASy online platform (https://www.expasy.org/, accessed on 19 September 2023) [38], providing detailed insights into their physicochemical properties. ProtComp 9.0 (http://linux1.softberry.com/berry.phtml?topic=protcomppl&group=programs&subgroup=proloc, accessed on 19 September 2023) was used to predict the subcellular localization of CeTPS proteins, providing insights into their cellular functions.

2.3. Chromosome Localization and Phylogenetic Analysis

Using the genome annotation data of *C. ensifolium*, Show Genes on Chromosome within TBtools was utilized to show the chromosomal positions of the *CeTPSs* in order to analyze their genomic distributions [36]. A phylogenetic tree was constructed using MEGA7.0 [39], based on TPS protein sequences from *C. ensifolium*, *A. thaliana*, *Oryza sativa*, *A. shenzhenica*, and *P. equestris*, to examine evolutionary relationships among TPS genes.

ClustalW was utilized for multiple sequence alignment, and 1000 bootstrap repetitions of the Neighbor-Joining method were used to generate a phylogenetic tree [39]. We employed Evolview v2 (https://evolgenius.info/, accessed on 22 September 2023) for the enhancement and visualization of the phylogenetic tree [40].

2.4. Conserved Motifs, Gene Structure, and Synteny Analysis

MEME (http://meme-suite.org/tools/meme, accessed on 26 September 2023) was used to analyze conserved motifs in CeTPS protein sequences in order to find common sequence patterns and functional domains [41]. The analysis was conducted with a maximum of 20 predicted motifs, with the settings maintained at default. TBtools' Gene Structure View was used to demonstrate the conserved motifs and structural details of the *CeTPSs*. One Step MCScanX in TBtools was used to investigate *TPS* gene duplications in *C. ensifolium*. To assess evolutionary selection pressures on *TPS* genes, we utilized TBtools' Simple Ka/Ks Calculator to calculate the non-synonymous (Ka) and synonymous (Ks) substitution rates, as well as their ratio. We conducted an analysis of duplication events involving *TPS* genes across three distinct species: *C. ensifolium*, *D. chrysotoxum*, and *C. goeringii*. Multiple Synteny Plot in TBtools was used to visualize the duplication patterns [36].

2.5. Cis-Acting Regulatory Elements Analysis

Promoter sequences 2000 bp upstream of the *CeTPS* genes were obtained using TBtools to analyze regulatory elements controlling gene expression [36]. In order to investigate the regulation mechanisms of *CeTPSs*, PlantCARE (http://bioinformatics.psb.ugent.be/webtools/plantcare/html/, accessed on 28 September 2023) was used for predicting the number and categories of *cis*-acting elements present in the promoter sequences [42]. Subsequently, the data were subjected to analysis and visualization utilizing Excel 2019 and TBtools [36].

2.6. Expression Patterns and qRT-PCR Analysis

This study initially identified potential *CeTPSs* involved in terpene biosynthesis using RNA-seq data from different organs. Samples of different organs (leaf, Le; root, Ro; pseudobulb, Ps; flower, Fl) of *C. ensifolium* were used for transcriptome analysis [35]. Calculations were performed to determine the Fragments per Kilobase per Million mapped reads (FPKM). Subsequently, we used the TBtools software to generate a comprehensive heatmap, illustrating the expression patterns of *CeTPSs* (FPKM values are in Supplementary Table S1) [36].

Three stages of flower development were examined for *CeTPS* gene expression patterns using qRT-PCR experiments. Total RNA was extracted from *C. ensifolium* flowers at three developmental stages using the Rapure Total RNA Plus Kit (Megan Biotechnology, Guangzhou, China). cDNA was synthesized using the Hifair® III 1st Strand cDNA Synthesis SuperMix kit (gDNA digester plus) from Yeasen Biotechnology (Shanghai, China). The qRT-PCR primers specific to *CeTPSs* were synthesized by Shanghai Sangon Biotech Company (Shanghai, China). Primer sequences used in this experiment are listed in Supplementary Table S2. qRT-PCR experiments were conducted on an ABI QuantStudio 6 Flex system (Applied Biosystems, Foster City, CA, USA). GAPDH was selected as the internal reference gene for accurate normalization of gene expression levels in the qRT-PCR experiments. To guarantee data reliability and robustness, the studies included three biological replicates. Using the $2^{-\Delta\Delta CT}$ formula, relative gene expression levels were calculated. The one-way ANOVA was conducted using SPSS 26.0 to assess gene expression differences across developmental stages, with results visualized using GraphPad Prism 9.5.

2.7. Subcellular Localization Analysis

We cloned the coding sequences (CDS) of *CeTPS1* and *CeTPS18* employing PCR amplification with primers detailed in Supplementary Table S3. These sequences were efficiently cloned into the pCAMBIA1302-35S-GFP vector. The 35S promoter-driven *CeTPS*-GFP vectors were successfully transformed into *Agrobacterium tumefaciens* strain GV3101

(Weidi Biotechnology, Shanghai, China). *Agrobacterium* strains carrying the *CeTPSs* were used to infect *N. benthamiana* leaves for transient expression experiments, with an empty vector as the control. After 48 h, CeTPS1 and CeTPS18 protein subcellular localization was observed utilizing the LSM880 confocal laser microscope (Carl Zeiss, Jena, Germany).

3. Results

3.1. Identification and Physicochemical Properties of CeTPS

An in-depth analysis of BlastP and HMMER search results was conducted. We identified 30 *CeTPS* genes, named *CeTPS1* to *CeTPS30*, based on their respective chromosomal positions. A thorough analysis was conducted to investigate the physicochemical properties of the 30 *CeTPS* genes, with the results summarized in Table 1, providing insights into their roles and potential applications in *C. ensifolium*. The lengths of the CeTPS protein sequences exhibited considerable variation, ranging from a minimum of 194 amino acids (aa) (CeTPS17) to a maximum of 807 aa (CeTPS19), measuring 477 aa on average. The substantial variation in protein lengths suggests the potential presence of pseudogenes. The molecular weights (MW) of CeTPS proteins varied widely, from a minimum of 22.24 kDa (CeTPS17) to a maximum of 91.98 kDa (CeTPS19). This variation in MW reflects the different lengths and compositions of the amino acid sequences in these proteins. The average theoretical isoelectric point (pI) of CeTPS proteins was 5.76, with all pI values below 7, indicating their acidic nature. The instability index (II), indicating protein stability, varied widely among CeTPS proteins, from a stable 29.24 in CeTPS22 to an unstable 61.68 in CeTPS3. The grand average of hydropathicity (GRAVY) for all CeTPS proteins was below 0, indicating their hydrophilic nature. Furthermore, the chloroplasts and cytoplasm were the targets of the subcellular localization prediction results, as shown by the ProtComp 9.0 database.

Table 1. Physicochemical properties of the CeTPS proteins.

Name	ID	AA (aa)	Mw (kDa)	pI	II	GRAVY	Subcellular Localization
CeTPS1	JL015507	413	48.61	5.97	57.61	−0.500	Chloroplast.
CeTPS2	JL027127	378	44.71	6.04	33.18	−0.273	Chloroplast. Cytoplasm.
CeTPS3	JL025499	218	25.05	5.59	61.68	−0.405	Chloroplast.
CeTPS4	JL026288	552	63.55	5.36	49.99	−0.142	Chloroplast.
CeTPS5	JL028400	549	64.38	5.62	52.65	−0.194	Chloroplast. Cytoplasm.
CeTPS6	JL028504	549	64.60	5.62	50.51	−0.210	Chloroplast. Cytoplasm.
CeTPS7	JL024344	408	47.55	5.28	48.98	−0.106	Chloroplast. Cytoplasm.
CeTPS8	JL026235	364	42.91	5.99	51.56	−0.126	Chloroplast. Cytoplasm.
CeTPS9	JL028294	424	50.01	6.38	52.50	−0.258	Chloroplast. Cytoplasm.
CeTPS10	JL024700	436	51.15	5.46	46.80	−0.280	Chloroplast. Cytoplasm.
CeTPS11	JL024633	498	58.14	5.46	47.35	−0.231	Chloroplast. Cytoplasm.
CeTPS12	JL008157	329	37.99	6.52	40.16	−0.107	Chloroplast. Cytoplasm.
CeTPS13	JL026427	703	82.25	6.15	38.04	−0.383	Chloroplast.
CeTPS14	JL027333	446	53.01	6.17	37.95	−0.358	Chloroplast.
CeTPS15	JL025789	446	52.80	6.16	37.71	−0.345	Chloroplast.
CeTPS16	JL022471	703	82.20	6.14	41.84	−0.395	Chloroplast.
CeTPS17	JL024194	194	22.24	5.36	42.68	−0.185	Chloroplast. Cytoplasm.
CeTPS18	JL012496	602	70.07	5.99	49.94	−0.271	Chloroplast.
CeTPS19	JL004677	807	91.98	5.95	48.28	−0.286	Chloroplast.
CeTPS20	JL027803	535	63.06	5.67	33.24	−0.211	Chloroplast. Cytoplasm.
CeTPS21	JL001410	287	33.47	5.73	32.94	−0.175	Chloroplast. Cytoplasm.
CeTPS22	JL026564	508	59.71	6.22	29.24	−0.233	Chloroplast. Cytoplasm.
CeTPS23	JL027817	529	62.46	5.74	33.84	−0.208	Chloroplast. Cytoplasm.
CeTPS24	JL028318	528	62.47	5.36	39.28	−0.192	Chloroplast. Cytoplasm.
CeTPS25	JL028319	535	63.31	5.60	33.94	−0.227	Chloroplast. Cytoplasm.
CeTPS26	JL027638	485	57.50	5.45	37.85	−0.161	Chloroplast. Cytoplasm.
CeTPS27	JL028133	320	36.79	5.14	31.87	−0.372	Chloroplast.

Table 1. *Cont.*

Name	ID	AA (aa)	Mw (kDa)	pI	II	GRAVY	Subcellular Localization
CeTPS28	JL027059	535	63.11	5.83	34.91	−0.219	Chloroplast. Cytoplasm.
CeTPS29	JL028844	501	59.04	5.46	37.25	−0.153	Chloroplast. Cytoplasm.
CeTPS30	JL028984	528	62.54	5.35	39.80	−0.158	Chloroplast. Cytoplasm.

3.2. Chromosome Localization and Phylogenetic Analysis of CeTPS

Chromosomal mapping revealed an uneven distribution of 28 *CeTPS* genes across the seven chromosomes of *C. ensifolium*, except for *CeTPS29* and *CeTPS30*, which were not localized on the chromosomes (Figure 2). Chromosomes 8 and 16 in *C. ensifolium* showed a significant accumulation of *CeTPSs*, each containing up to eight genes due to the presence of two separate gene clusters. In contrast, chromosomes 2 and 12 each hold only one *CeTPS* gene. The clustering of most *CeTPSs* on chromosomes suggests they likely originated from tandem or segmental duplications. No correlation was found between the size of the chromosomes and the distribution of the *CeTPSs*. Based on the CeTPS and TPS protein sequences from other species, a phylogenetic tree was constructed using the Neighbor-Joining method (Figure 3). All TPS protein sequences were classified into five subfamilies: TPS-a, TPS-b, TPS-c, TPS-e/f, and TPS-g. The CeTPS family was categorized into three subfamilies: TPS-b (17 members), TPS-a (8 members), and TPS-c (5 members) (Figure 3). TPS proteins from monocot and dicot species established separate subgroups in the TPS-a subfamily. This observation aligns with the findings of prior investigations [11]. In addition, CeTPS members were extremely similar to the TPS protein of *A. shenzhenica* and *P. equestris* from the Orchidaceae.

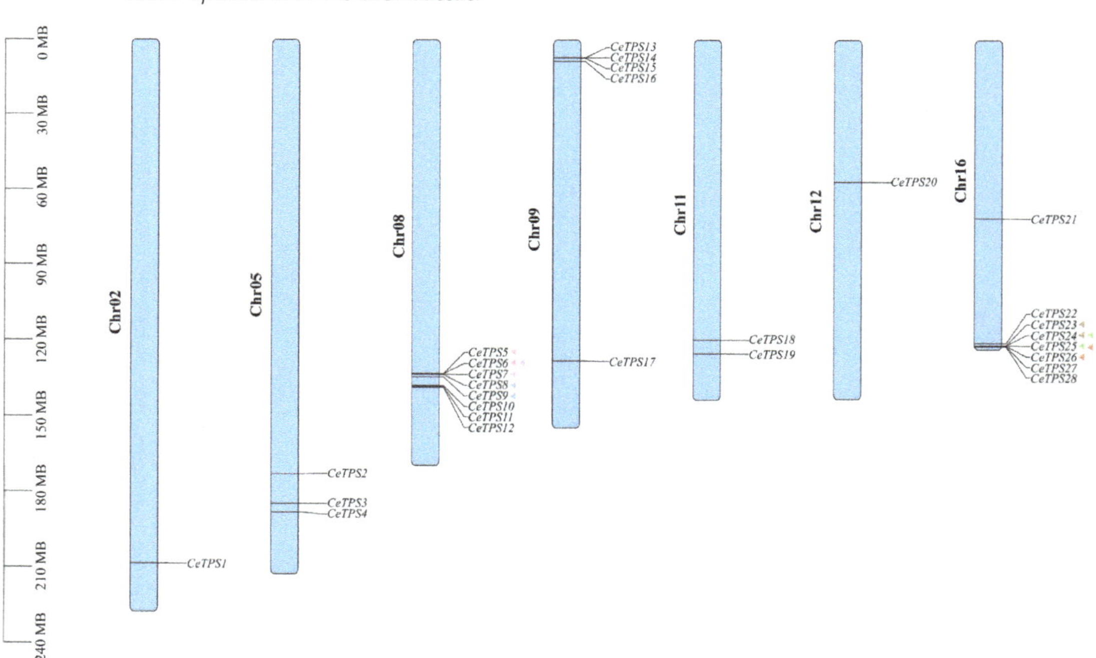

Figure 2. Distribution of the *CeTPSs* on *C. ensifolium* chromosomes. The scale shows chromosomal distances in megabases (MB). The colored triangle icons indicate tandem repeat relationships between *CeTPSs*, and triangle icons of the same color *CeTPSs* have tandem repeats.

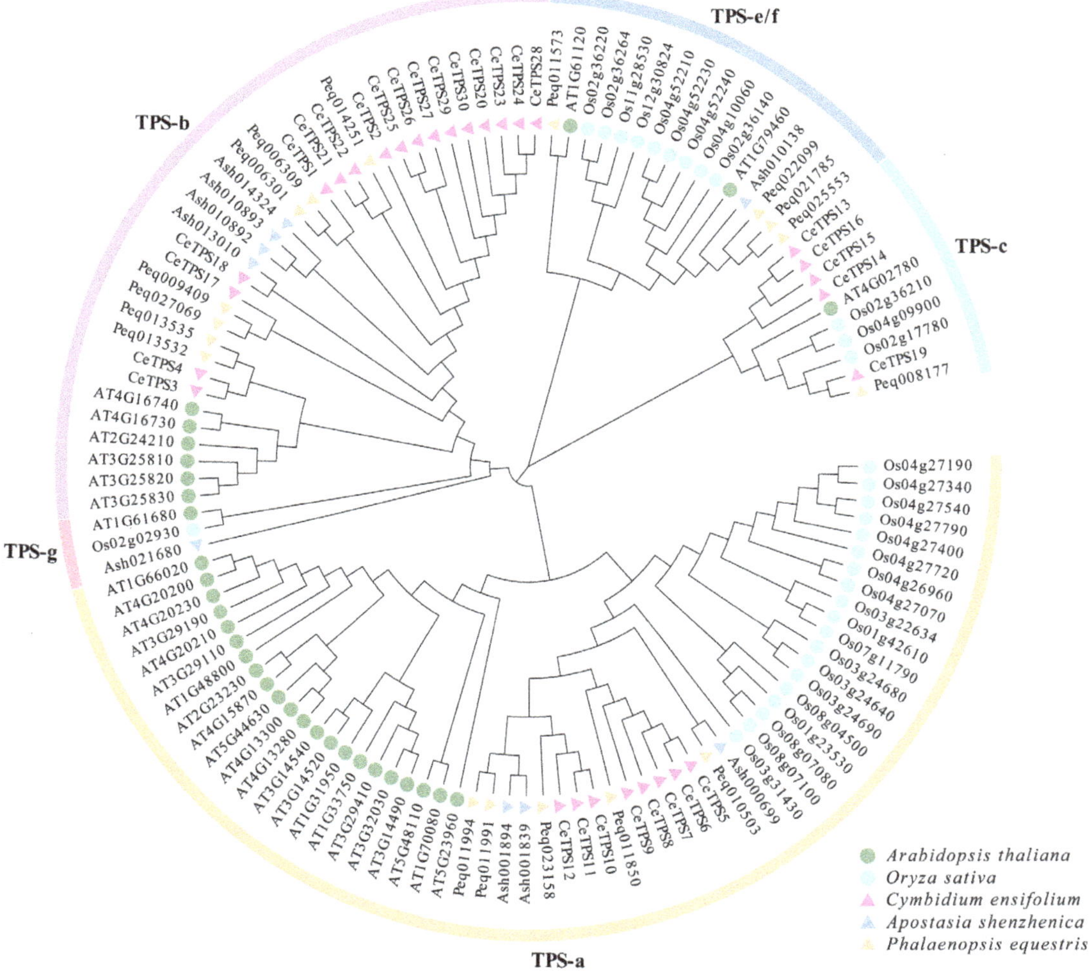

Figure 3. Phylogenetic analysis of TPS protein sequences from five plant species. Members of the TPS subfamilies from different species are marked with different colors and shapes. Green and cyan circles represent the TPS proteins from *A. thaliana* and *O. sativa*, respectively. Purple, blue, and yellow triangles represent the TPS proteins from *C. ensifolium*, *A. shenzhenica*, and *P. equestris*, respectively.

3.3. Analysis of CeTPS Conserved Motifs, Gene Structure, and Synteny

Utilizing the web program MEME, we identified 20 conserved motifs within the CeTPS protein sequences, providing valuable insights into their structural and functional conservation. The findings demonstrated that CeTPSs had between three and sixteen motifs (Figure 4B). The most conserved motifs were discovered in CeTPS20, CeTPS23, CeTPS25, and CeTPS28, whereas CeTPS3 and CeTPS17 only had three motifs. All CeTPS sequences contained Motif 2. Twenty-three CeTPS protein sequences contained the DDxxD (Motif 1) and RRX_8W (Motif 3) motif, respectively. NSE/DTE motifs (Motif 8) were found in 18 CeTPS genes (Figure S1). In the TPS-a subfamily, all members except CeTPS12 contain the RRX_8W motif. CeTPS27 of the TPS-b subfamily and all TPS-c subfamily members lack the conserved RRX_8W motif, suggesting functional and structural differences within the TPS gene family in *C. ensifolium*. The RRX_8W motif is essential for the cyclization process of monoterpenes [43]. The DDxxD and NSE/DTE motifs are essential for the enzymatic cleavage of prenyl diphosphate substrates, playing critical roles in substrate binding and catalysis in terpene synthase enzymes. These motifs facilitate the coordination of Mg^{2+} or Mn^{2+} ions at the C-terminus [4,14]. Results showed that members of the same subfamily had highly similar motif compositions, while distinct subfamilies were characterized by unique motifs, indicating conserved functions within subfamilies and functional divergence between them.

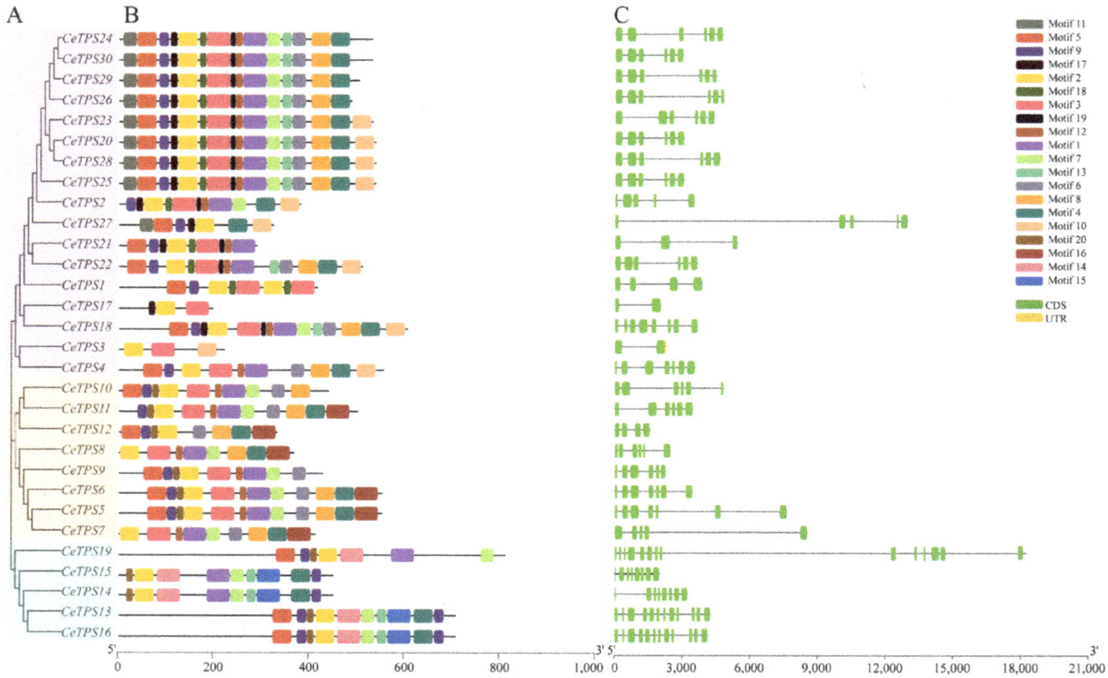

Figure 4. Phylogenetic analysis, motif conservation, and gene structure of CeTPS. (**A**) Phylogenetic tree of CeTPS proteins. (**B**) Conserved motifs in CeTPS proteins. (**C**) Gene structure of *CeTPSs*.

The *CeTPSs* displayed variability in exon count, spanning from two to fourteen exons (Figure 4C). The *CeTPS19* gene had the most exons, while *CeTPS3* and *CeTPS17* had only two exons. Members of the same subfamily share comparable intron–exon structures. By analyzing the collinearity of the *CeTPSs* within *C. ensifolium*, we identified six pairs of tandem duplicates and one pair of segmental duplicates (Figures 2 and 5). A Ka/Ks value of 1 indicates neutral selection, values greater than 1 signify positive selection, and values less than 1 suggest purifying selection [44]. The segmentally duplicated gene pair had a Ka/Ks ratio of 0.4476, while the tandem duplicates ranged from 0.5462 to 0.8025 (Table 2) These results, with all Ka/Ks values below 1, suggest that purifying selection has been applied to the *CeTPS* gene family, indicating that these genes are conserved and likely essential for maintaining specific functions in *C. ensifolium*. To explore the evolutionary relationships among Orchidaceae's TPS family, we analyzed the collinearity of *TPSs* among *C. ensifolium*, *D. chrysotoxum*, and *C. goeringii*, revealing gene duplication events. *CeTPSs* had nine and eight pairs of homologous genes with *C. goeringii* and *D. chrysotoxum*, respectively (Figure 6).

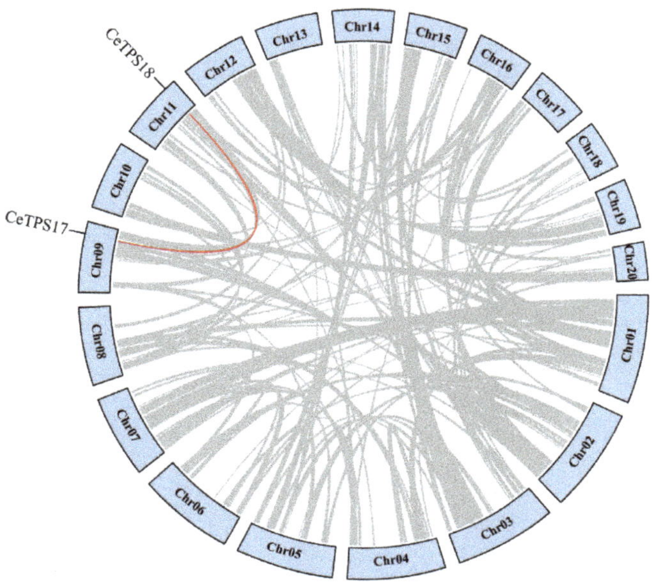

Figure 5. Collinear relationships of *CeTPSs* on chromosomes. The red line represents the *CeTPSs* with segmental duplications in *C. ensifolium*.

Table 2. Ka/Ks analysis of *CeTPS* genes.

Gene Pairs		Ka	Ks	Ka/Ks	Duplication Type	Purify Selection
CeTPS5	*CeTPS6*	0.017950909	0.032074859	0.559656681	Tandem	Yes
CeTPS6	*CeTPS7*	0.029937944	0.054806694	0.546246113	Tandem	Yes
CeTPS8	*CeTPS9*	0.146080662	0.22792656	0.6409111	Tandem	Yes
CeTPS17	*CeTPS18*	0.442114425	0.987765176	0.447590618	Segmental	Yes
CeTPS23	*CeTPS24*	0.0254454	0.037389324	0.68055256	Tandem	Yes
CeTPS24	*CeTPS25*	0.027606516	0.040278134	0.685397089	Tandem	Yes
CeTPS25	*CeTPS26*	0.002618184	0.003262648	0.802472277	Tandem	Yes

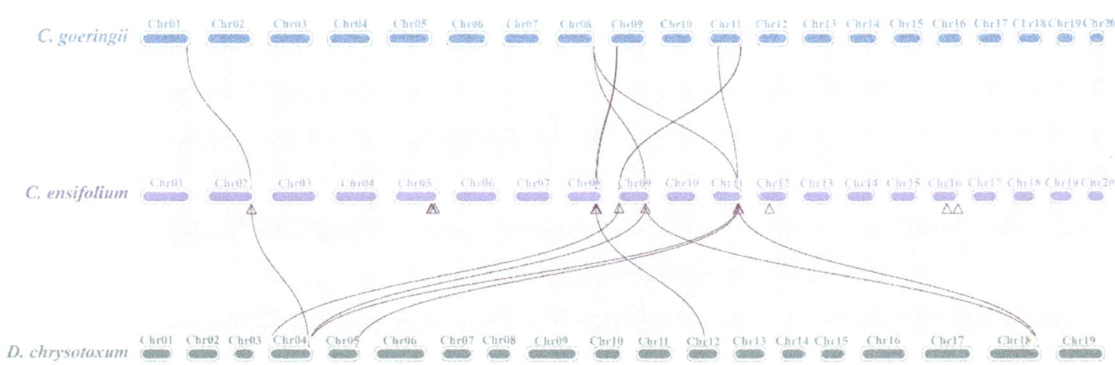

Figure 6. Collinearity analysis of *TPS* genes in three orchid species. The red triangles indicate the positions of the *CeTPSs* on chromosomes. The red lines highlight the *TPS* genes with collinear relationships between three orchid species.

3.4. *Cis*-Elements Analysis of *CeTPS*

Analysis of promoter sequences 2000 bp upstream of the *CeTPSs* identified 641 *cis*-acting elements, highlighting the complex regulatory mechanisms controlling gene expression and terpenoid biosynthesis in *C. ensifolium*. These *cis*-acting elements were categorized into three functional groups (Figure 7A,B). There were 135 *cis*-acting elements related to plant growth and development which were classified into 10 groups: As-1 element (33%), AAGAA-motif (21%), MRE (17%), AT-rich element (8%), CCAAT-box (5%), etc. In the *CeTPS* gene promoters, 8 categories were created from the identification of 155 *cis*-acting elements associated with plant stress response, including anaerobic induction (ARE, 43%), stress response (STRE, 18%), drought response (MBS, 15%), and wounding response (WUN-motif, 8%), etc. The phytohormone responsiveness category contained the most *cis*-acting elements (351/641), primarily responsive to MeJA (MYC, 31%; TGACG-motif, 13%; CGTCA-motif, 12%), ethylene (ERE, 23%), abscisic acid (ABRE, 5%), and auxin (TGA-element, 5%) (Figure 7C). Our findings revealed that MeJA serves as the primary regulator of *CeTPS* gene expression, corroborating previous research in plants [20,29].

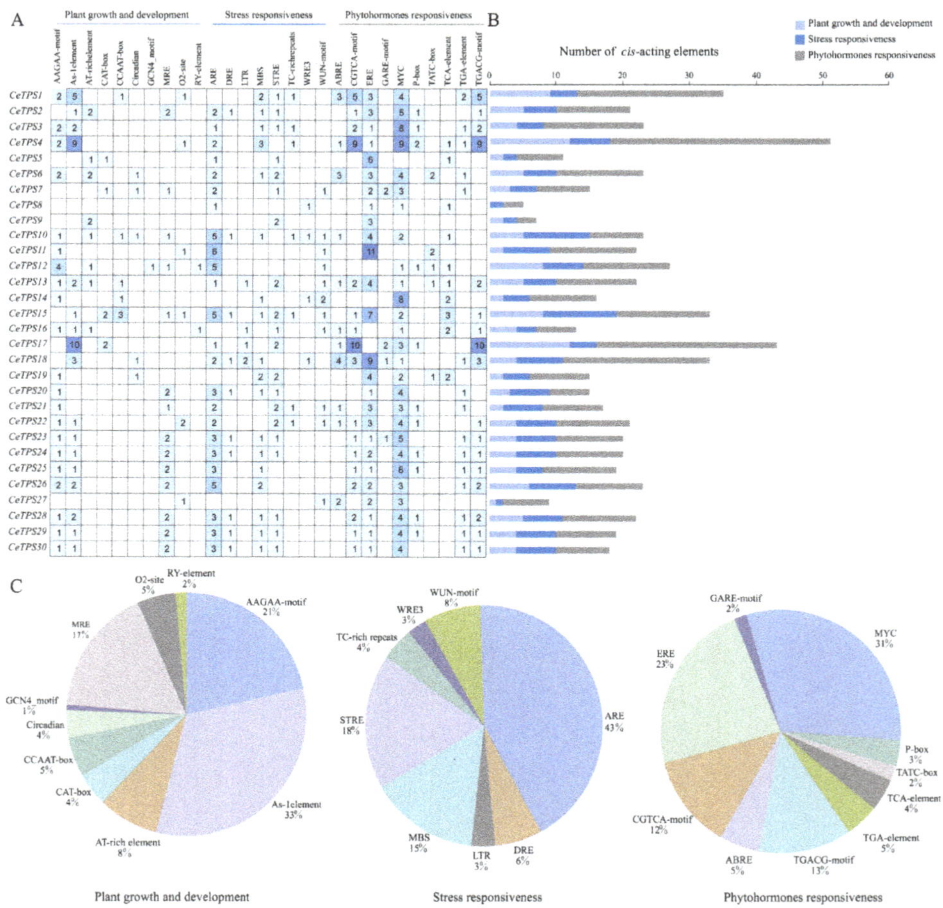

Figure 7. Promoter analysis of *CeTPSs*. (**A**) The count of distinct *cis*-elements identified in each *CeTPS* gene. (**B**) Total *cis*-regulatory elements related to plant growth and development, stress responsiveness, and phytohormone responsiveness. (**C**) The proportion of *cis*-elements related to plant growth and development, stress responsiveness, and phytohormone responsiveness.

3.5. Expression Patterns and qRT-PCR Analysis of CeTPS

Based on the transcriptome expression data of *CeTPSs* in various organs, we found that *CeTPSs* have tissue-specific expression characteristics (Figure 8A). Specifically, *CeTPS14* and *CeTPS16* showed comparatively high expression levels in the roots and leaves, while *CeTPS23* and *CeTPS28* showed significant expression levels in pseudobulbs. Additionally, *CeTPS1* and *CeTPS18* displayed high transcript abundance in flowers. We conducted qRT-PCR experiments to quantify the abundance of *CeTPS1* and *CeTPS18* transcripts at different developmental stages of the flower (Figure 8B). These genes showed the highest expression levels during blooming. It is hypothesized that these genes contribute to the biosynthesis of terpenoids in the flowers of *C. ensifolium*.

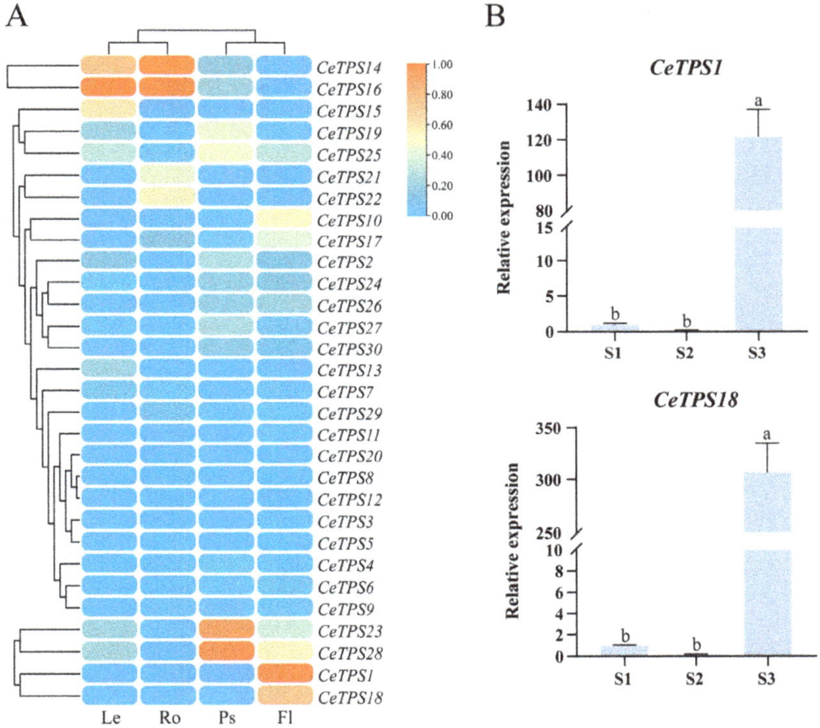

Figure 8. Heatmap and qRT-PCR analysis of *CeTPSs* expression in *C. ensifolium*. (**A**) Expression patterns of *CeTPSs* in different organs (FPKM values) (Le: leaf; Ro: root; Ps: pseudobulb; Fl: flower). (**B**) Expression in flowers at three stages (S1: small bud; S2: large bud; S3: flowering). Bars display biological replicate mean values ± SE; significant statistical differences are shown by unique letters ($p < 0.05$, Duncan).

3.6. Subcellular Localization of CeTPS

To further validate the functions of the CeTPS1 and CeTPS18 proteins, we created expression vectors (35S: *CeTPS1*-GFP and 35S: *CeTPS18*-GFP) and transiently expressed them in *N. benthamiana* leaves. This method allowed us to track gene expression and visualize subcellular localization. Strong GFP fluorescence signals of *CeTPS1*-GFP and *CeTPS18*-GFP were detected in the cytoplasm, while non-overlapping punctate red spots were observed in the corresponding chlorophyll autofluorescence (Chl) images (Figure 9). The experimental results demonstrated that both CeTPS1 and CeTPS18 proteins localized in chloroplasts (Figure 9).

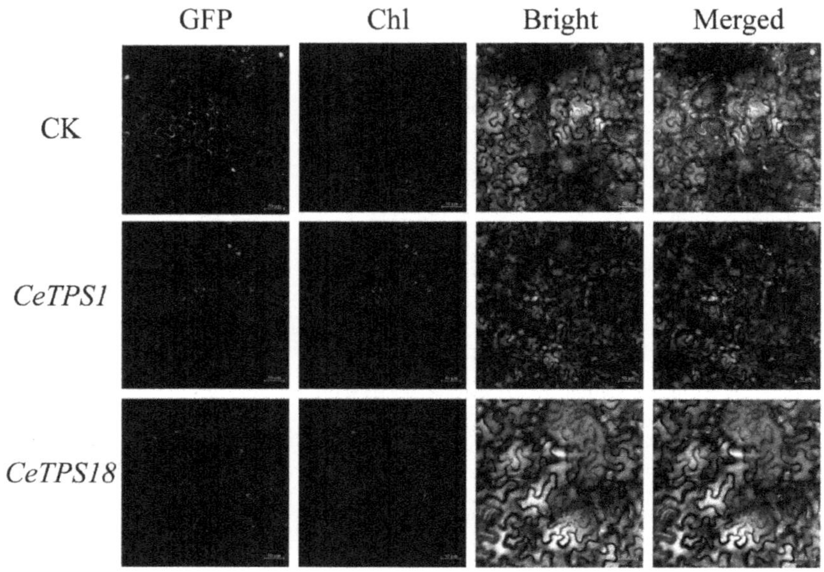

Figure 9. CeTPS proteins' subcellular localization in *N. benthamiana* leaves (GFP: fluorescence image of green fluorescent protein; Chl: chlorophyll autofluorescence image; Bright: brightfield image; Merged: all channels combined).

4. Discussion

Floral scent is crucial for the diverse evolution of orchids and is a key ornamental trait. Terpenes, which are primary components of orchid floral scents, are synthesized by *TPS* genes, whose roles in terpene production have been confirmed in various orchid species. *C. ensifolium*, a traditional Chinese orchid, has been cherished for its fresh and elegant fragrance [35]. However, comprehensive studies on the identification of the TPS gene family in *C. ensifolium* remain scarce, limiting our understanding of terpenoid biosynthesis and its potential applications. Therefore, analyzing and categorizing the *TPS* genes in *C. ensifolium* holds significant importance from a theoretical standpoint.

To accurately identify *CeTPS* genes, we combined HMMER search and BlastP results, followed by validation using CDD and SMART to exclude incomplete sequences, ultimately identifying 30 TPS gene family members in *C. ensifolium*. A comparative analysis of *C. ensifolium* TPS family members with other orchid species, including *D. chrysotoxum* (48) [32], *D. officinale* (34) [29], *C. faberi* (32) [30], and *C. goeringii* (40) [31], reveals that *C. ensifolium* has a medium-sized TPS gene family. Phylogenetic analysis of CeTPS proteins in *C. ensifolium* identified three unique subfamilies: TPS-a, TPS-b, and TPS-c (Figure 3), which differ from classifications in other Orchidaceae species, suggesting unique evolutionary adaptations [28–30,34]. Evidence suggests that *C. ensifolium* has lost TPS-e/f and TPS-g subfamily members over its evolutionary history. The CeTPS gene family comprised 8 TPS-a, 17 TPS-b, and 5 TPS-c subfamily members. Unlike the proportion of TPS-a subfamily in *A. thaliana* (68.75%), *C. faberi* (43.33%), and *D. chrysotoxum* (41.17%), *C. ensifolium* has only 26.67% of TPS-a subfamily members, and 56.67% of CeTPSs are distributed in the TPS-b subfamily. *TPS* genes in plants usually cluster together to generate functional gene clusters [45]. The TPS functional gene clusters exist in numerous plants, including *A. thaliana* [17], *Solanum lycopersicum* [45], *Daucus carota* [46], *A. shenzhenica*, *V. planifolia*, *D. catenatum*, and *P. equestris* [28]. The TPS gene clusters are highly conserved in plants, contributing to the stability and diversity of terpene biosynthesis pathways [45]. Chromosomal analysis of *CeTPSs* in *C. ensifolium* revealed that TPS-a, TPS-c, and TPS-b subfamily members were clustered on chromosomes Chr08, Chr09, and Chr16, respectively (Figure 2).

Gene duplication is crucial for gene family expansion, diversification, and understanding evolutionary relationships among homologous genes. Gene duplications are classified into tandem, segmental, and genome-wide types. Tandem duplications place homologous genes on the same chromosome, while segmental duplications occur when duplicated genes are located on different chromosomes [47]. Seven duplication events, including six tandem and one segmental, were identified in the *CeTPSs* (Figures 2 and 5), with Ka/Ks ratios ranging from 0.4476 to 0.8025 (Table 2), indicating purifying selection and suggesting evolutionary conservation. The findings indicate that tandem and segmental duplications have contributed to expanding the TPS gene family in *C. ensifolium*, driving genetic diversity and enhancing terpenoid biosynthesis essential for ecological interactions. Through further collinearity analysis of *C. ensifolium*, *C. goeringii*, and *D. chrysotoxum*, multiple TPS homologous genes were identified among these three species (Figure 6). The detection of these homologous genes highlights the evolutionary conservation among these orchid species and implies potential similarities in gene function or regulatory mechanisms. This knowledge lays a critical foundation for further exploration of the evolutionary processes and gene functions of the TPS gene in orchids.

Numerous *cis*-regulatory elements regulate phytohormone signaling and stress responses in *C. ensifolium*, including MeJA regulation, ERE regulation, anaerobic induction, drought response, etc. (Figure 7). The findings show that hormones and stress may modulate *CeTPSs* expression. Several environmental factors regulate the release of volatile terpenoids from plants, including light intensity, circadian clock, ambient temperature, and relative humidity [48]. MeJA generates geraniol by promoting the expression of the *D. officinale* TPS gene *DoGES1* [33]. Some TPS genes in *C. sinensis* are suppressed or promoted in expression after treatments with MeJA and various stressors [21]. Genes responsible for MeJA biosynthesis in *C. ensifolium* are highly expressed in mature flower buds and fully bloomed flowers [35], implying that the MeJA signaling pathway may influence the expression of *CeTPSs*. Various transcription factors regulate the TPS gene in plants, playing a crucial role in modulating terpenoid biosynthesis by controlling gene expression in response to development and the environment. HcMYBs in *Hedychium coronarium* regulate *HcTPS* expression by binding to core MYB-binding sites [49]. Transient expression of the *PbbHLH4* increases the content of monoterpenes in scentless *phalaenopsis* orchid flowers by 950-fold [50]. DobHLH4 interacts with DoJAZ1 and binds to the *DoTPS10* promoter's G-box, upregulating *DoTPS10* expression, crucial for MeJA-mediated linalool accumulation in *D. officinale* [51]. Various transcription factors are thought to regulate the expression of *CeTPS* genes.

CeTPSs exhibit organ-specific expression patterns according to the transcriptome expression data. This observation suggests that members from different subfamilies may fulfill distinct roles in various plant organs, contributing to a diversity of biological processes. For example, *CeTPS14* and *CeTPS16* are predominantly expressed in roots and leaves, whereas *CeTPS23* and *CeTPS28* show significant expression in pseudobulbs (Figure 8A). This pattern indicates that these genes might be involved in stress responses or other physiological functions related to plant defense mechanisms. Furthermore, the elevated expression of *CeTPS1* and *CeTPS18* in flowers suggests their involvement in floral scent biosynthesis (Figure 8A). These findings provide essential theoretical foundations and valuable insights for future research on the specific roles of *CeTPSs* in plant growth, development, and their underlying molecular mechanisms across different organs. Floral organs primarily emit volatile organic compounds (VOCs). The release of floral scents follows a distinct pattern across the stages of flower development. Flowers emit more VOCs when they prepare for pollination [25]. We examined gene expression patterns in *C. ensifolium* at various stages of flower development, revealing key regulatory genes involved in scent production. Then we identified a peak in *CeTPS* transcript abundance, specifically during the flowering period. The present results align with earlier investigations [27]. Gene expression patterns indicated that *CeTPS1* and *CeTPS18* exhibited the highest expression levels during flowering (Figure 8B). CeTPS1 and CeTPS18 belong to the TPS-b subfamily. These genes are considered essential in the monoterpene biosynthesis pathways of *C. ensifolium*. These findings offer essential insights and establish a foundation for future research on floral scent biosynthesis and molecular breeding in orchids.

5. Conclusions

In this research, we identified 30 *CeTPS* genes within the *C. ensifolium* genome and performed an extensive analysis of their physicochemical properties. The CeTPS proteins were classified into three TPS subfamilies. Genes within the same subfamily shared remarkably comparable structures and conserved motifs, indicating functional conservation, whereas different subfamilies had distinct features, demonstrating functional differentiation. The genome of *C. ensifolium* contained seven pairs of duplicated genes with purifying selection, suggesting that the evolution of *CeTPSs* was conserved. *Cis*-element analysis of *CeTPS* gene promoters indicated regulation primarily by plant hormones and stress. Transcriptome analysis revealed the *CeTPSs* demonstrated organ-specific expression patterns. The qRT-PCR experiments further validated two functional genes that were highly expressed during flowering and analyzed their subcellular localization. These findings provide valuable insights and data, laying the groundwork for future research on floral scent biosynthesis mechanisms and molecular breeding in orchids.

Supplementary Materials: The following supporting information can be downloaded at: https://www.mdpi.com/article/10.3390/horticulturae10101015/s1, Figure S1: Motif logo of Motif1, Motif2, Motif3, and Motif8. Table S1: The FPKM values of *CeTPSs* of different organs of *C. ensifolium*; Table S2: Primers of qRT-PCR experiments; Table S3: Primers of subcellular localization.

Author Contributions: The experiments were devised and conceived by C.F. (Chuanming Fu) and Y.A.; B.L. conducted the bioinformatics analysis; the experiments and data analysis were carried out by M.W.; M.W. and B.L. drafted the manuscript; J.L., N.H. and Y.T. were responsible for the visualization and preparation of the figures; L.G. and C.F. (Caiyun Feng) prepared the plant materials. All authors contributed to revising the manuscript. M.W. and B.L. contributed equally to this work. All authors have read and agreed to the published version of the manuscript.

Funding: This research was supported by the Guilin Innovation Platform and Talent Plan Project (No. 20210218-10), the Regional Fund of the National Natural Science Foundation of China (No. 32160096), the Basal Research Fund of Guangxi Academy of Sciences (No. CQZ-E-1910), and the Science and Technology Innovation Special Fund Project of Fujian Agricultural and Forestry University (No. KFb22057XA).

Data Availability Statement: The original contributions presented in the study are included in the article/Supplementary Materials, further inquiries can be directed to the corresponding authors.

Conflicts of Interest: The authors declare no conflict of interest.

References

1. Yan, X.M.; Zhou, S.S.; Liu, H.; Zhao, S.W.; Tian, X.C.; Shi, T.L.; Bao, Y.T.; Li, Z.C.; Jia, K.H.; Nie, S.; et al. Unraveling the evolutionary dynamics of the *TPS* gene family in land plants. *Front. Plant Sci.* **2023**, *14*, 1273648. [CrossRef]
2. Zhou, F.; Pichersky, E. More is better: The diversity of terpene metabolism in plants. *Curr. Opin. Plant Biol.* **2020**, *55*, 1–10. [CrossRef] [PubMed]
3. Gershenzon, J.; Dudareva, N. The function of terpene natural products in the natural world. *Nat. Chem. Biol.* **2007**, *3*, 408–414. [CrossRef] [PubMed]
4. Tholl, D. Biosynthesis and biological functions of terpenoids in plants. *Adv. Biochem. Eng. Biotechnol.* **2015**, *148*, 63–106. [CrossRef] [PubMed]
5. Spréa, R.M.; Fernandes, Â.; Calhelha, R.C.; Pereira, C.; Pires, T.; Alves, M.J.; Canan, C.; Barros, L.; Amaral, J.S.; Ferreira, I. Chemical and bioactive characterization of the aromatic plant *Levisticum officinale* W.D.J. Koch: A comprehensive study. *Food Funct.* **2020**, *11*, 1292–1303. [CrossRef] [PubMed]
6. Aprotosoaie, A.C.; Hăncianu, M.; Costache, I.I.; Miron, A. Linalool: A review on a key odor-ant molecule with valuable biological properties. *Flavour Fragr. J.* **2014**, *29*, 193–219. [CrossRef]
7. Bohlmann, J.; Meyer-Gauen, G.; Croteau, R. Plant terpenoid synthases: Molecular biology and phylogenetic analysis. *Proc. Natl Acad. Sci. USA* **1998**, *95*, 4126–4133. [CrossRef]

8. Vranová, E.; Coman, D.; Gruissem, W. Network analysis of the MVA and MEP pathways for isoprenoid synthesis. *Annu. Rev. Plant Biol.* **2013**, *64*, 665–700. [CrossRef]
9. Pichersky, E.; Noel, J.P.; Dudareva, N. Biosynthesis of plant volatiles: Nature's diversity and ingenuity. *Science* **2006**, *311*, 808–811. [CrossRef]
10. Tholl, D. Terpene synthases and the regulation, diversity and biological roles of terpene metabolism. *Curr. Opin. Plant Biol.* **2006**, *9*, 297–304. [CrossRef]
11. Chen, F.; Tholl, D.; Bohlmann, J.; Pichersky, E. The family of terpene synthases in plants: A mid-size family of genes for specialized metabolism that is highly diversified throughout the kingdom. *Plant J.* **2011**, *66*, 212–229. [CrossRef] [PubMed]
12. Boutanaev, A.M.; Moses, T.; Zi, J.; Nelson, D.R.; Mugford, S.T.; Peters, R.J.; Osbourn, A. Investigation of terpene diversification across multiple sequenced plant genomes. *Proc. Natl. Acad. Sci. USA* **2015**, *112*, E81–E88. [CrossRef] [PubMed]
13. Reed, J.; Osbourn, A. Engineering terpenoid production through transient expression in *Nicotiana Benthamiana*. *Plant Cell Rep.* **2018**, *37*, 1431–1441. [CrossRef] [PubMed]
14. Jiang, S.Y.; Jin, J.; Sarojam, R.; Ramachandran, S. A comprehensive survey on the terpene synthase gene family provides new insight into its evolutionary patterns. *Genome Biol. Evol.* **2019**, *11*, 2078–2098. [CrossRef]
15. Karunanithi, P.S.; Zerbe, P. Terpene synthases as metabolic gatekeepers in the evolution of plant terpenoid chemical diversity. *Front. Plant Sci.* **2019**, *10*, 1166. [CrossRef]
16. Li, G.; Köllner, T.G.; Yin, Y.; Jiang, Y.; Chen, H.; Xu, Y.; Gershenzon, J.; Pichersky, E.; Chen, F. Nonseed plant *Selaginella Moellendorffii* has both seed plant and microbial types of terpene synthases. *Proc. Natl. Acad. Sci. USA* **2012**, *109*, 14711–14715. [CrossRef]
17. Aubourg, S.; Lecharny, A.; Bohlmann, J. Genomic analysis of the terpenoid synthase (AtTPS) gene family of *Arabidopsis thaliana*. *Mol. Genet. Genom.* **2002**, *267*, 730–745. [CrossRef]
18. Martin, D.M.; Aubourg, S.; Schouwey, M.B.; Daviet, L.; Schalk, M.; Toub, O.; Lund, S.T.; Bohlmann, J. Functional annotation, genome organization and phylogeny of the grapevine (*Vitis vinifera*) terpene synthase gene family based on genome assembly, FLcDNA cloning, and enzyme assays. *BMC Plant Biol.* **2010**, *10*, 226. [CrossRef]
19. Zhang, A.; Xiong, Y.; Fang, J.; Jiang, X.; Wang, T.; Liu, K.; Peng, H.; Zhang, X. Diversity and functional evolution of terpene synthases in Rosaceae. *Plants* **2022**, *11*, 736. [CrossRef]
20. Yang, Z.; Zhan, T.; Xie, C.; Huang, S.; Zheng, X. Genome-wide analyzation and functional characterization on the TPS family provide insight into the biosynthesis of mono-terpenes in the camphor tree. *Plant Physiol. Biochem.* **2023**, *196*, 55–64. [CrossRef]
21. Zhou, H.C.; Shamala, L.F.; Yi, X.K.; Yan, Z.; Wei, S. Analysis of terpene synthase family genes in *Camellia sinensis* with an emphasis on abiotic stress conditions. *Sci. Rep.* **2020**, *10*, 933. [CrossRef] [PubMed]
22. Zhang, D.; Zhao, X.; Li, Y.; Ke, S.; Yin, W.; Lan, S.; Liu, Z. Advances and prospects of orchid research and industrialization. *Hortic. Res.* **2022**, *9*, uhac220. [CrossRef] [PubMed]
23. Schiestl, F.P.; Ayasse, M.; Paulus, H.F.; Löfstedt, C.; Hansson, B.S.; Ibarra, F.; Francke, W. Sex pheromone mimicry in the early spider orchid (*Ophrys sphegodes*): Patterns of hydrocarbons as the key mechanism for pollination by sexual deception. *J. Comp. Physiol. A* **2000**, *186*, 567–574. [CrossRef] [PubMed]
24. Hsiao, Y.; Pan, Z.; Hsu, C.; Yang, Y.; Hsu, Y.; Chuang, Y.; Shih, H.; Chen, W.; Tsai, W.; Chen, H. Research on orchid biology and biotechnology. *Plant Cell Physiol.* **2011**, *52*, 1467–1486. [CrossRef] [PubMed]
25. Ramya, M.; An, H.R.; Baek, Y.S.; Reddy, K.E.; Park, P.H. Orchid floral volatiles: Biosynthesis genes and transcriptional regulations. *Sci. Hortic.* **2018**, *235*, 62–69. [CrossRef]
26. Hsiao, Y.Y.; Tsai, W.C.; Kuoh, C.S.; Huang, T.H.; Wang, H.C.; Wu, T.S.; Leu, Y.L.; Chen, W.H.; Chen, H.H. Comparison of transcripts in *Phalaenopsis bellina* and *Phalaenopsis equestris* (Orchidaceae) flowers to deduce monoterpene biosynthesis pathway. *BMC Plant Biol.* **2006**, *6*, 14. [CrossRef]
27. Ramya, M.; Park, P.H.; Chuang, Y.; Kwon, O.K.; An, H.R.; Park, P.M.; Baek, Y.S.; Kang, B.; Tsai, W.; Chen, H. RNA sequencing analysis of *Cymbidium goeringii* identifies floral scent biosynthesis related genes. *BMC Plant Biol.* **2019**, *19*, 337. [CrossRef]
28. Huang, L.M.; Huang, H.; Chuang, Y.C.; Chen, W.H.; Wang, C.N.; Chen, H.H. Evolution of terpene synthases in Orchidaceae. *Int. J. Mol. Sci.* **2021**, *22*, 6947. [CrossRef]
29. Yu, Z.; Zhao, C.; Zhang, G.; Teixeira, D.S.J.; Duan, J. Genome-wide identification and expression profile of tps gene family in *Dendrobium officinale* and the role of *DoTPS10* in linalool biosynthesis. *Int. J. Mol. Sci.* **2020**, *21*, 5419. [CrossRef]
30. Wang, Q.Q.; Zhu, M.J.; Yu, X.; Bi, Y.Y.; Zhou, Z.; Chen, M.K.; Chen, J.; Zhang, D.; Ai, Y.; Liu, Z.J.; et al. Genome-wide identification and expression analysis of terpene synthase genes in *Cymbidium faberi*. *Front. Plant Sci.* **2021**, *12*, 751853. [CrossRef]
31. Sun, Y.; Chen, G.; Huang, J.; Liu, D.; Xue, F.; Chen, X.; Chen, S.; Liu, C.; Liu, H.; Ma, H.; et al. The *Cymbidium goeringii* genome provides insight into organ development and adaptive evolution in orchids. *Ornam. Plant Res.* **2021**, *1*, 10. [CrossRef]
32. Zhang, Y.; Zhang, G.; Zhang, D.; Liu, X.; Xu, X.; Sun, W.; Yu, X.; Zhu, X.; Wang, Z.; Zhao, X.; et al. Chromosome-scale assembly of the *Dendrobium chrysotoxum* genome enhances the understanding of orchid evolution. *Hortic. Res.* **2021**, *8*, 183. [CrossRef] [PubMed]
33. Zhao, C.; Yu, Z.; Silva, J.; He, C.; Wang, H.; Si, C.; Zhang, M.; Zeng, D.; Duan, J. Functional characterization of a *Dendrobium officinale* geraniol synthase DOGES1 involved in floral scent formation. *Int. J. Mol. Sci.* **2020**, *21*, 7005. [CrossRef] [PubMed]
34. Huang, H.; Kuo, Y.W.; Chuang, Y.C.; Yang, Y.P.; Huang, L.M.; Jeng, M.F.; Chen, W.H.; Chen, H.H. *Terpene synthase-b* and *terpene synthase-e/f* genes produce monoterpenes for *Phalaenopsis bellina* floral scent. *Front. Plant Sci.* **2021**, *12*, 700958. [CrossRef] [PubMed]

35. Ai, Y.; Li, Z.; Sun, W.H.; Chen, J.; Zhang, D.; Ma, L.; Zhang, Q.H.; Chen, M.K.; Zheng, Q.D.; Liu, J.F.; et al. The *Cymbidium* genome reveals the evolution of unique morphological traits. *Hortic. Res.* **2021**, *8*, 255. [CrossRef] [PubMed]
36. Chen, C.; Chen, H.; Zhang, Y.; Thomas, H.R.; Frank, M.H.; He, Y.; Xia, R. TBtools: An integrative toolkit developed for interactive analyses of big biological data. *Mol. Plant.* **2020**, *13*, 1194–1202. [CrossRef]
37. Mistry, J.; Chuguransky, S.; Williams, L.; Qureshi, M.; Salazar, G.A.; Sonnhammer, E.; Tosat-to, S.; Paladin, L.; Raj, S. Richardson, L.J.; et al. Pfam: The protein families database in 2021. *Nucleic. Acids. Res.* **2021**, *49*, D412–D419. [CrossRef]
38. Duvaud, S.; Gabella, C.; Lisacek, F.; Stockinger, H.; Ioannidis, V.; Durinx, C. Expasy, the Swiss bioinformatics resource portal, as designed by its users. *Nucleic. Acids. Res.* **2021**, *49*, W216–W227. [CrossRef]
39. Kumar, S.; Stecher, G.; Tamura, K. MEGA7: Molecular evolutionary genetics analysis version 7.0 for bigger datasets. *Mol. Biol. Evol.* **2016**, *33*, 1870–1874. [CrossRef]
40. He, Z.; Zhang, H.; Gao, S.; Lercher, M.J.; Chen, W.H.; Hu, S. Evolview v2: An online visualization and management tool for customized and annotated phylogenetic trees. *Nucleic. Acids. Res.* **2016**, *44*, W236–W241. [CrossRef]
41. Bailey, T.L.; Boden, M.; Buske, F.A.; Frith, M.; Grant, C.E.; Clementi, L.; Ren, J.; Li, W.W.; Noble, W.S. MEME SUITE: Tools for motif discovery and searching. *Nucleic. Acids. Res.* **2009**, *37*, W202–W208. [CrossRef] [PubMed]
42. Lescot, M.; Déhais, P.; Thijs, G.; Marchal, K.; Moreau, Y.; Van de Peer, Y.; Rouzé, P.; Rom-bauts, S. PlantCARE, a database of plant *cis*-acting regulatory elements and a portal to tools for in silico analysis of promoter sequences. *Nucleic. Acids. Res.* **2002**, *30*, 325–327. [CrossRef]
43. Dudareva, N.; Martin, D.; Kish, C.M.; Kolosova, N.; Gorenstein, N.; Fäldt, J.; Miller, B.; Bohlmann, J. (*E*)-beta-ocimene and myrcene synthase genes of floral scent biosynthesis in snapdragon: Function and expression of three terpene synthase genes of a new terpene synthase subfamily. *Plant. Cell.* **2003**, *15*, 1227–1241. [CrossRef] [PubMed]
44. Nekrutenko, A.; Makova, K.D.; Li, W.H. The K(A)/K(S) ratio test for assessing the protein-coding potential of genomic regions: An empirical and simulation study. *Genome Res.* **2002**, *12*, 198–202. [CrossRef] [PubMed]
45. Matsuba, Y.; Nguyen, T.T.; Wiegert, K.; Falara, V.; Gonzales-Vigil, E.; Leong, B.; Schäfer, P.; Kudrna, D.; Wing, R.A.; Bolger, A.M.; et al. Evolution of a complex locus for terpene biosynthesis in *solanum*. *Plant. Cell.* **2013**, *25*, 2022–2036. [CrossRef]
46. Reichardt, S.; Budahn, H.; Lamprecht, D.; Riewe, D.; Ulrich, D.; Dunemann, F.; Kopertekh, L. The carrot monoterpene synthase gene cluster on chromosome 4 harbours genes encoding flavour-associated sabinene synthases. *Hortic. Res.* **2020**, *7*, 190. [CrossRef]
47. Qiao, X.; Li, Q.; Yin, H.; Qi, K.; Li, L.; Wang, R.; Zhang, S.; Paterson, A.H. Gene duplication and evolution in recurring polyploidization-diploidization cycles in plants. *Genome Biol.* **2019**, *20*, 38. [CrossRef]
48. Qiao, Z.; Hu, H.; Shi, S.; Yuan, X.; Yan, B.; Chen, L. An update on the function, biosynthesis and regulation of floral volatile terpenoids. *Horticulturae* **2021**, *7*, 451. [CrossRef]
49. Abbas, F.; Ke, Y.; Zhou, Y.; Yu, Y.; Waseem, M.; Ashraf, U.; Wang, C.; Wang, X.; Li, X.; Yue, Y.; et al. Genome-wide analysis reveals the potential role of MYB transcription factors in floral scent formation in *Hedychium coronarium*. *Front. Plant Sci.* **2021**, *12*, 623742. [CrossRef]
50. Chuang, Y.; Hung, Y.; Tsai, W.; Chen, W.; Chen, H. PbbHLH4 regulates floral monoterpene biosynthesis in *phalaenopsis* orchids. *J. Exp. Bot.* **2018**, *69*, 4363–4377. [CrossRef]
51. Yu, Z.; Zhang, G.; Teixeira Da Silva, J.A.; Zhao, C.; Duan, J. The methyl jasmonate-responsive transcription factor DobHLH4 promotes *DoTPS10*, which is involved in linalool biosynthesis in *Dendrobium officinale* during floral development. *Plant Sci.* **2021**, *309*, 110952. [CrossRef] [PubMed]

horticulturae

MDPI

Article

Comparative Transcriptomic Analysis of Pyrethrin and EβF Biosynthesis in *Tanacetum cinerariifolium* Stems and Flowers

Tuo Zeng [1,2,†], Jiawen Li [2,†], Caiyun Wang [2] and Jinjin Li [2,*]

[1] School of Life Sciences, Guizhou Normal University, Guiyang 550025, China; zengtuo@gznu.edu.cn
[2] National Key Laboratory for Germplasm Innovation & Utilization of Horticultural Crops, College of Horticulture & Forestry Sciences, Huazhong Agricultural University, Wuhan 430070, China; lijiawen211@pku.edu.cn (J.L.); wangcy@mail.hzau.edu.cn (C.W.)
* Correspondence: jinjin2023ucr@gmail.com
† These authors contributed equally to this work.

Abstract: *Tanacetum cinerariifolium*, a perennial Asteraceae plant, is renowned for its ornamental value and natural insecticidal compounds, especially pyrethrins. These compounds, primarily stored in flower heads, are highly effective as insecticides with low toxicity to mammals, making them crucial for organic agriculture, along with the sesquiterpene (E)-β-farnesene (EβF), play critical roles in *T. cinerariifolium* defense mechanisms. However, the spatiotemporal patterns of these secondary metabolites in stems and flower heads, as well as their regulatory mechanisms, remain unclear. This study investigated the biosynthesis and regulation of pyrethrins and EβF across developmental stages (S1–S4) in flowers and stems using GC-MS and transcriptomics. Transcriptome analysis revealed that the expression of pyrethrin biosynthetic genes was not synchronized with pyrethrin accumulation. The main pyrethrin biosynthetic genes exhibited coordinated expression patterns, peaking during early flowering stages (S1–S2), while pyrethrin accumulation was primarily observed during mid-flower development. In contrast, the biosynthetic genes of EβF showed synchronized expression with EβF accumulation, with the highest activity observed in stems and early flowers. WGCNA identified jasmonic acid signaling, trichome differentiation, and terpene transport pathways as potentially associated with pyrethrin biosynthesis. Hub genes including *MYC2* were identified as playing pivotal roles in regulating secondary metabolite biosynthesis. These findings provide new insights into the regulation and biosynthesis of pyrethrins and EβF, offering a foundation for optimizing bioactive compound production and advancing sustainable pest management strategies.

Keywords: *Tanacetum cinerariifolium*; pyrethrins; transcriptomics; flower development; (E)-β-farnesene

1. Introduction

Tanacetum cinerariifolium, commonly known as pyrethrum, is a perennial plant in the Asteraceae family, valued for its ornamental application and its role as a source of natural insecticides. The plant is characterized by white ray florets surrounding a dense center of yellow disc florets and is particularly renowned for the high concentration of natural pyrethrins in its flower heads. These pyrethrins, comprising pyrethrin I, cinerin I, jasmolin I, pyrethrin II, cinerin II, and jasmolin II, provide significant economic benefits [1]. Pyrethrins are highly effective insecticides that incapacitate pests quickly, exhibit minimal toxicity to mammals, and are resistant to pest tolerance. Moreover, they are fully biodegradable, making them environmentally friendly [2,3]. These properties have ensured their long-standing use in household and agricultural pest control [4]. Due to their potent insecticidal properties, pyrethrins have made pyrethrum a globally cultivated plant, not only for ornamental purposes but also as a key resource for bio-pesticides [5,6]. This makes pyrethrum an excellent choice for organic farming, home gardening, and agro-industrial applications.

Recent advancements in plant physiological ecology and secondary metabolite research have expanded interest in *T. cinerariifolium*, focusing on its primary insecticidal compounds and other bioactive substances. Pyrethrum flower extracts are recognized for their potent insecticidal properties and repellent effects against aphids and mosquitoes. Notably, approximately 70% of the volatile compounds in pyrethrum are composed of (*E*)-β-farnesene (EβF) and germacrene D (GD) [7]. The efficacy of these compounds is attributed to the synergistic action between the highly volatile EβF and the less volatile pyrethrins [7,8].

EβF plays a crucial role as a key component of alarm pheromones in many aphid species, attracting predatory insects, such as ladybugs, in response to predator attacks [9,10]. This pheromone is primarily released from a viscous droplet secreted from the aphid's dorsal abdominal tubes during attacks. At high concentrations and purity, EβF can trigger an alarm response, causing conspecifics to disperse from the feeding site. Moreover, this signaling compound is likely intercepted by predatory insects, serving as an important cue in their search for food [9,11]. Pyrethrins and EβF together form the defensive arsenal of *T. cinerariifolium*. Notably, pyrethrins predominantly accumulate in the ovary walls of the flower heads, accounting for over 94% of the plant's total pyrethrins [12,13]. Both the flower heads and stems contain high levels of EβF. These compounds exhibit significant spatiotemporal variations in their accumulation.

While pyrethrins are monoterpenoid derivatives, EβF is classified as a sesquiterpene. Both compounds are synthesized via the isoprenoid metabolic pathway, which involves a complex interplay of competition and enhancement. Volatile substances dominated by EβF can even promote the synthesis of pyrethrins [14,15]. EβF is synthesized from the precursor molecules isopentenyl pyrophosphate (IPP) and its isomer dimethylallyl pyrophosphate (DMAPP) through the mevalonate (MVA) pathway. Farnesyl diphosphate synthase (FPS) catalyzes the formation of farnesyl diphosphate (FPP), which is then converted to EβF by sesquiterpene synthase (E)-beta-farnesene synthase (EbF) [16]. In *T. cinerariifolium*, pyrethrins are primarily biosynthesized in the flower heads through the esterification of a monoterpenoid acyl moiety (pyrethric or chrysanthemic acid) with an alcohol moiety (pyrethrolone, jasmolone, and cinerolone). The acid component originates from the methylerythritol-4-phosphate (MEP) pathway within plastids, a key branch of the terpene biosynthetic network [17], whereas the alcohol component is derived from jasmonates [2]. Initial studies have explored the metabolic roles and biosynthetic pathways of these compounds in pyrethrum. However, detailed information regarding their expression patterns, regulatory mechanisms, and ecological functions during different growth stages, particularly during key phases of flower development, remains limited. Systematic omics approaches are needed to further investigate the biological and ecological significance of pyrethrin and EβF synthesis and release during these critical stages.

In *T. cinerariifolium*, the flowering phase is characterized by elongated peduncles that are typically erect or slightly curved, serving to support the capitulum and facilitate adaptation to the pollination environment. Flower head development proceeds through eight distinct stages: S1, a well-developed closed bud; S2, with ray flowers positioned vertically; S3, where ray flowers become horizontal, and the first row of disc flowers opens; S4, with three rows of disc flowers open; S5, when all disc flowers are open; S6, the early senescent stage marked by fading disc flower color while ray flowers remain intact; S7, the late senescent stage with minimal disc flower coloration and dried ray flowers; and S8, when ray flowers are shed, thereby enabling wind-driven seed dispersal (Figure S1a). Notably, during the S2 stage, pyrethrum flowers are rich in volatile secondary metabolites and contain moderate levels of pyrethrins (further, GC–MS analyses for additional stages are provided in the Supplementary Information (Figure S1b).

This study employs high-throughput transcriptomics and GC-MS technologies to systematically analyze the expression and regulation of pyrethrins and EβF during stem and flower development stages S1 to S4 in *T. cinerariifolium*. The goal is to elucidate the metabolic pathways and regulatory networks controlling the biosynthesis and accumulation of these secondary metabolites during critical developmental stages. This comprehensive analysis is expected to provide new insights into plant secondary metabolite regulatory mechanisms and their potential applications in pest management.

2. Results

2.1. The Content of Pyrethrin and EβF in T. cinerariifolium

To characterize the tissue-specific biosynthesis patterns of key defensive metabolites, we quantified pyrethrins and EβF concentrations across vegetative stems and flowering developmental stages (S1–S5) using GC-MS. Details of identified compounds are listed in Table S1. The total pyrethrin content is lowest in the stem and increases progressively from S1, peaking during stages S4 and S5. No significant difference in pyrethrin content was observed between S4 and S5 (Figure 1a; Table S2). For optimal production efficiency and large-scale mechanical harvesting, pyrethrum is typically collected during stages S4 and S5 [6]. In contrast, EβF content is highest in the stem, relatively high at S1, and gradually declines and stabilizes between S3 and S5 (Figure 1b; Table S2). This distribution pattern indicates that EβF may serve primarily in early floral developmental stages and the stem for signaling and defensive purposes, complementing the role of pyrethrins, which are predominantly involved in defense during the flower mature stages.

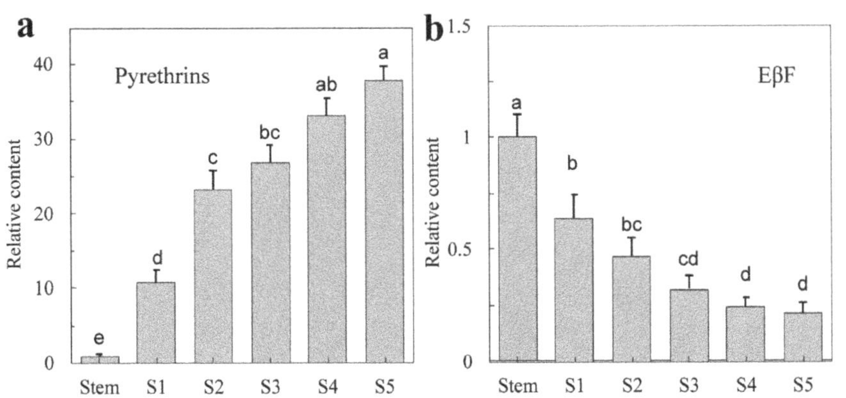

Figure 1. Comparison of pyrethrins and EβF levels in *T. cinerariifolium*. (**a**) Relative content of pyrethrins at different flower developmental stages and in stems. (**b**) Relative content of EβF at different flower developmental stages and in stems. Error bars represent mean ± SD. Different lower-case letters indicate a significant difference ($p < 0.05$) from one-way ANOVA followed by a post hoc Tukey test. The values are normalized based on the stem sample.

2.2. Transcriptome Sequencing, Assembly, and Annotation of T. cinerariifolium

A total of 15 sequencing libraries were constructed from *T. cinerariifolium* flower stems and flowering stages (S1–S4), yielding 101.5 Gb of clean data. Each sample produced more than 6.09 Gb of clean data, with a Q30 base percentage exceeding 93.63% (Table S3), which met the quality standards for subsequent analysis. The clean data from all 15 libraries were assembled using Trinity v2.6.6 [18], resulting in 100,099 unigenes. The average read length was 977 bp, with a GC content of 38.67%. Clean reads were aligned to the reference sequences derived from the Trinity assembly. BUSCO analysis confirmed that the majority of genes were well represented in the BUSCO library (Figure 2a), demonstrating the high quality of the sequencing data for downstream analyses.

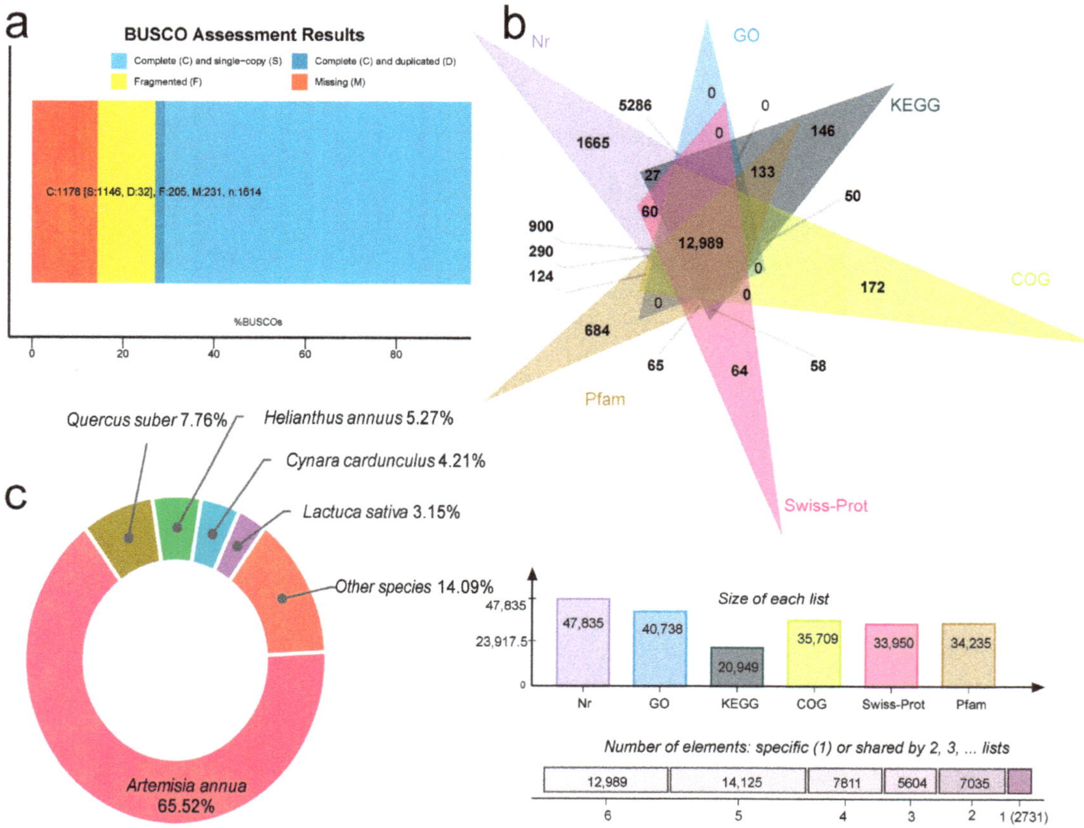

Figure 2. Transcriptome assembly and gene functional annotation. (**a**) BUSCO assessment results. (**b**) The Venn diagram shows the total number of functional annotations in databases. (**c**) Species distribution of the BlastX result against the Nr database.

To gain comprehensive functional information, gene annotation was performed using Blast v2.10 and Diamond v5.10 software against seven major databases: UniProt, KEGG, GO, Nr, Pfam, COG, and KEGG Pathway. In total, 50,295 unigenes (50.25%) were annotated, with 12.98% of the unigenes annotated across all databases. The number of unigenes annotated in each database was as follows: Nr (47,835; 47.79%), Pfam (34,235; 34.20%), Swiss-Prot (33,950; 33.92%), GO (40,738; 40.70%), COG (35,709; 35.67%), and KEGG (20,949; 20.93%) (Figure 2b). Species annotation from the Nr database revealed that the top five annotated species were *Artemisia annua* (65.52%), *Quercus suber* (7.76%), *Helianthus annuus* (5.27%), *Cynara cardunculus* (4.21%), and *Lactuca sativa* (3.15%) (Figure 2c).

To evaluate the consistency among samples, Pearson correlation and principal component analysis (PCA) were performed. The Pearson correlation results revealed minimal variation within each sample group, indicating a high degree of consistency within groups (Figure 3a). The PCA results further showed clear separation between different sample groups, suggesting distinct clustering patterns, the observed separation between groups provides strong visual evidence of group differentiation, confirming the consistency of the data for transcriptome analysis (Figure 3b). Differentially expressed genes (DEGs) were identified using the criteria of p-adj < 0.05 and $|\log2\ \text{FoldChange}| > 1$. Compared to the stem, a total of 13,239 DEGs were detected in S1 (6008 upregulated and 7231 downregulated). Between S1 and S3, 12,160 DEGs were identified (6635 upregulated and 5525 downregulated), and a similar number of DEGs (12,160) was observed between S1 and S4 (7139 upregulated and 4768 downregulated) (Figure 3c). These findings highlight marked differences between the stem and the flowering stages, while expression patterns were more similar between adjacent stages, such as S1 and S2, or S3 and S4 (Figure 3d).

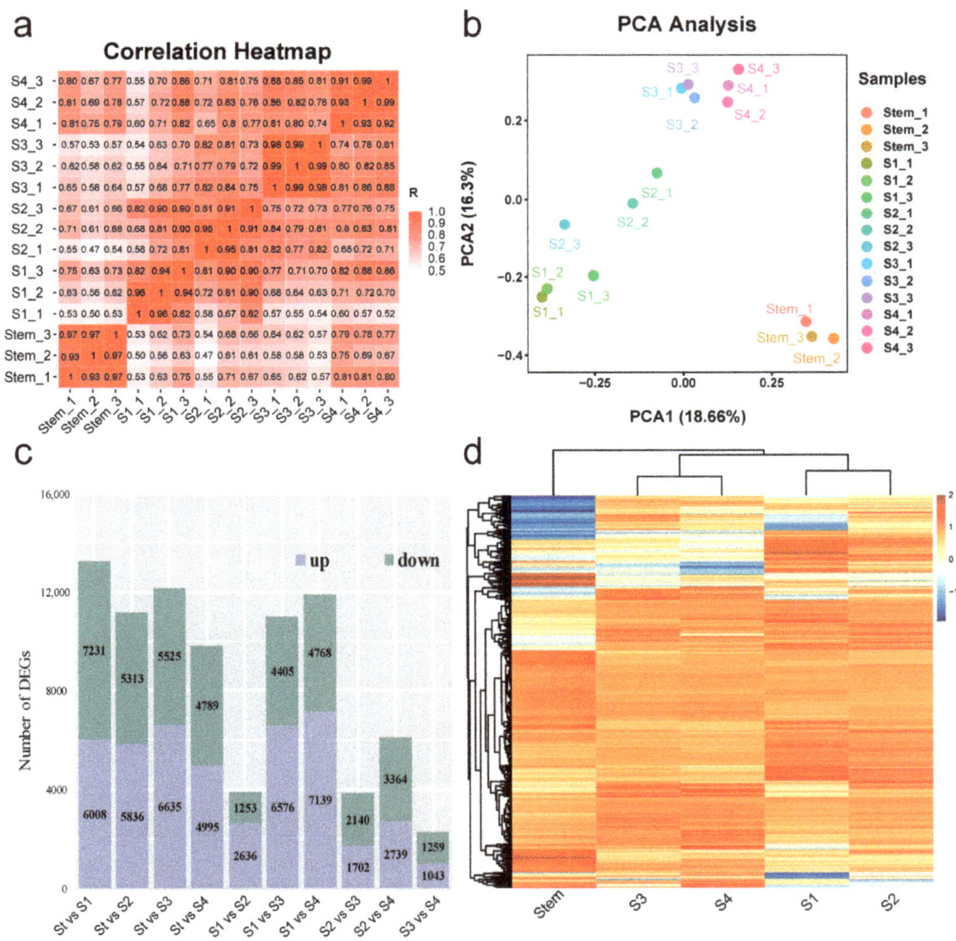

Figure 3. The quantification and DEGs of the samples. (**a**) Pearson correlation analysis of samples. (**b**) Principal Component Analysis of samples. (**c**) Distribution of DEGs between different periods. (**d**) Heat map of DEG clusters. The x-axis represents hierarchical clustering of samples based on gene expression similarity.

2.3. Differential Gene Expression in JA Biosynthesis

JA plays a critical role in flower development and serves as a precursor for the synthesis of pyrethrins. An analysis of DEGs involved in JA revealed notable expression patterns for key genes such as *SPLA2* (*Secretory Phospholipase A2*, 1 DEG), *LOX* (*Lipoxygenase*, 8 DEGs), *OPR3* (*12-Oxophytodienoate Reductase*, 8 DEGs), *AOS* (*Allene Oxide Synthase*, 1 DEG), *AOC* (*Allene Oxide Cyclase*, 1 DEG), *ACX* (*Acyl-CoA Oxidase*, 2 DEGs), and *MFP* (*Multifunctional Protein*, 2 DEGs) (Figure 4). These genes are expressed at very low levels in the stem but are highly expressed during the early flowering stages (S1 and S2). Subsequently, their expression levels decline, with a particularly pronounced decrease in *AOC*, a key gene in JA biosynthesis. qRT-PCR results confirm that the expression of *LOX*, *AOS*, and *AOC* increases from the stem to the early flower stages (S1 and S2). *LOX* peaks at S3 and S4, while *AOC* and *AOS* show their highest expression at S1 and S2 before progressively declining through S3 and S4. These trends indicate that JA biosynthesis is most active during the early to mid-flower stages, supporting its role in initiating defense and developmental processes.

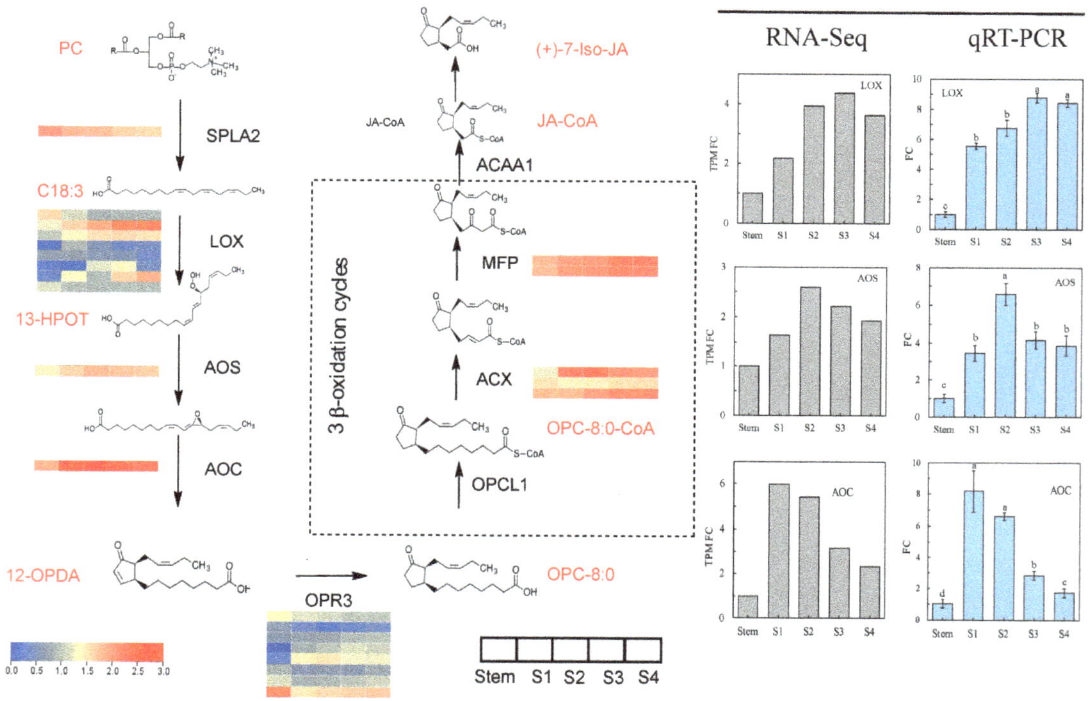

Figure 4. DEGs assigned to JAs biosynthesis pathway. The colors in the heatmaps represent the normalized expression levels of DEGs, expressed as log10(TPM + 1). RNA-Seq (gray bars) and qRT-PCR (blue bars) data are shown. FC values are presented for expression level and normalized to the stem samples. Error bars represent mean ± SD. Different lower-case letters indicate a significant difference ($p < 0.05$) from one-way ANOVA followed by a post hoc Tukey test.

2.4. Differential Gene Expression in Pyrethrins Biosynthesis

While some pathways in the synthesis of pyrethrins remain unclear, most have been well elucidated. The DEGs selected for expression analysis have been previously validated in *T. cinerariifolium* as key regulators or enzymes involved in the biosynthesis of pyrethrins and EβF. DEG analysis reveals that the expression of pyrethrin biosynthetic genes is extremely low in the stem but significantly elevated during the early flowering stages (S1 and

S2). Their expression is subsequently downregulated during the mid-flowering stages (S3 and S4). qRT-PCR results show that genes such as *CDS* (*Chrysanthemyl Diphosphate Synthase*), *ADH* (*Alcohol Dehydrogenase*), *ALDH* (*Aldehyde Dehydrogenase*) and *GLIP* (*GDSL-like Lipase*) are highly expressed during the early to mid-flowering stages (S1 and S2). Notably, *JMH* (*Jasmonate Hydroxylase*) maintains consistently high expression levels throughout all flowering stages. In contrast, EβF expression is highest in the stem and at the early flowering stage (S1) (Figure 5).

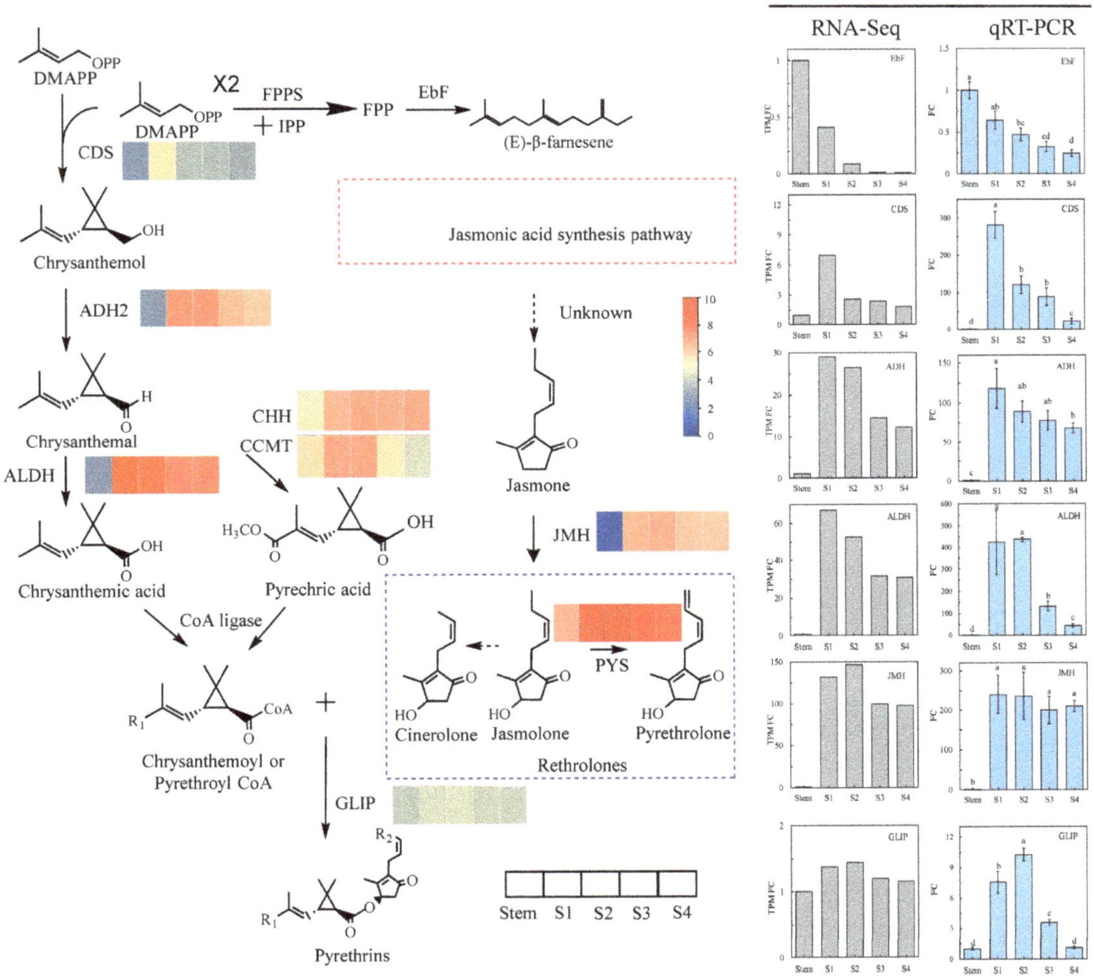

Figure 5. The biosynthesis of pyrethrin and gene expression. The color shows the normalized expression score from $\log_2$(TPM + 1) of DEGs. Solid arrows indicate known steps in the pathway, dotted arrows indicate steps not yet elucidated, red dotted box indicates JAs synthesis pathway, blue dotted box indicates rethrolones, a pyrethrin molecule incorporates one of three rethrolones, right diagram: grey color represents RNA-Seq, blue color represents qRT-PCR. Error bars represent mean ± SD. Different lower-case letters indicate a significant difference ($p < 0.05$) from one-way ANOVA followed by a post hoc Tukey test.

2.5. WGCNA Co-Expression Analysis

To further investigate the biosynthesis of pyrethrins in *T. cinerariifolium*, a filtering threshold was applied, selecting genes with an average TPM value greater than 1 and a coefficient of variation exceeding 0.2. Using the filtered expression matrix containing 28,749 genes, co-expression modules were constructed with a fitting index and optimal connectivity, using a soft threshold power of 11. A merging threshold of 0.3 yielded 66 modules, whereas a threshold of 0.2 reduced the number to 22 modules (Figure 6a,b). Among these, the darkolivegreen module (2856 genes), honeydew1 module (1091 genes), and coral1 module (3026 genes) were identified as being highly correlated with pyrethrin biosynthesis (Figure 6c).

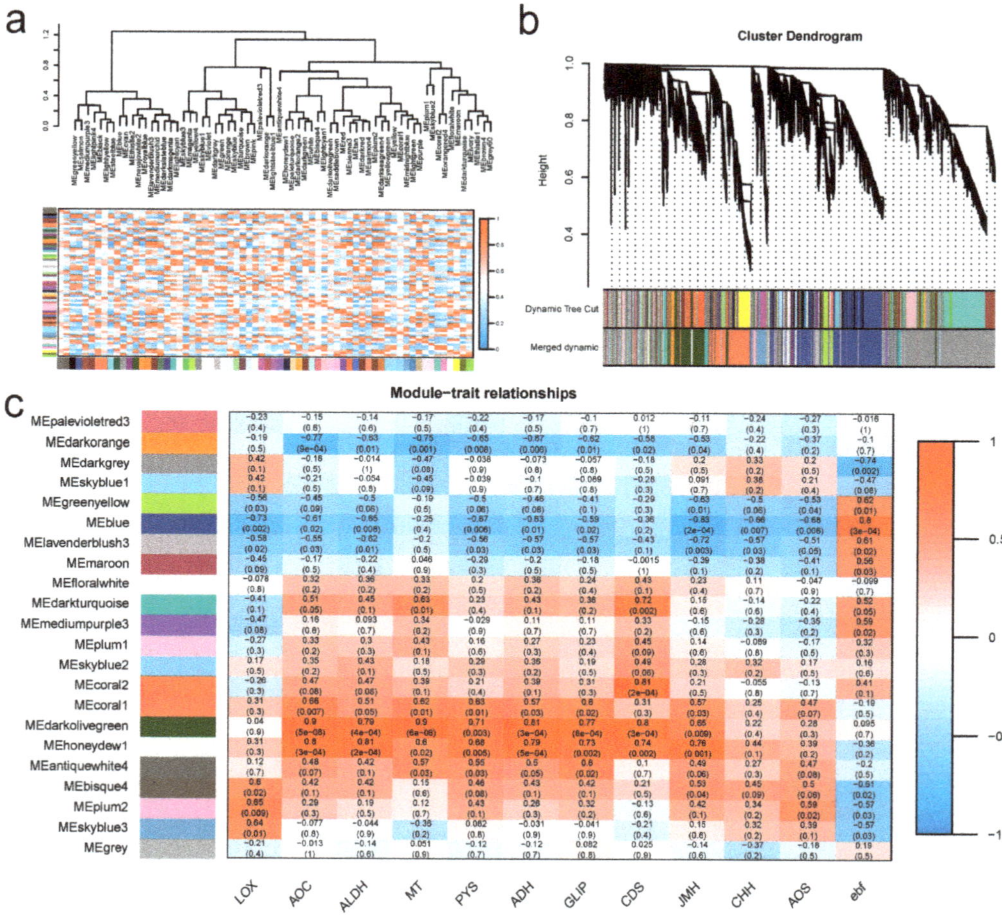

Figure 6. Co-expression modules determined by weighted gene co-expression network analysis. (**a**) The cluster dendrogram groups genes into modules based on their co-expression patterns. Each color below the dendrogram represents a distinct module. (**b**) Co-expression module expression spectrum diagram. Modules were identified using the dynamic tree cut method and subsequently merged based on their similarity. (**c**) Heatmap of correlations between module eigengenes (rows) and sample traits (columns). Red denotes a positive correlation, blue indicates a negative correlation, and the intensity of the color reflects the strength of the correlation. The numbers in each cell indicate the correlation coefficient and the p-value (in parentheses). Values are presented using scientific notation, such as "5e-06", which denotes a value of 5×10^{-6}.

To further investigate the association between these three modules and pyrethrin biosynthesis, KEGG enrichment analysis was performed, and the top 20 pathways were visualized based on their significance (Figure S2a–c). The major enriched pathways in the coral1, honeydew1, and darkolivegreen modules were largely consistent, including metabolic pathways (222, 218, and 73 genes, respectively) and biosynthesis of secondary metabolites (102, 132, and 73 genes, respectively). Additionally, the honeydew1 module showed specific enrichment in pathways related to monoterpenoid biosynthesis (4 genes) and plant–pathogen interactions (30 genes), highlighting potential links to plant defense mechanisms. In contrast, the darkolivegreen module was enriched in plant hormone signal transduction pathways (31 genes) and α-linolenic acid metabolism pathways (14 genes), both associated with JA biosynthesis. These results suggest that hormonal regulation, particularly involving JA, plays a pivotal role in pyrethrin biosynthesis in *T. cinerariifolium*.

GO enrichment further confirmed these results (Figure S2d–f). In the honeydew1 module, the main enrichments included 104 protein bindings, 4 trichome differentiations, and 3 ABC-type xenobiotic transporter activities. The synthesis of pyrethrins is highly related to gland differentiation. During the flowering stage, it is synthesized by the glands and then transported to the ovary wall outside the glands. The pathways associated with the activity of these ABC-type transporters may play a significant role in pyrethrin synthesis and transport. In the darkolivegreen module, the most significant enrichment was observed in 297 plasma membrane proteins, as well as numerous pathways related to hormones, including 32 responses to jasmonic acid, 56 responses to abscisic acid, 25 responses to ethylene, 11 jasmonic acid biosynthetic processes, and 14 jasmonic acid-mediated signaling pathways. This module also included 34 defense responses to fungus and 118 DNA-binding transcription factor activities. Additionally, 28 responses to wounding pathways were identified, indicating that genes involved in these pathways may play a crucial role in enhancing pyrethrin content in response to damage. In the coral1 module, the most significant enrichments were 328 plasma membrane proteins, 35 responses to wounding, 31 defense responses to fungus, and 21 responses to jasmonic acid. These pathways related to damage response, antifungal activity, and response to MeJA are highly associated with pyrethrin synthesis.

The enrichment of JA-related pathways across multiple modules further supports the role of JA as a central regulator in pyrethrin biosynthesis. Given that JA is well known for its function in plant defense, it is plausible that pyrethrins and JA share a coordinated regulatory mechanism. The identification of genes involved in ABC-type transporters, trichome differentiation, and defense responses suggests that both transport and glandular structures are crucial for efficient pyrethrin accumulation. Moreover, the observed enrichment in wounding response pathways reinforces the idea that pyrethrin biosynthesis may be inducible by biotic or abiotic stressors, similar to other secondary metabolites in plants.

To identify key regulatory genes in these modules, protein–protein interaction (PPI) networks were constructed using the STRING database, and hub genes were identified with the Cytohubba plugin in Cytoscape. In the honeydew1 module, hub genes included *AP2*, which is involved in transcriptional regulation, *TTG1* (trichome development), *TPS11* (terpene synthesis), *PLA2A* (lipid hydrolysis), and transcription factors *MYB73* and *MYB12* (Figure 7a). The darkolivegreen module featured hub genes such as *ALDH7B4*, *ALDH3H1*, and *ALDH10A8*, which are known to play roles in pyrethrin biosynthesis. Additionally, this module included *EIN3* (ethylene-insensitive 3), which interacts with *MYC2*, *MYC3*, and *MYC4* to inhibit the expression of jasmonic acid-induced wound response genes and herbivore defense genes (Figure 7b). In the coral1 module, hub genes included *COI1* (*CORONATINE INSENSITIVE 1*), *MYC2*, and *JAZ1* (Figure 7c). Among these, the *MYC2* gene corresponds to the *TRINITY_DN73856_c0_g1* gene. Amplification of the *MYC2* interaction network revealed that *MYC2* plays a central role in the jasmonic acid pathway, interacting with multiple JAZ proteins to mediate responses in the JA signaling pathway (Figure 7d).

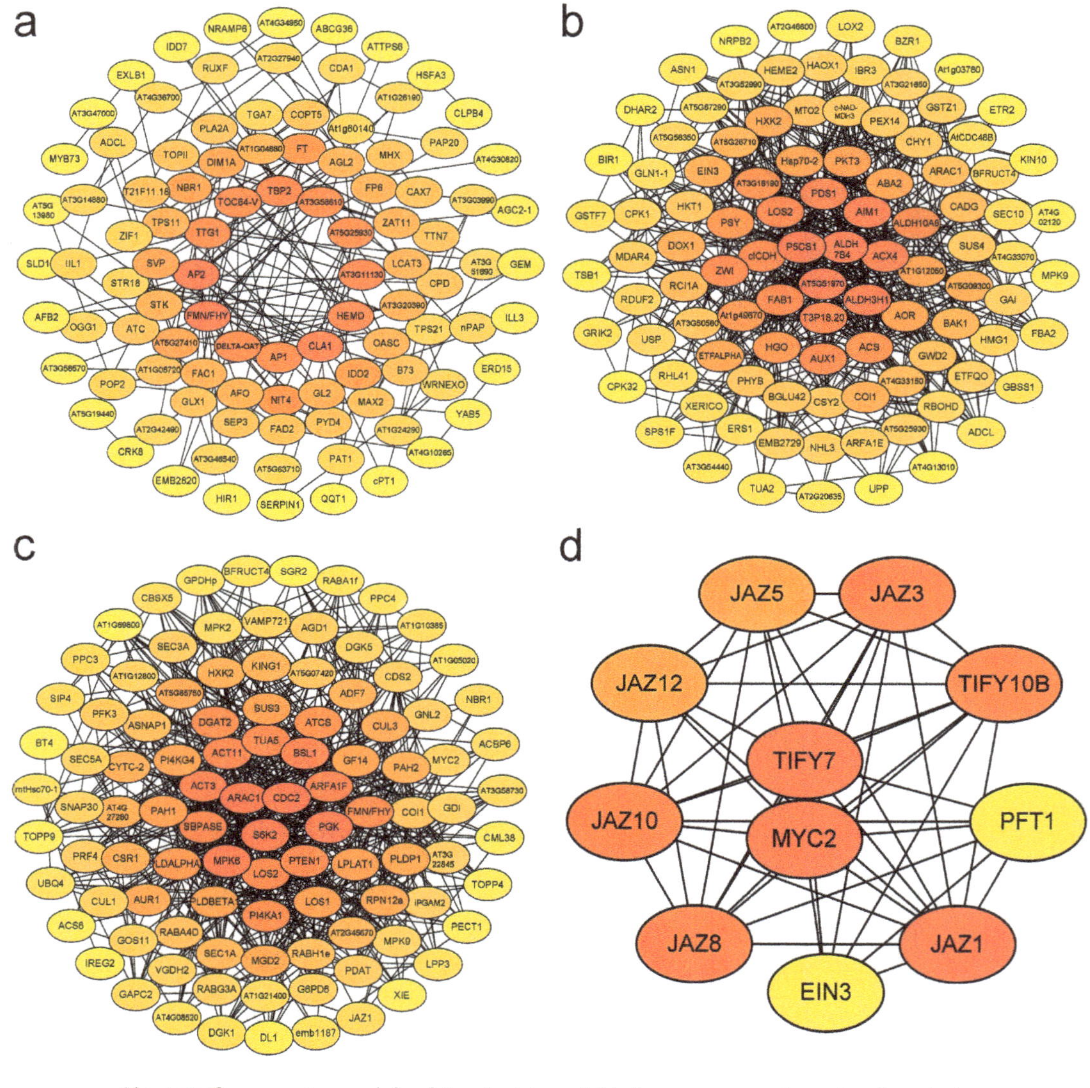

Figure 7. Co-expression modules determined by weighted gene co-expression network analysis. (**a**–**c**) The predicted protein–protein interaction network of the honeydew1, darkolivergreen, and coral1 modules, constructed based on the Arabidopsis protein database using STRING. (**d**) Correlation network of *MYC2*.

3. Discussion

3.1. Dynamics of Pyrethrin and EβF Biosynthesis in T. cinerariifolium

This study provides a comprehensive analysis of the biosynthesis and regulation of pyrethrins and EβF in *T. cinerariifolium*, elucidating their accumulation patterns and the gene expression profiles. Pyrethrin content increases from S1 to S5, with relatively high levels observed during the S4–S5, aligning with its critical role in plant defense and representing a common harvest time for industrial applications [6]. Conversely, EβF levels are elevated in the stem and early floral stages, suggesting its function in early defense signaling and ecological interactions, a phenomenon observed in fields where pyrethrum attracts numerous predatory ladybugs during the S1–S2 stages [8]. The biosynthetic pathways of pyrethrins and EβF, both stemming from isoprenoid metabolic precursors but diverging into monoterpenoid and sesquiterpenoid routes, respectively, exemplify a sophisticated ecological defense strategy. Pyrethrins, synthesized via the MEP pathway [17], are crucial for deterring pests in later floral stages. EβF, synthesized from the MVA pathway [16], acts as an alarm pheromone to attract predatory insects during the early flowering stages [8]. In addition, EβF enhances the pyrethrin's effectiveness [19]. This establishes a nuanced regulatory network that orchestrates pyrethrin biosynthesis and supports the plant's adaptive responses to environmental cues.

Endogenous JAs and pyrethrin synthesis-related genes exhibited similar expression patterns: they were expressed at low levels during the stem elongation stage, increased during the S1 and S2 stages of flowering, and were downregulated in S3 and S4. This co-expression pattern, particularly the simultaneous upregulation of JA and pyrethrin-related genes during early flowering, suggests a coordinated regulation of these pathways. Such synchronization indicates that other co-expressed genes during these peak periods could play roles in pyrethrin biosynthesis, providing potential targets for identifying new genes involved in this synthetic pathway.

Notably, while our study highlights the coordinated expression of JA and pyrethrin biosynthesis genes in flower heads, previous studies have reported contrasting findings regarding the role of MeJA treatment. For example, in leaves, MeJA treatment strongly induced early-stage pyrethrin biosynthesis genes but downregulated late-stage genes encoding enzymes such as GLIP, JMH, and PYS, ultimately slowing pyrethrin production [20]. In flower heads, however, pyrethrin synthesis exhibited a diminished response to MeJA treatment, with no significant impact on pyrethrin content [21]. These earlier findings suggest substantial tissue-specific regulatory differences in pyrethrin biosynthesis.

3.2. Insights from Co-Expression Analysis

WGCNA is a robust tool for studying gene interactions and identifying modules associated with secondary metabolite pathways. It has been widely used to integrate transcriptomic and metabolomic data, revealing key regulatory mechanisms. For instance, WGCNA effectively identified genes and pathways associated with high sugar, high acid, and high anthocyanin traits in *Vaccinium duclouxii* [22], flavonoids linked to cold tolerance in chrysanthemum [23], and the effects of forchlorfenuron and thidiazuron on flavonoid biosynthesis in table grape skins [24].

However, WGCNA has limitations, especially when gene expression and metabolite synthesis or accumulation are asynchronous. qRT-PCR analysis revealed that genes related to pyrethrin biosynthesis are highly expressed during the S1–S2 stages of flowering but exhibit significant lag in pyrethrin accumulation, which becomes substantial only during the S4–S5 stages. In contrast, EβF shows synchronous patterns of gene expression and accumulation. This difference can be attributed to the contrasting biosynthetic pathways and storage characteristics of the two compounds. Pyrethrin biosynthesis involves the complete MEP pathway and parts of the JA synthesis pathway, making it a complex, multi-step process influenced by numerous regulatory factors [2,17]. Additionally, pyrethrins are primarily stored in intercellular spaces or the fruit pericarp tissues, which likely reduces their phytotoxic effects but also delays their accumulation [25]. This storage mechanism further contributes to the diminished responsiveness of pyrethrin synthesis in flower heads to JA signal [20,21]. In contrast, EβF synthesis is a simpler, single-step process catalyzed by sesquiterpene synthase, and it is localized in glandular trichomes and head cells, facilitating faster synthesis and transport [11,26].

The asynchronous nature of pyrethrin gene expression and metabolite accumulation underscores the limitations of using pyrethrin content as a direct trait for WGCNA analysis. Therefore, in this study, analyzing the co-expression patterns of known pyrethrin biosynthetic genes represents a more reliable approach for identifying regulatory networks and candidate genes involved in pyrethrin synthesis.

Through WGCNA, three key modules, honeydew1, darkolivegreen and coral1, were identified as strongly associated with pyrethrin biosynthesis and defense responses. Each module revealed distinct but interconnected pathways contributing to secondary metabolite production. The honeydew1 module was enriched in genes related to trichome differentiation, plant–pathogen interactions, and monoterpenoid biosynthesis, underscoring its relevance to glandular structures where pyrethrins are synthesized. The darkolivegreen module highlighted pathways involving JA biosynthesis and signaling, while the coral1 module included key JA pathway components such as *MYC2*, *JAZ1*, and *COI1*, which are critical regulators of JA responses. The presence of ABC transporters and plasma membrane proteins across these modules suggests their roles in metabolite synthesis and transport. Collectively, these modules provide a comprehensive regulatory map integrating pyrethrin biosynthesis with broader plant defense mechanisms.

Our analysis also revealed modules associated with trichome differentiation. Laboratory experiments demonstrated that MeJA treatment enhances trichome density in seedlings, correlating directly with increased pyrethrin content [20]. Similarly, field studies reported a positive correlation between trichome density and pyrethrin accumulation [27]. Supporting this, experiments on *T. cinerariifolium* callus and CRISPR-Cas9-edited hairy roots devoid of glandular trichomes failed to produce pyrethrins, even under induced differentiation conditions [28]. These observations emphasize that glandular trichomes are essential for pyrethrin biosynthesis and may represent a critical rate-limiting factor in its production.

Among the key genes identified, *TcTTG1*, a WD40-repeat protein, was co-expressed with pyrethrin biosynthetic genes. *TcTTG1* shares homology with *TTG1* in *A. thaliana* and *A. annua*, both of which are involved in trichome development [29]. In *A. thaliana*, *TTG1* interacts with the MYB-family gene *GL1* and bHLH-family gene *GL3*, forming a MYB-bHLH-WD40 ternary complex that governs trichome formation [30,31]. Similarly, *A. annua* homolog AaGSW2, a WRKY transcription factor, positively regulates glandular development and enhances artemisinin synthesis [32]. Although functional details of TcTTG1 require further validation, its co-expression with pyrethrin biosynthetic genes suggests it may play a critical role in glandular trichome formation and pyrethrin biosynthesis.

Furthermore, transcription factors such as *MYB73*, *MYC2*, and JAZ1 emerged as key regulatory genes. Studies have shown that *MYC2*, enriched in MeJA-treated leaf transcriptomes, acts as a central regulator of JA-mediated pathways [20], while *MYB73* in apples regulates vacuolar transport systems to influence malic acid accumulation [33]. Additionally, ABC transporters, identified within these modules, are critical for transporting pyrethrin precursors and products between plastids, cytoplasm, and extracellular spaces. These transport systems, along with key transcription factors, represent integral components of the pyrethrin biosynthetic pathway and promising targets for metabolic engineering to enhance pyrethrin production.

3.3. Implications for Pest Management and Future Directions

The elucidation of pyrethrin and EβF biosynthesis pathways provides valuable insights into how *T. cinerariifolium* produces and deploys these compounds as defense mechanisms. The distinct roles of pyrethrins in direct insecticidal activity and EβF in predator attraction and alarm signaling highlight a synergistic defense strategy that can be exploited in integrated pest management. By enhancing the natural production of these compounds, either through genetic engineering or optimized cultivation practices, it may be possible to develop bio-based solutions that reduce reliance on synthetic pesticides, aligning with global goals for sustainable agriculture.

The co-expression modules identified through WGCNA, particularly the honeydew1, darkolivegreen, and coral1 modules, point to key regulatory nodes that could be targeted for metabolic engineering. For example, manipulating MYC2, EIN3, and ABC transporters may enhance pyrethrin accumulation without compromising plant growth or other metabolic functions. Moreover, understanding the interplay between the JA pathway and other hormonal pathways, such as ABA and ethylene provides a framework for designing strategies that optimize plant resilience under various environmental conditions while maintaining high secondary metabolite yields.

Future research should focus on the application of these findings to improve the commercial viability of *T. cinerariifolium* as a bio-pesticide resource. This includes developing molecular tools to precisely regulate pyrethrin and EβF biosynthesis during key growth stages, as well as exploring how environmental factors, such as light, temperature, and biotic stress, influence these pathways. Additionally, investigating the role of glandular trichomes and transport mechanisms in pyrethrin localization and secretion may reveal further targets for enhancing yield and efficiency. Advancing transcriptomic and metabolomic techniques will allow for a more comprehensive understanding of how secondary metabolism is dynamically regulated in response to environmental stimuli, paving the way for innovations in plant science and sustainable agriculture.

4. Materials and Methods

4.1. Plant Materials

The *T. cinerariifolium* plants, derived from tissue culture (clonal progeny), were cultivated in the greenhouse at Huazhong Agricultural University. Flowers at different developmental stages (S1–S4) and flower stalks were collected according to the stages defined by Ramirez's study [34], and the flower stalks were collected as described in [8]. At each stage, tissues were randomly sampled from different plants within the same stage and then pooled to ensure representativeness. A total of 15 samples were collected, including 3 stem samples and 3 samples from each flowering stage (S1–S4). All samples were replicated in triplicate and immediately transferred to liquid nitrogen for rapid freezing and further analysis.

4.2. GC-MS Analysis

For the stored samples, 100 mg of the thoroughly ground powder was transferred to a 2 mL vial containing n-hexane with 8.7 ng/µL methyl laurate as an internal standard (chromatography grade). The mixture was vortexed for 15 s, centrifuged at 5000× g for 10 min, and the supernatant was passed through a drying column packed with anhydrous sodium sulfate. For tissues too small to be ground, n-hexane extraction was performed directly, and the extracts were transferred to 1.5 mL autosampler vials and stored at −20 °C until analysis.

The GC/MS analysis was performed using a GC/MS-QP2010 Ultra system (Shimadzu Corporation, Kyoto, Japan) equipped with an HP-5 MS column. The column temperature was maintained at 40 °C for 3 min, then ramped at 10 °C/min to 280 °C and held for 2 min. The ion source and transfer line temperatures were set at 230 °C and 280 °C, respectively. Electron ionization was conducted at 70 eV, with a mass scan range of 45 to 450 m/z and a scanning rate of 5 scans per second. Data analysis was carried out using Shimadzu GC solutions software v4.20, with qualitative and quantitative assessments based on the NIST2011 and the PESTEI3 pesticide databases. Each sample was analyzed in triplicate with chrysanthemum flower head extract (Sigma-Aldrich, Saint Louis, MO, USA) used as the standard reference material.

4.3. Total RNA Extraction, cDNA Library Construction and Sequencing

Total RNA was extracted from the samples using the phenol/chloroform method. RNA concentration and integrity were assessed using the NanoDrop 2000 spectrophotometer (Thermo Scientific, Waltham, MA, USA) and the Agilent Bioanalyzer 2100 System (Agilent Technologies, Santa Clara, CA, USA). The NEBNext® Ultra™ RNA Library Prep Kit for Illumina® (NEB, Ipswich, MA, USA) was used to generate sequencing libraries. The PCR products were purified and the library quality was assessed using the Agilent Bioanalyzer 2100 system. Sequencing was performed by a commercial sequencing company (MGI, Wuhan, China). The raw sequences described in this article were submitted and released to The National Genomics Data Center (NGDC, https://ngdc.cncb.ac.cn) under BioProject PRJCA033633 (accessed on 13 December 2024).

4.4. Transcript Splicing, Annotation, and Quality Assessment

Raw reads were processed for quality control using Fastp v0.20.1 software with default parameters to remove low-quality values, adapter sequences, and poly-N sequences [35]. Reads with low-quality values, adapter sequences, and poly-N sequences were filtered out. Clean reads were then assembled using Trinity v2.6.6 [18]. The longest transcript from the assembly was designated as the unigene for subsequent analyses. The assembly's accuracy and completeness were assessed using N50 and BUSCO v3.0.2 metrics [36].

The TransDecoder v5.5.0 software was used to predict the reading frames of unigenes and translate the coding sequences (https://github.com/TransDecoder/TransDecoder, accessed on 13 December 2024, parameter: -m 50). Transcription factors were identified using the PlantTFdb (http://planttfdb.gao-lab.org/, accessed on 13 December 2024) database [37]. Alignment was performed using Diamond v0.9.24 (parameters: -max-target-seqs 1, -evalue 1×10^{-5}) [38], aligning unigenes to the NCBI non-redundant (Nr) database (https://ftp.ncbi.nlm.nih.gov/blast/db/, accessed on 13 December 2024), COG database (http://www.ncbi.nlm.nih.gov/COG, accessed on 13 December 2024), Swiss-Prot database (http://www.expasy.ch/sprot), KEGG database (http://www.genome.jp/kegg, accessed on 13 December 2024), and Gene Ontology (GO) database (http://www.geneontology.org/, accessed on 13 December 2024). Based on the alignment results against the Nr database, the species distribution of the aligned sequences was statistically analyzed and plotted. The Venn diagram was generated using Evenn (http://www.ehbio.com/test/venn/, accessed on 13 December 2024) [39].

The reads were aligned back to the assembled transcripts using Bowtie2 v2.3.4 (parameters: mismatch 0) software [40]. The alignment results were analyzed using RSEM v1.3.1 [41], to obtain the read count of each sample mapped to the unigenes, which were then normalized to TPM values. For quality assessment, Pearson correlation coefficients and PCA were performed using R v3.6.1 programming language to evaluate the correlation and reproducibility among samples.

4.5. Screening and Enrichment Analysis of Differential Genes

Differential gene expression analysis between samples was performed using the R package DESeq2 v1.10 [42], with a threshold set at $|\log2FoldChange| > 1$ and an adjusted p-value (Padj) < 0.05. The DEGs (differentially expressed genes) were subjected to

K-means hierarchical clustering. The R package ClusterProfiler facilitated GO and KEGG enrichment analysis of these genes [43].

4.6. WGNCA Analysis

Co-expression network analysis was performed using the R package WGCNA v1.703 in R v3.6.3 to assess modules of genes with high correlation [44]. Transcripts with an average expression below 1 were excluded. Modules linked to phenotypic traits were pinpointed in the network by converting the adjacency matrix into a topological overlap matrix using WGCNA. Modules were clustered based on similar expression patterns. The obtained modules were subjected to KEGG and GO enrichment analysis using KOBAS v3.0 [45]. Protein–protein interaction networks were retrieved from the STRING v11.5 database (http://string-db.org/). Hub genes were selected using the CytoHubba plugin [46], and visualization analysis was performed using Cytoscape v3.8.2 [47].

4.7. Transcriptome qRT-PCR Validation

To validate the accuracy of RNA-Seq data, pyrethrins synthesis-related genes were selected for qRT-PCR. The RNA samples were reverse transcribed using a reverse transcription kit (Toyobo, Osaka, Japan). Specific primers were designed (Table S4), and qRT-PCR was performed using the SYBR preMix Ex Taq Kit (Takara, Kusatsu, Japan) and the Roche LightCycler® 96 System (Roche, Basel, Switzerland) with the local transcriptome library validated using Tbtools v1.075 [48]. The qRT-PCR reaction system, with a total volume of 20 μL, included 10 μL of SYBR mixture, 0.4 μL of the upstream primer, 0.4 μL of the downstream primer, and 0.5 ng of the template. The amplification program consisted of pre-denaturation at 94 °C for 30 s, followed by 40 cycles of denaturation at 94 °C for 5 s, annealing at 60 °C for 15 s, and extension at 72 °C for 10 s. The relative expression levels were calculated using the $2-\Delta\Delta CT$ method with the GADPH gene as the reference gene [49]. Each sample was analyzed by qRT-PCR with three independent biological replicates and two technical replicates.

5. Conclusions

This study provides novel insights into the dynamic biosynthesis and regulation of pyrethrins and EβF in *T. cinerariifolium*, highlighting the complexity of their spatiotemporal expression and functional roles in plant defense. The observed asynchrony between pyrethrin gene expression and metabolite accumulation underscores the intricate regulation of monoterpenoid pathways, while the synchronization of EβF biosynthesis points to its distinct ecological functions during early flower development. By identifying key pathways such as JA signaling, trichome differentiation, and terpene transport as potentially associated with pyrethrin biosynthesis, and pinpointing hub genes like *MYC2*, this study lays the groundwork for future research into metabolic engineering and targeted breeding strategies. Furthermore, the results emphasize the significance of integrating transcriptomic data with functional studies to unravel the regulation of secondary metabolite pathways. These findings not only advance our understanding of *T. cinerariifolium* biology but also offer practical applications in optimizing bio-pesticide production for sustainable agriculture.

Supplementary Materials: The following supporting information can be downloaded at: https://www.mdpi.com/article/10.3390/horticulturae11020201/s1, Figure S1: Developmental stages of *T. cinerariifolium* flowers and GC-MS analysis of key metabolites in S2. Figure S2: WGCNA modules gene functional enrichment analysis. Table S1: The mass spectrograms of *T. cinerariifolium* S2 flowers. Table S2: GCMS Analysis of Pyrethrins and EβF in Different Samples. Table S3: Summary of Illumina HiSeq sequencing data, Table S4: Primers used in experiment.

Author Contributions: Conceptualization, T.Z. and J.L. (Jiawen Li); investigation, T.Z. and J.L. (Jiawen Li); methodology, T.Z. and J.L. (Jiawen Li); software, T.Z. and J.L. (Jiawen Li); validation, T.Z. and J.L. (Jiawen Li); data curation, J.L. (Jinjin Li); writing—original draft preparation, T.Z. and J.L. (Jinjin Li); writing—review and editing, T.Z., J.L. (Jinjin Li) and C.W.; visualization, T.Z.; supervision, J.L. (Jinjin Li) and C.W. All authors have read and agreed to the published version of the manuscript.

Funding: This work was supported by the National Natural Science Foundation of China (grant no. 32160718): the Guizhou Normal University QSXM[2022]19, the college student innovation and entrepreneurship training program project (202310663009), and Guizhou Science and technology innovation team project (Molecular biology of stress resistance in crop).

Data Availability Statement: The raw sequences described in this article were submitted and released to The National Genomics Data Center (NGDC, https://ngdc.cncb.ac.cn) under BioProject PRJCA033633 (accessed on 13 December 2024).

Conflicts of Interest: The authors declare no conflicts of interest.

References

1. Mossa, A.H.; Mohafrash, S.M.M.; Chandrasekaran, N. Safety of natural insecticides: Toxic effects on experimental animals. *BioMed Res. Int.* **2018**, *2018*, 4308054. [CrossRef] [PubMed]
2. Lybrand, D.B.; Xu, H.; Last, R.L.; Pichersky, E. How plants synthesize pyrethrins: Safe and biodegradable insecticides. *Trends Plant Sci.* **2020**, *25*, 1240–1251. [CrossRef] [PubMed]
3. Matsuda, K. Chemical and biological studies of natural and synthetic products for the highly selective control of pest insect species. *Biosci. Biotechnol. Biochem.* **2022**, *86*, 1–11. [CrossRef] [PubMed]
4. Grdisa, M.; Carovic-Stanko, K.; Kolak, I.; Satovic, Z. Morphological and biochemical diversity of dalmatian pyrethrum (*Tanacetum cinerariifolium* (Trevir.) Sch Bip.). *Agric. Conspec. Sci.* **2009**, *74*, 73–80.
5. Jeran, N.; Grdiša, M.; Varga, F.; Šatović, Z.; Liber, Z.; Dabić, D.; Biošić, M. Pyrethrin from Dalmatian pyrethrum (*Tanacetum cinerariifolium*/Trevir./Sch. Bip.): Biosynthesis, biological activity, methods of extraction and determination. *Phytochem. Rev.* **2021**, *20*, 875–905. [CrossRef]
6. Zeng, T.; Li, J.; Li, J.; Hu, H.; Zhu, L.; Liu, K.; Bai, J.; Jiang, Q.; Wang, C. Pyrethrins in *Tanacetum cinerariifolium*: Biosynthesis, regulation, and agricultural application. *Ornam. Plant Res.* **2024**, *4*, e015. [CrossRef]
7. Li, J.J.; Hu, H.; Chen, Y.; Xie, J.; Li, J.; Zeng, T.; Wang, M.Q.; Luo, J.; Zheng, R.R.; Jongsma, M.A.; et al. Tissue specificity of (E)-beta-farnesene and germacrene D accumulation in pyrethrum flowers. *Phytochemistry* **2021**, *187*, 112768. [CrossRef]
8. Li, J.J.; Hu, H.; Mao, J.; Yu, L.; Stoopen, G.; Wang, M.Q.; Mumm, R.; de Ruijter, N.C.A.; Dicke, M.; Jongsma, M.A.; et al. Defense of pyrethrum flowers: Repelling herbivores and recruiting carnivores by producing aphid alarm pheromone. *New Phytol.* **2019**, *223*, 1607–1620. [CrossRef] [PubMed]
9. Vandermoten, S.; Mescher, M.C.; Francis, F.; Haubruge, E.; Verheggen, F.J. Aphid alarm pheromone: An overview of current knowledge on biosynthesis and functions. *Insect Biochem. Mol. Biol.* **2012**, *42*, 155–163. [CrossRef] [PubMed]
10. Zhang, R.; Wang, B.; Grossi, G.; Falabella, P.; Liu, Y.; Yan, S.; Lu, J.; Xi, J.; Wang, G. Molecular basis of alarm pheromone detection in aphids. *Curr. Biol.* **2017**, *27*, 55–61. [CrossRef]
11. Li, J.; Hu, H.; Ren, S.; Yu, L.; Luo, Y.; Li, J.; Zeng, T.; Wang, M.; Wang, C. Aphid alarm pheromone mimicry in transgenic *Chrysanthemum morifolium*: Insights into the potential of (E)-β-farnesene for aphid resistance. *Front. Plant Sci.* **2024**, *15*, 1373669. [CrossRef] [PubMed]
12. Bhat, B.K.; Menary, R.C. Scanning electron microscopic study of oil glands in pyrethrum flowers. *Pyrethrum Post* **1975**, *15*, 11–15.
13. Head, S.W. A study of the insecticidal constituents in *Chrysanthemum cinerariaefolium.* (1) Their development in the flower head (2) Their distribution in the plant. *Pyrethrum Post* **1966**, *8*, 32–37.
14. Kikuta, Y.; Ueda, H.; Nakayama, K.; Katsuda, Y.; Ozawa, R.; Takabayashi, J.; Hatanaka, A.; Matsuda, K. Specific regulation of pyrethrin biosynthesis in *Chrysanthemum cinerariaefolium* by a blend of volatiles emitted from artificially damaged conspecific plants. *Plant Cell Physiol.* **2011**, *52*, 588–596. [CrossRef] [PubMed]
15. Ueda, H.; Kikuta, Y.; Matsuda, K. Plant communication: Mediated by individual or blended VOCs? *Plant Signal. Behav.* **2012**, *7*, 222–226. [CrossRef] [PubMed]
16. Liu, Y.; Wang, Z.; Cui, Z.; Qi, Q.; Hou, J. Progress and perspectives for microbial production of farnesene. *Bioresour. Technol.* **2022**, *347*, 126682. [CrossRef]
17. Matsuda, K.; Kikuta, Y.; Haba, A.; Nakayama, K.; Katsuda, Y.; Hatanaka, A.; Komai, K. Biosynthesis of pyrethrin I in seedlings of *Chrysanthemum cinerariaefolium*. *Phytochemistry* **2005**, *66*, 1529–1535. [CrossRef]

18. Grabherr, M.G.; Haas, B.J.; Yassour, M.; Levin, J.Z.; Thompson, D.A.; Amit, I.; Adiconis, X.; Fan, L.; Raychowdhury, R.; Zeng, Q.; et al. Full-length transcriptome assembly from RNA-Seq data without a reference genome. *Nat. Biotechnol.* **2011**, *29*, 644–652. [CrossRef]
19. Liu, F.; Wang, Q.; Xu, P.; Andreazza, F.; Valbon, W.R.; Bandason, E.; Chen, M.; Yan, R.; Feng, B.; Smith, L.B.; et al. A dual-target molecular mechanism of pyrethrum repellency against mosquitoes. *Nat. Commun.* **2021**, *12*, 2553. [CrossRef] [PubMed]
20. Zeng, T.; Li, J.W.; Xu, Z.Z.; Zhou, L.; Li, J.J.; Yu, Q.; Luo, J.; Chan, Z.L.; Jongsma, M.A.; Hu, H.; et al. *TcMYC2* regulates pyrethrin biosynthesis in *Tanacetum cinerariifolium*. *Hortic. Res.* **2022**, *9*, uhac178. [CrossRef]
21. Zito, S.W.; Zieg, R.G.; Staba, E.J. Distribution of pyrethrins in oil glands and leaf tissue of *Chrysanthemum cinerariaefolium*. *Planta Medica* **1983**, *47*, 205–207. [CrossRef] [PubMed]
22. Zeng, T.; He, Z.J.; He, J.F.; Lv, W.; Huang, S.X.; Li, J.W.; Zhu, L.Y.; Wan, S.; Zhou, W.F.; Yang, Z.S.; et al. The Telomere-to-telomere gap-free reference genome of wild blueberry (*Vaccinium duclouxii*) provides its high soluble sugar and anthocyanin accumulation. *Hortic. Res.* **2023**, *10*, uhad209. [CrossRef] [PubMed]
23. Wu, D.; Wu, Y.; Gao, R.; Zhang, Y.; Zheng, R.; Fang, M.; Li, Y.; Zhang, Y.; Guan, L.; Gao, Y. Integrated metabolomics and transcriptomics reveal the key role of flavonoids in the cold tolerance of chrysanthemum. *Int. J. Mol. Sci.* **2024**, *25*, 7589. [CrossRef] [PubMed]
24. Ting, Z.; Pengcheng, Z.; Jiang, X.; Lingzhu, W.; Wanting, S.; Jiang, W.; Jianhui, C. Integrated transcriptomic and metabolomic analysis reveals the effects of forchlorfenuron and thidiazuron on flavonoid biosynthesis in table grape skins. *Curr. Plant Biol.* **2024**, *40*, 100417.
25. Li, J.J.; Hu, H.; Fu, H.S.; Li, J.; Zeng, T.; Li, J.W.; Wang, M.Q.; Jongsma, M.A.; Wang, C.Y. Exploring the co-operativity of secretory structures for defense and pollination in flowering plants. *Planta* **2024**, *259*, 41. [CrossRef] [PubMed]
26. Li, J.; Luo, Y.; Li, M.; Li, J.; Zeng, T.; Luo, J.; Chang, X.; Wang, M.; Jongsma, M.A.; Hu, H.; et al. Nocturnal burst emissions of germacrene D from the open disk florets of pyrethrum flowers induce moths to oviposit on a nonhost and improve pollination success. *New Phytol.* **2024**, *244*, 2036–2048. [CrossRef] [PubMed]
27. Suraweera, D.; Groom, T.; Nicolas, M. Pattern of pyrethrin accumulation, achene and trichome development in relation to pattern of flower development in pyrethrum. *II Int. Symp. Pyrethrum* **2015**, *1169*, 93–100. [CrossRef]
28. Li, J.W.; Zeng, T.; Xu, Z.Z.; Li, J.J.; Hu, H.; Yu, Q.; Zhou, L.; Zheng, R.R.; Luo, J.; Wang, C.Y. Ribozyme-mediated CRISPR/Cas9 gene editing in pyrethrum (*Tanacetum cinerariifolium*) hairy roots using a RNA polymerase II-dependent promoter. *Plant Methods* **2022**, *18*, 32. [CrossRef] [PubMed]
29. Tian, H.; Wang, S. TRANSPARENT TESTA GLABRA1, a key regulator in plants with multiple roles and multiple function mechanisms. *Int. J. Mol. Sci.* **2020**, *21*, 4881. [CrossRef] [PubMed]
30. Pesch, M.; Dartan, B.; Birkenbihl, R.; Somssich, I.E.; Hülskamp, M. *Arabidopsis TTG2* regulates *TRY* expression through enhancement of activator complex-triggered activation. *Plant Cell* **2014**, *26*, 4067–4083. [CrossRef]
31. Zhao, M.; Morohashi, K.; Hatlestad, G.; Grotewold, E.; Lloyd, A. The TTG1-bHLH-MYB complex controls trichome cell fate and patterning through direct targeting of regulatory loci. *Development* **2008**, *135*, 1991–1999. [CrossRef] [PubMed]
32. Xie, L.H.; Yan, T.X.; Li, L.; Chen, M.H.; Ma, Y.N.; Hao, X.L.; Fu, X.Q.; Shen, Q.; Huang, Y.W.; Qin, W.; et al. The WRKY transcription factor *AaGSW2* promotes glandular trichome initiation in *Artemisia annua*. *J. Exp. Bot.* **2020**, *72*, 1691–1701. [CrossRef]
33. Hu, D.G.; Li, Y.Y.; Zhang, Q.Y.; Li, M.; Sun, C.H.; Yu, J.Q.; Hao, Y.J. The *R2R3-MYB* transcription factor *MdMYB73* is involved in malate accumulation and vacuolar acidification in apple. *Plant J.* **2017**, *91*, 443–454. [CrossRef] [PubMed]
34. Ramirez, A.M. Pyrethrum Secondary Metabolism: Biosynthesis, Localization and Ecology of Defence Compounds. Ph.D. Thesis, Wageningen University, Wageningen, The Netherlands, 2013.
35. Chen, S.F.; Zhou, Y.Q.; Chen, Y.R.; Gu, J. Fastp: An ultra-fast all-in-one FASTQ preprocessor. *Bioinformatics* **2018**, *34*, 884–890. [CrossRef] [PubMed]
36. Simão, F.A.; Waterhouse, R.M.; Ioannidis, P.; Kriventseva, E.V. BUSCO: Assessing genome assembly and annotation completeness with single-copy orthologs. *Bioinformatics* **2015**, *19*, 3210–3212. [CrossRef]
37. Tian, F.; Yang, D.C.; Meng, Y.Q.; Jin, J.P.; Gao, G. PlantRegMap: Charting functional regulatory maps in plants. *Nucleic Acids Res.* **2020**, *48*, D1104–D1113. [CrossRef] [PubMed]
38. Buchfink, B.; Xie, C.; Huson, D.H. Fast and sensitive protein alignment using DIAMOND. *Nat. Methods* **2015**, *12*, 59–60. [CrossRef] [PubMed]
39. Chen, T.; Zhang, H.; Liu, Y.; Liu, Y.X.; Huang, L. EVenn: Easy to create repeatable and editable Venn diagrams and Venn networks online. *J. Genet. Genom.* **2021**, *48*, 863–866. [CrossRef]
40. Langmead, B.; Salzberg, S.L. Fast gapped-read alignment with Bowtie 2. *Nat. Methods* **2012**, *9*, 357–359. [CrossRef]
41. Li, B.; Dewey, C.N. RSEM: Accurate transcript quantification from RNA-Seq data with or without a reference genome. *BMC Bioinform.* **2011**, *12*, 16. [CrossRef]
42. Anders, S.; Huber, W. Differential expression analysis for sequence count data. *Genome Biol.* **2010**, *11*, R106. [CrossRef] [PubMed]
43. Yu, G.; Wang, L.G.; Han, Y.; He, Q.Y. ClusterProfiler: An R package for comparing biological themes among gene clusters. *Omics J. Integr. Biol.* **2012**, *16*, 284–287. [CrossRef]
44. Zhang, B.; Horvath, S. A general framework for weighted gene co-expression network analysis. *Stat. Appl. Genet. Mol. Biol.* **2005**, *4*, 1. [CrossRef] [PubMed]

45. Bu, D.; Luo, H.; Huo, P.; Wang, Z.; Zhang, S.; He, Z.; Wu, Y.; Zhao, L.; Liu, J.; Guo, J.; et al. KOBAS-i: Intelligent prioritization and exploratory visualization of biological functions for gene enrichment analysis. *Nucleic Acids Res.* **2021**, *49*, W317–W325 [CrossRef] [PubMed]
46. Chin, C.H.; Chen, S.H.; Wu, H.H.; Ho, C.W.; Ko, M.T.; Lin, C.Y. CytoHubba: Identifying hub objects and sub-networks from complex interactome. *BMC Syst. Biol.* **2014**, *8*, S11. [CrossRef] [PubMed]
47. Shannon, P.; Markiel, A.; Ozier, O.; Baliga, N.S.; Wang, J.T.; Ramage, D.; Amin, N.; Schwikowski, B.; Ideker, T. Cytoscape: A software environment for integrated models of biomolecular interaction networks. *Genome Res.* **2003**, *13*, 2498–2504. [CrossRef]
48. Chen, C.J.; Chen, H.; Zhang, Y.; Thomas, H.R.; Frank, M.H.; He, Y.H.; Xia, R. TBtools: An integrative toolkit developed for interactive analyses of big biological data. *Mol. Plant* **2020**, *13*, 1194–1202. [CrossRef]
49. Livak, K.J.; Schmittgen, T.D. Analysis of relative gene expression data using real-time quantitative PCR and the $2^{-\Delta\Delta CT}$ method *Methods* **2001**, *408*, 402–408. [CrossRef]

horticulturae

MDPI

Article

The Comprehensive Root Metabolite–Rhizomicrobiota Response Patterns of *Rhododendron delavayi* (*R. delavayi*) to Waterlogging Stress and Post–Waterlogging Recovery

Jing Tang [1,2,3,†], Qingqing Huang [1,2,†], Qian Wang [1], Fei Shan [2], Shaolong Wu [1], Ximin Zhang [1,2], Ming Tang [1,3] and Yin Yi [1,3,*]

1 Key Laboratory of Plant Physiology and Development Regulation, School of Life Sciences, Guizhou Normal University, Guiyang 550025, China; tangjing2016@gznu.edu.cn (J.T.); hqq477455@163.com (Q.H.); wangqian202511@163.com (Q.W.); m16685518259@163.com (S.W.); zhxm409@outlook.com (X.Z.); mingtang@gznu.edu.cn (M.T.)

2 Key Laboratory of Environment Friendly Management on Alpine Rhododendron Diseases and Pests of Institutions of Higher Learning in Guizhou Province, Guizhou Normal University, Guiyang 550025, China; sdft368@163.com

3 Key Laboratory of State Forestry Administration on Biodiversity Conservation in Karst Area of Southwest, Guizhou Normal University, Guiyang 550025, China

* Correspondence: gzklppdr@gznu.edu.cn; Tel.: +86-0851-8322-7099

† These authors contributed equally to this work.

Abstract: Waterlogging is a critical abiotic stressor that significantly impacts plant growth. Plants under waterlogging stress release metabolic signals that recruit rhizosphere microorganisms and enhance stress resistance. However, the mechanisms through which the non-adaptive species *R. delavayi* responds to waterlogging stress via the synergistic interaction between root metabolites and rhizosphere microbiota remain poorly elucidated. Here, we employed pot experiments to characterize the responses of the root metabolite–microbiota complex in *R. delavayi* during waterlogging stress and subsequent recovery. Our results revealed that waterlogging altered the root morphology, the root metabolite profile, rhizosphere microbial diversity and network complexity, and these effects persisted during recovery. A significant correlation between root metabolites and the rhizosphere microbial community structure during waterlogging stress and recovery. Importantly, some differentially accumulated metabolites had significant effects on the assembly of rhizosphere microbes. Most of the core microbes in the rhizosphere microbial community under waterlogging and post–waterlogging recovery treatment were likely beneficial bacteria. Based on these findings, we propose a model for how root metabolites and rhizosphere microbes interact to help *R. delavayi* cope with waterlogging and recover. Based on these findings, we propose a possible response pattern of root metabolites and rhizosphere microbiota complex in *R. delavayi* under waterlogging stress and recovery. This work provides new insights into the synergistic mechanisms enhancing plant waterlogging tolerance and highlights the potential of harnessing rhizosphere microbiota to improve resilience in rhododendrons.

Keywords: rhododendron; waterlogging stress; root–rhizomicrobiota complex; beneficial bacteria; multi–omics

1. Introduction

Due to the marked increase in heavy precipitation and recurrent precipitation events, the response mechanisms of plants to waterlogging stress have become a key research focus.

 https://doi.org/10.3390/horticulturae11070770

Rhododendrons, both evergreen and deciduous woody plants, have high horticultural value and are cultivated worldwide [1]. In the northwest of Guizhou Province, the Baili Azalea Nature Reserve (BANR) contains approximately 40 species of Rhododendron and serves as an important rhododendron germplasm bank. *Rhododendron delavayi* (*R. delavayi*) is a crucial member of the BANR [2]. With global climate change, the frequency and intensity of waterlogging events in August in Guizhou have increased significantly [2]. Unfortunately, waterlogging hinders rhododendron growth and development and can be fatal. However, research on their response patterns to waterlogging stress and post-waterlogging recovery remains limited.

The rhizosphere—the soil region within 2 mm of the root system—is a hotspot for microbial activity [3]. Rhizosphere microbiotas are highly sensitive to environmental fluctuations, particularly changes in soil moisture [4]. Growing evidence underscores their critical role in plant health and adaptation to waterlogging stress [5,6]. For instance, waterlogging alters spring wheat root microbiota, depleting beneficial bacteria [7], while enriching *Geobacter* in soybeans [8] and *Pseudomonas* in sugar beets [6]. Hypoxia during waterlogging imposed severe impacts on both plants and their microbiota, favoring anaerobic taxa (e.g., *Firmicutes, Fusobacteria*) [9]. Although the effects of waterlogging on rhizosphere communities have been documented [10], post–waterlogging recovery—particularly whether microbiota could restore functional communities to support plant resilience—remains poorly understood [11]. Waterlogging stress modulates rhizosphere microbiota assembly through root-mediated metabolites, altering plant–microbe interactions. Waterlogging-induced shifts in root metabolites, including mucilage composition, phytohormone fluxes such as strigolactones [12], and anaerobic byproducts (ethanol, aldehydes) [13,14], drive microbial community restructuring [15,16]. Following waterlogging, recovering plants face multiple challenges, including the detoxification of phytotoxins accumulated during inundation, nutrient deficiencies, and increased susceptibility to biotic stress [17,18]. To overcome these challenges, recovering plants rely on root metabolites to recruit specific rhizosphere microbial communities that aid in stress mitigation and recovery [9,15]. Consequently, understanding the responses of the root metabolite–rhizosphere microbiota complex to waterlogging stress and subsequent recovery is essential for elucidating the mechanisms by which waterlogging impacts rhododendrons.

In this study, a pot water control experiment was conducted to investigate the effects of waterlogging and post–waterlogging recovery on 3–year–old *R. delavayi* seedlings. The responses of root metabolites and rhizosphere microbial communities to waterlogging stress and recovery were analyzed using 16S rRNA gene sequencing (targeting the V4 region) and ultrahigh-performance liquid chromatography–mass spectrometry (UPLC–MS/MS). The objectives of this study were to (i) elucidate the responses of rhizosphere microbiotas and root metabolites in rhododendrons to waterlogging stress; (ii) assess whether rhizosphere microbiotas and root metabolites return to their pre-waterlogging state during the recovery phase; and (iii) explore how *Rhododendron* root metabolites unite specific rhizosphere microorganisms to enhance resistance to waterlogging stress. Insights into the roles of microorganisms and root metabolites under waterlogging stress may provide a valuable reference for understanding the mechanisms underlying the effects of waterlogging on rhododendrons from the perspective of plant–microbe interactions.

2. Result

2.1. Generating Metabolites and Rhizosphere Microbiotas for *R. delavayi* Root During Waterlogging and Recovery

To survey the metabolic and microbial characteristics of *R. delavayi* roots during waterlogging and recovery, we collected the root tissues and rhizosphere soil of 21 three-year-old

R. delavayi seedlings (Figure 1a). The seedlings of *R. delavayi* were exposed to waterlogging stress for 0, 10, 20, or 30 days (W0, W10, W20, W30). In the post–waterlogging recovery treatments, waterlogging was performed for 10, 20, or 30 days, followed by an equal recovery period of 10, 20, or 30 days, designated as WR10, WR20, and WR30, respectively. In addition, we performed 16s rRNA sequencing on microbes in the rhizosphere soils (Figure 1b). We used UPLC–MS/MS to determine the contents of metabolites in each root sample (Figure 1c). Through the experimental design, we identified 693 metabolites in the root system of *R. delavayi* and classified a total of 6808 operational taxonomic units (OTUs) from its rhizosphere microbial community, covering 81 phyla and 687 genera.

2.2. Effects of Waterlogging Stress and Restoration on R. delavayi Seedling Roots

Waterlogging stress significantly reduced fresh root weight (Supplementary Figure S1). This decline persisted during recovery, exacerbated by prolonged prior stress durations. Adventitious roots emerged following 20 days of waterlogging but diminished during recovery beyond 20 days. By WR30, all adventitious roots were absent, with the remaining roots exhibiting darkening and necrosis alongside a pronounced reduction in root biomass. These results demonstrated persistent and irreversible root damage in *R. delavayi* under extended waterlogging stress and recovery.

2.3. Changes in Root Metabolites During Waterlogging Stress and Recovery Period

Root metabolomic profiling of *R. delavayi* via UPLC–MS/MS identified 693 metabolites (Supplementary Figure S2), predominantly lipids (17.5%), phenolic acids (17.0%), flavonoids (14.9%), and others (50.6%). Principal component analysis (PCA) revealed a marked divergence between waterlogging and recovery root metabolite profiles (Supplementary Figure S3A). Bray–Curtis dissimilarity was greatest between W0 and W30, exceeding all other group comparisons except W0 vs. WR30 (Supplementary Figure S3B). Both extended waterlogging (30 days) and prolonged recovery (30 days) induced significantly stronger root metabolite changes than shorter durations (10/20 days), demonstrating time-effect metabolic changes.

Using orthogonal partial least squares discriminant analysis (OPLS–DA), a total of 470 differentially accumulated metabolites (DAMs) were identified across waterlogging–recovery phases (Supplementary Figure S4A). Venn analysis showed that less than 30% of the DAMs reversed their changes within the recovery period after experiencing waterlogging (Supplementary Figure S4B). Temporal profiling via Short Time series Expression Miner (STEM) revealed five significant expression patterns encompassing 97 DAMs (Supplementary Figure S4C), which were predominantly enriched in flavonoid biosynthesis, purine metabolism, linoleic acid metabolism, and alpha-linolenic acid metabolism. A total of 22 flavonoid metabolites were identified in the metabolome of *R. delavayi* roots under waterlogging stress and recovery, including 13 DAMs (Supplementary Figure S4D). Most of these differential flavonoid metabolites exhibited significant upregulation during the waterlogging stress period, such as Quercetin, Hesperetin, and Phloretin.

Through a comprehensive literature review of metabolites, we identified 48 out of the 97 DAMs that played a critical role in plant resistance to adverse conditions (Supplementary Table S2). Overall, the functions of these 48 DAMs were categorized into four main groups: (1) scavenging reactive oxygen species (ROS) and maintaining redox balance (e.g., L–Proline, α–Linolenic Acid, Quercetin), (2) antioxidation (e.g., Protocatechualdehyde, Phloretin, Dihydrokaempferol), (3) antifungal, antibacterial, and antiviral effects (e.g., Hesperetin, 9,10,13–Trihydroxy–11–Octadecenoic Acid), and (4) regulation of root growth (e.g., Isorhamnetin, Quercetin–3–O–sophoroside). These results suggested that the root metabolism of *R. delavayi* actively participated in both waterlogging stress and recovery.

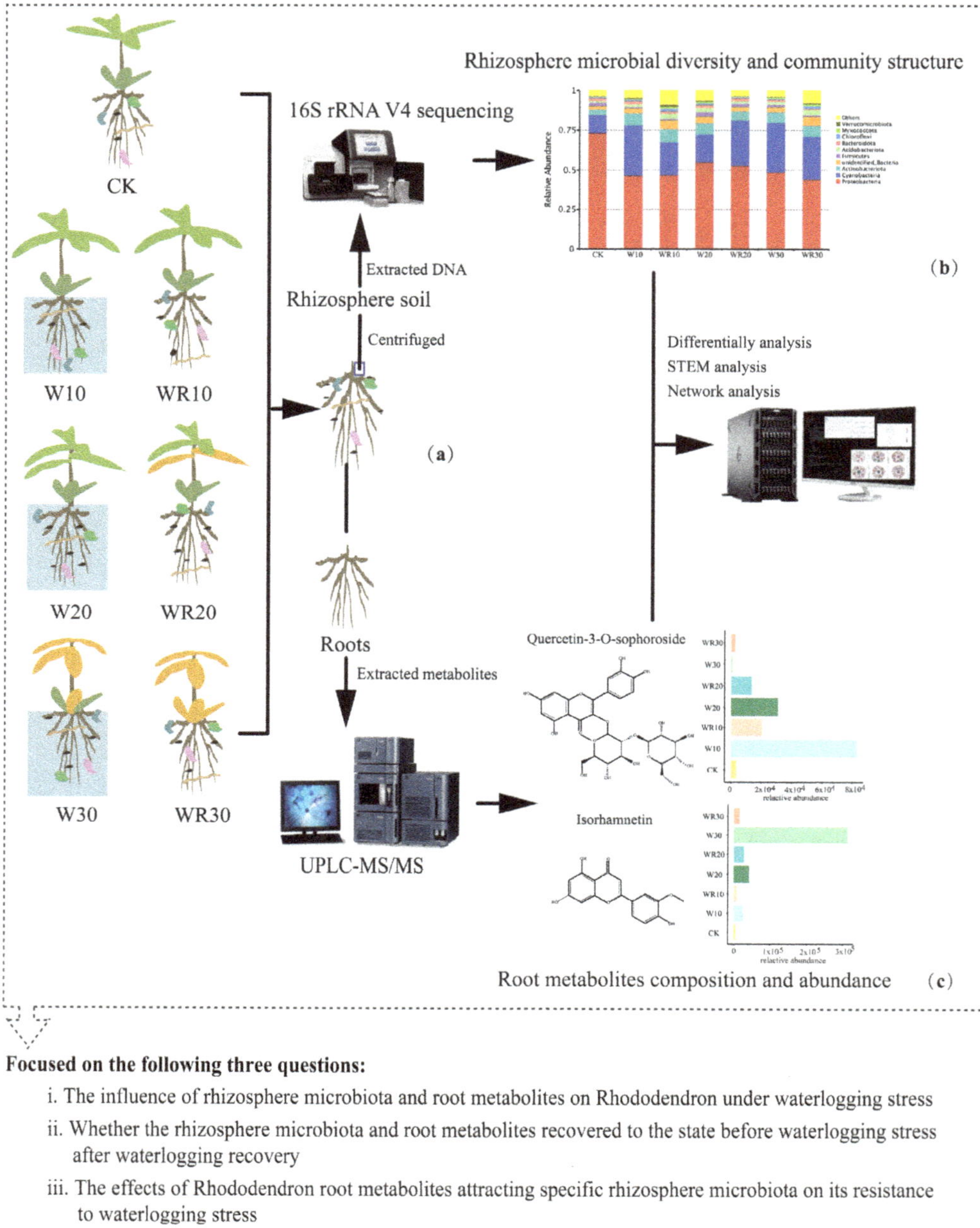

Figure 1. Overview of the experimental design. (**a**) 21 samples were retrieved for normal (CK), waterlogging (W), and recovery (WR) on 3-year-old the *R. delavayi* seedlings. The root of *R. delavayi* seedling was collected and divided into two parts. (**b**) The root part was used to analyze the root metabolites by UPLC-MS/MS. (**c**) The rhizosphere soil was used to analyze the microbiota by sequencing.

2.4. Effects of Waterlogging and Post–Waterlogging Recovery on Rhizosphere Microbial Diversity and Community Structure

Utilizing sequencing of the V4 region of the 16S rRNA gene and subsequent analysis, a total of 6808 OTUs were classified in the rhizosphere microbiotas of *R. delavayi* under waterlogging stress and restoration. Only 17% (998/6068) constituted the consistently presenting microbiotas across all treatments (Supplementary Figure S5A). Waterlogging induced pronounced community turnover, retaining merely 40.5% of W0 OTUs (1265/3124) during stress and 40.5% (1267) post-recovery. Likelihood ratio testing identified 363 and 376 differentially abundant OTUs during waterlogging and recovery, respectively (Supplementary Table S3). These results revealed substantial rhizosphere microbiome restructuring in *R. delavayi* under waterlogging–recovery cycles. Notably, during waterlogging and the subsequent recovery period, a large number of OTUs in the rhizosphere exhibit decreased or increased abundance. Then, α-diversity analyses revealed a time-dependent shift in the microbial community of the *R. delavayi* rhizosphere under waterlogging–recovery cycles. The recovery phases (WR10, WR30) exhibited elevated Shannon diversity ($p < 0.05$) compared to W0, while waterlogging duration differentially impacted richness: PD whole tree and Chao1 indices peaked at W20 before declining at W30 (Supplementary Table S4). Non–Metric Multidimensional Scaling (NMDS) ordination of Bray–Curtis distances demonstrated a marked divergence between waterlogged and recovery communities (PERMANOVA $R^2 = 0.32$, $p = 0.001$) (Supplementary Figure S5B). Unweighted UniFrac distances showed maximal structural shifts between the W0–W30 (waterlogging climax) and W20–WR20 (recovery transition) groups (Supplementary Figure S5C).

The 6808 OTUs were mostly divided into 81 Phyla (Supplementary Table S5). The highest relative abundances at the phylum level were for Proteobacteria (44.38~73.29%), followed by Cyanobacteria (11.29~31.50%) and Actinobacteriota (3.29~8.31%). Following the waterlogging and post–waterlogging recovery of the rhizosphere microbiota, there was a notable reduction in the relative abundances of Proteobacteria, while Cyanobacteria and Actinobacteriota exhibited marked increases. Interestingly, there was a substantial number of unidentified_bacteria (2.23~5.68%), and their abundance shows a significant increase following waterlogging and subsequent recovery. Moreover, to evaluate variations in the rhizosphere microbiota during waterlogging and post–waterlogging recovery, we evaluated differentially abundant genera in the waterlogging and recovery groups. A total of 687 genera were annotated in the *R. delavayi* rhizosphere microbiota, of which 187 genera were affected by waterlogging stress and restoration (Supplementary Figure S5D). Most genera (115) were significantly affected by waterlogging stress. Thirty-three genera only significantly changed during the post-waterlogging recovery period but experienced no significant changes during waterlogging. All of the results suggested that waterlogging and subsequent recovery exert substantial influence on rhizosphere microbiotas.

2.5. Effects of Waterlogging and Post–Waterlogging Recovery on Rhizosphere Microbial Co–Occurrence Network Complexity and Core Microbes

To evaluate the influence of waterlogging and the subsequent recovery period on the complexity of rhizosphere microbial co-occurrence networks, six distinct networks were constructed. We calculated various network topological parameters, including the total number of nodes, number of links, network diameter, relative modularity (RM), average clustering coefficient (average CC), and average degree. Analysis of network topological parameters indicated that the rhizosphere microbial co-occurrence networks manifested divergent trends in response to differing waterlogging durations and recovery periods (Figure 2a, Table 1). Prolonged waterlogging progressively increased the complexity and density of the overall structure of the co-occurrence network (nodes, links, average degree, and diameter increased from W10 to W30), with elevated modularity (RM increased) and reduced clustering (average CC reduced), indicating a transition toward compartmentalized communities. Intriguingly, the network structure within post–waterlogging recovery exhibited phase–dependent variation. During a shorter post-waterlogging recovery period (10 days), the network structure underwent simplification and loss (nodes, links, and average degree reduced, but diameter increased), concurrent with the decline in compartmentalization (RM and average CC reduced). Conversely, with prolonged waterlogging and subsequent restoration (20 or 30 days), the structure of the co-occurrence network increased again in complexity and density, with a higher degree of compartmentalization. Furthermore, we evaluated the robustness (resistance to node loss), cohesion, and negative co-occurrence ratio of rhizosphere microbial co-occurrence networks to examine how different durations of waterlogging and subsequent recovery periods affected network interaction and stability (Figure 2b). The robustness of waterlogging groups increased with stress duration, whereas recovery phases exhibited temporal heterogeneity. The 10 d and 30 d recovery groups showed reduced robustness compared to their corresponding waterlogging groups, while the 20 d recovery group displayed higher robustness than its waterlogging counterpart. In addition, we assessed the cohesion to measure the microbial interactions [19]. By analyzing the positive cohesion and negative cohesion, we found that the microbial interactions in the 10 d and 30 d waterlogging groups were higher than in their corresponding post–waterlogging recovery groups. Interestingly, the 20 d group exhibited the opposite trend. Notably, although robustness and cohesion exhibited fluctuations, these variations did not reach statistical significance. Under disturbances, the presence of negative interactions enhanced network stability [20]. Although the negative co-occurrence ratio was higher in the 10 d waterlogging group, it remained notably low across all networks following both waterlogging and recovery periods (ratio < 2.0%). The effects of waterlogging and subsequent recovery on microbial interactions and network stability were limited.

Table 1. Topological properties of rhizosphere bacterial networks.

Network	Nodes	Links	Average Degree	Diameter	Average CC	RM
W10	327	1253	7.66	6.91	0.82	1.74
W20	334	1390	8.85	12.86	0.80	1.74
W30	449	4417	19.67	16.57	0.76	3.49
WR10	283	865	6.11	9.88	0.73	1.18
WR20	486	4758	19.58	14.61	0.71	3.11
WR30	475	5976	25.16	17.51	0.75	4.04

Node: the total number of nodes; Links: the total number of links; Diameter: network diameter; Average CC: average clustering coefficient; RM: relative modularity.

Native core microbiomes associated with the plant rhizoplane hold the key to harnessing the full potential of soil microbiomes for supporting plant growth [21]. Here, two methods were used to jointly define the core microbes. Firstly, by calculating the within-module connectivity (Zi) and among-module connectivity (Pi), network hub microbiotas were obtained. The W10 network contained one module hub microbiotas, while W20, W30, WR10, and WR20 harbored one, seven, one, and three hub microbiota, respectively (Figure 2c). Secondly, to identify specialist microbiotas associated with each treatment, we selected OTUs with specificity and occupancy greater than or equal to 0.7 in the specificity–occupancy (SPEC–OCCU) plots. Here, a total of 49 specialist OTUs were found (Figure 2d, Supplementary Table S6). A total of 59 core microbes were found using the two methods, of which only *Rubrobacter* fulfilled the criteria for hub microbiotas and specialist microbiotas. (Figure 2e).

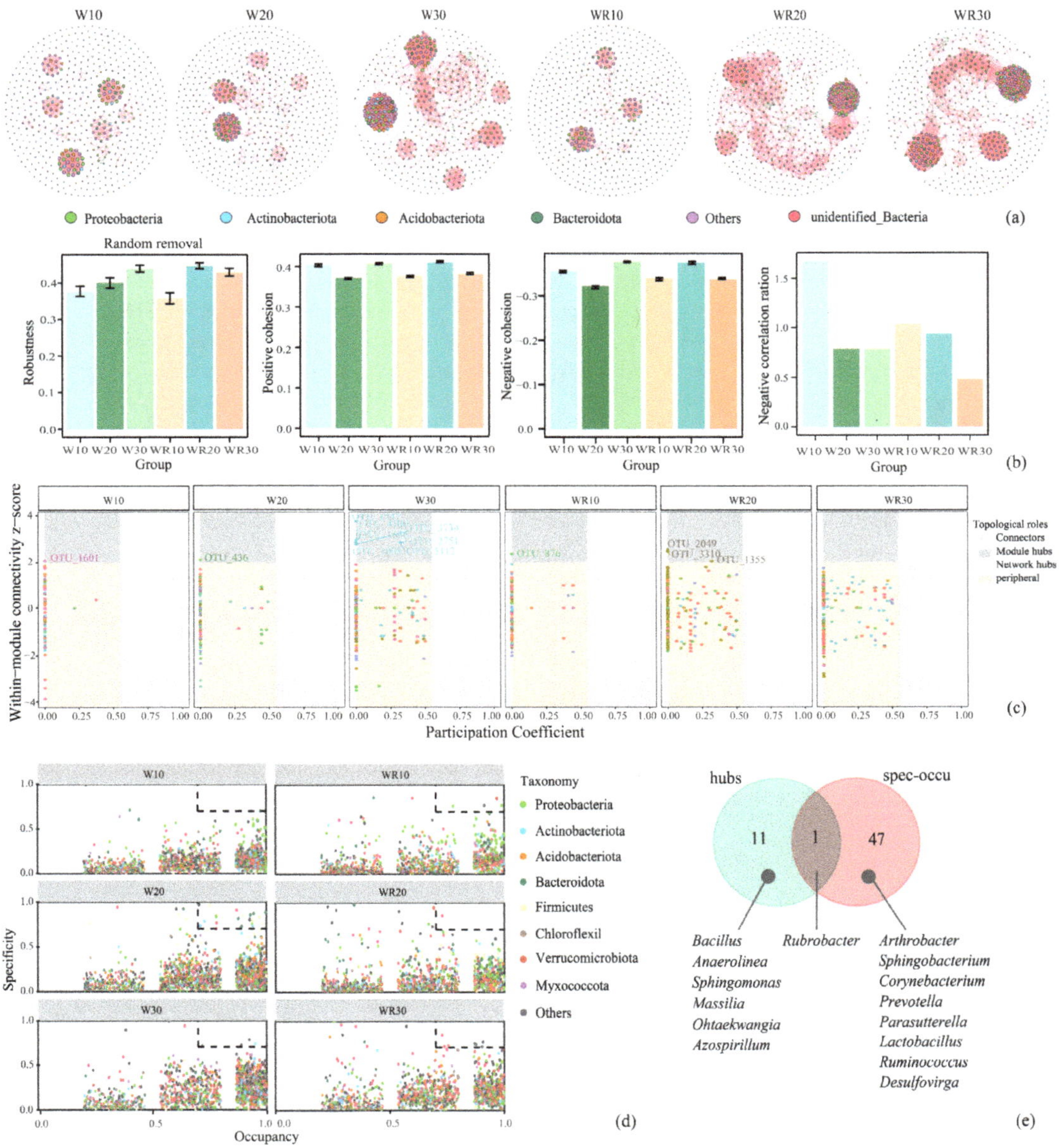

Figure 2. Rhizosphere microbial correlation networks and core microbiotas with different waterlogging and recovery periods. (**a**) Visualization of constructed networks in 6 different waterlogging and recovery period treatments. (**b**) The robustness, positive and negative cohesion, and negative correlation ratio of the networks. (**c**) Hub microbiotas with 6 rhizosphere microbial correlation networks. (We identified module hubs (Zi ≥ 2.0, Pi < 0.55), connectors (Zi < 2.0, Pi ≥ 0.55), and network hubs (Zi ≥ 2.0, Pi ≥ 0.55), which are referred to as keystone nodes). (**d**) The SPEC-OCCU plots of 6 different waterlogging and recovery period treatment groups. (**e**) Venn analysis between the hub microbiotas and specialist microbiotas.

2.6. Co-Occurrence Between Root Metabolites and Rhizosphere Microbial Communities and Microbial Assembly Processes Under Waterlogging and Post-Waterlogging Recovery

Plants selectively recruit specific rhizosphere microbial communities through root metabolites, thereby establishing plant–rhizosphere microbial complexes that synergistically enhance plant resistance to stressors [22,23]. Root metabolites were significantly correlated with the composition of the rhizosphere microbial community under waterlogging and post–waterlogging recovery, based on a Mantel test ($p = 0.021$). We propose that root metabolites during waterlogging and recovery correlate with dynamic shifts in the rhizosphere microbiota. Hence, we investigated the potential driving role of metabolites in the microbial community assembly process under these conditions. By calculating the β–nearest taxon index (βNTI) between sample pairs, we found that deterministic and stochastic assembly processes were observed across all the rhizosphere bacterial communities of waterlogging and post–waterlogging recovery (W10, W30, WR10 of βNTI > 2, W20, WR20, WR30 of |βNTI| < 2) (Figure 3a). Furthermore, dispersal limitation (RCbray > 0.95) was found to be dominant among the stochastic processes, and deterministic processes were dominated by homogeneous selection. Finally, correlation analyses integrated βNTI with 48 DAMs identified through OPLS–DA, STEM clustering, and literature validation. A total of 16 metabolites (including L–valine, L–citrulline, N–glycyl–L–leucine, etc.) demonstrated significant associations with microbial community assembly processes (Figure 3b). To explore the root metabolite–rhizomicrobiota response patterns under waterlogging stress and post–waterlogging recovery, cross–domain network pipeline analysis was conducted between 48 different metabolites screened above and the rhizosphere microbiota of the network. It was found that 24 metabolites were significantly correlated with 57 microorganisms (Figure 3c). Quercetin–3–O–sophoroside, L–Citrulline, L–Valine, N–Glycyl–L–leucine, Guanine, and L–Glycyl–L–phenylalanine exhibited significant co–occurrences with a greater number of rhizosphere microbiotas, most of which were positive.

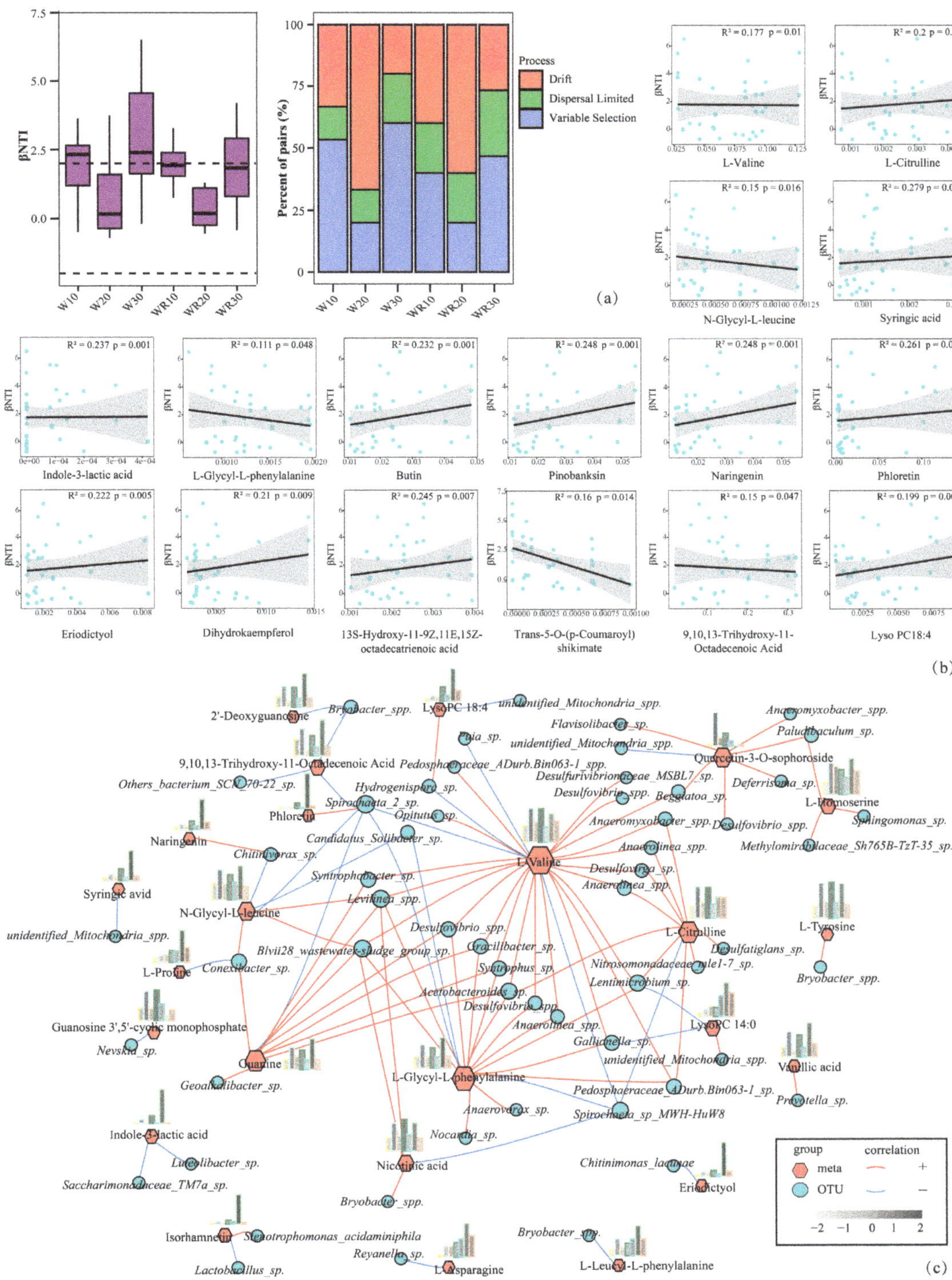

Figure 3. Correlation analysis of root metabolites with rhizosphere microbial community and microbial assembly processes under waterlogging and post-waterlogging recovery. (**a**) Contributions

of deterministic and stochastic processes to the aggregation of rhizosphere microbial communities were analyzed by βNTI, and each community assembly process was defined by the percentage of site pairs. (**b**) Scatter plots illustrating the correlation between βNTI and the relative abundance of the 16 root differential metabolites. Adjusted R2 and p values from linear regressions are shown. (**c**) Correlation network of metabolite production and microbial community of *R. delavayi* root under waterlogging and post-waterlogging recovery. (The bar chart illustrated the variation of expression for corresponding metabolites). From left to right, each bar represented the following groups: CK, W10, WR10, W20, WR20, W30, and WR30).

3. Discussion

3.1. Effects of Waterlogging and Post–Waterlogging Recovery on Root Metabolites and Rhizosphere Microbes

Global climate change has increased waterlogging frequency over six decades, with partial/complete inundation damaging most terrestrial plants. While plant acclimation to waterlogging is well studied, post-stress recovery mechanisms remain underexplored [24]. Although rhizosphere microbiome involvement in waterlogging resistance is established [9], its role during recovery remains unclear. Prolonged waterlogging caused progressive root damage in *R. delavayi*, with incomplete restoration observed post-recovery. Root metabolite alterations intensified with treatment duration, peaking at 30 days of waterlogging/recovery. Consistent with reported waterlogging-induced microbial shifts [6], our data revealed persistent community restructuring during recovery. Rhizosphere microbial diversity and phylogenetic complexity increased during waterlogging and persisted through recovery. These findings provide further evidence of continuous waterlogging–hypoxia impacts during plant recovery [25,26], emphasizing the critical need to investigate post-stress microbial and metabolic dynamics for enhancing plant resilience.

Waterlogging and subsequent recovery exert substantial influence on the rhizosphere microbiota across taxonomic ranks. Previous studies of rhizosphere microorganisms in sugar beet and maize found that Proteobacteria and Firmicutes were significantly enriched under waterlogging treatment [6,27]. However, another research on spring wheat (*Triticum aestivum*) showed that the proportion of Actinobacteria and Proteobacteria decreased [7]. We also found a notable reduction in Proteobacteria abundance in the rhizosphere of *R. delavayi*. At the same time, Cyanobacteria and Actinobacteriota exhibited a marked increase. We think that this large variation may be related to differences in the soil in which plants grow. Soil type could also significantly influence microbial community structure [8]. Compared with the neutral soil of sugar beet and maize, the planting soil of *R. delavayi* has higher acidity (pH 4.8 $\pm$ 0.3). Some studies have suggested that waterlogging is less harmful to neutral soils than acidic soils, at least on the level of microbial functions [8]. The variability in rhizosphere microbial responses, attributed to the dissimilarities of soil and plants, suggests that the current research on the interaction between plants and rhizosphere microorganisms for resisting stress is still profoundly inadequate. Moreover, there was a substantial number of unidentified bacteria in the rhizosphere of *R. delavayi*, and their abundance shows a significant increase following waterlogging and subsequent recovery. These indicated that there are still many unknown bacteria in the rhizosphere microbes of *R. delavayi* that we need to isolate and study, which may bring us more challenges. Many studies have reported that host plants recruit rhizosphere beneficial bacteria to promote plant growth under various stresses, such as waterlogging, high salt, etc. [28,29]. Through database comparison and identification, 44 potentially rhizosphere-beneficial bacteria were identified from the 187 genera of rhizosphere microbiota affected by waterlogging stress and subsequent recovery (Supplementary Table S7) [30]. Nearly all potentially rhizosphere-beneficial bacteria demonstrated varying degrees of enrichment under distinct waterlogging and recovery periods.

3.2. Effects of Waterlogging and Post–Waterlogging Recovery on Core Microbes

Core microbes constitute an essential and interrelated group within the rhizosphere microbiome community, where changes in their presence and abundance can induce significant variations in microbiome structure and function [31,32]. A total of 59 core microbes were found using two methods, in which only *Rubrobacter* fulfilled the criteria for hub microbiotas and specialist microbiotas. (Figure 2e). *Rubrobacter* was present during the plant developmental processes and played a crucial role in promoting plant growth and nutrient cycling [33,34]. Interestingly, we found that the hub microbiotas were almost all potentially beneficial to plant growth or nutrient cycling. The genus *Bacillus* has long been recognized for its beneficial interactions with plants, enhancing growth, nutrient uptake, and stress resistance [35]. The genera *Anaerolinea* and *Ohtaekwangia* played a significant role in soil carbon and nitrogen cycles and promoted plant growth [36–38]. The genus *Sphingomonas* produced highly beneficial phytohormones and improved plant growth under stress conditions [39]. *Massilia* genera had rock phosphate solubilizing ability, improved plant phosphorus nutrition status, and driven plant growth [40,41]. *Azospirillum* genera was a nitrogen-fixing bacterium, which promoted plant growth and increased plant tolerance to water stress [42,43]. Some specialist microbiotas have also been shown to promote plant growth or nutrient cycling. The genus *Arthrobacter* is a nitrogen-fixing bacterium, which promotes plant growth. *Sphingobacterium*, *Rubrobacter*, and *Lactobacillus* are plant growth-promoting rhizobacteria [44]. In addition, we found that many of the specialist microbiotas were anaerobic bacteria or facultative anaerobes, which can adapt to anoxic conditions. Examples include the genera *Corynebacterium*, *Sphingobacterium*, *Prevotella*, *Parasutterella*, *Lactobacillus*, *Ruminococcus*, and *Desulfovirga* [45–49]. The results indicated that the core microbes in the rhizosphere microbial community under waterlogging and post–waterlogging recovery treatment may exhibit two key characteristics: (i) they are important for promoting plant growth and nutrition cycling; (ii) they are adapted to anoxic conditions. Nevertheless, the way in which the core potentially rhizosphere beneficial bacteria we discovered establish a defense mechanism against waterlogging damage in association with *R. delavayi* is still unclear. In the future, to address these issues, we can use synthetic communities to explore the extent to which they contribute to *R. delavayi* resistance to waterlogging stress [50,51].

3.3. Root Metabolites Assembled Rhizosphere Microbial Community to Resist Stress for Plant

Current research indicates that root metabolites are correlated with rhizosphere microbiota succession and might be particularly influential under stress conditions [52,53]. Plant roots under stress release metabolic signals to recruit rhizosphere microorganisms, which promote plant growth and contribute to stress resistance [29,54]. In our studies, the differentially accumulated metabolites were significantly correlated with the composition and assembly process of *R. delavayi* rhizosphere microbial communities under waterlogging and post–waterlogging recovery. Meanwhile, based on the existing literature on root metabolite function and rhizosphere beneficial bacteria, we found several interesting co–occurrences between metabolites and rhizosphere microbiota. L–Homoserine exhibited a significant positive co-occurrence with the beneficial bacterium *Sphingomonas* sp., which synthesizes highly advantageous plant hormones under adverse conditions, thereby promoting plant growth. The plant growth–promoting bacterium *Bryobacter* sp. correlated with Nicotinic acid, 9,10,13–Trihydroxy–11–Octadecenoic Acid, 2′–Deoxyguanosine, L–Leucyl–L–phenylalanine and L–Tyrosine. *Nocardia* sp. secreted Indoleacetic acid (IAA) and siderophores to promote plant growth, which correlated with L–Glycyl–L–phenylalanine. *Lactobacillus* sp. and *Stenotrophomonas acidaminiphila* demonstrated the ability to fix nitrogen, increase phosphorus, and increase potassium. Isorhamnetin was likely to concentrate *Lactobacillus* sp. and *Stenotrophomonas acidaminiphila* together to regulate the growth of both primary and lateral roots. Additionally, Quercetin–3–O–sophoroside, which similarly regulated the growth of both primary and lateral roots, enhanced the abundance of the plant growth–promoting bacterium *Flavisolibacter* sp. [30]. Hence, we suggested that enriching these metabolites may play a significant role in rhizosphere microbial community assembly and plant resistance to stress.

3.4. The Response Patterns of Root Metabolite–Rhizomicrobiota in R. delavayi Under Waterlogging Stress and Post-Waterlogging Recovery

This study described the responses of root metabolites and rhizosphere microorganisms to waterlogging stress and post–waterlogging recovery. By combining our results, we attempt to describe the possible response patterns of root metabolites and rhizosphere microbiota complexes in *R. delavayi* under waterlogging stress and recovery (Figure 4). Three response patterns were identified: (1) The effects of waterlogging stress on *R. delavayi* persisted throughout the corresponding recovery period, as evidenced by the root morphology and the composition of root metabolites and rhizosphere microorganisms. (2) When *R. delavayi* was subjected to waterlogging stress, the abundances of numerous root metabolites and rhizosphere microorganisms changed significantly. Some metabolites associated with both biotic and abiotic stress responses were enriched. The enrichment of these metabolites may be a common response of *R. delavayi* to waterlogging stresses. There are many potentially beneficial bacteria in the rhizosphere microbiota of *R. delavayi*, which may contribute to waterlogging stress resistance. However, the enrichment degree of potentially beneficial bacteria differed across different waterlogging and recovery periods. Core microbes play a pivotal role in organizing the assembly and function of plant-associated microbial communities [31]. Notably, most of the core microbiotas of *R. delavayi* responding to waterlogging stress were potentially beneficial bacteria. (3) The waterlogged *R. delavayi* assembled specific rhizosphere microbiotas by means of root metabolites resistant to the biological or abiotic stresses of host plants, within which some beneficial bacteria in the rhizosphere were recruited.

Nonetheless, to propose more detailed mechanisms underlying rhizosphere responses to waterlogging stress in the future, it is important to recognize that the patterns identified in this study require further experimental validation and a more comprehensive explanation of several key aspects. First, the enrichment of synergistic beneficial microbiota by specific root metabolites needs to be experimentally confirmed. Future studies should also focus on elucidating the succession dynamics of these metabolites and microorganisms during the post–waterlogging recovery phase. Finally, to deepen our understanding of the roles of root metabolites and rhizosphere microorganisms in the waterlogging resistance of *R. delavayi*, further investigations should incorporate more detailed time-series analyses.

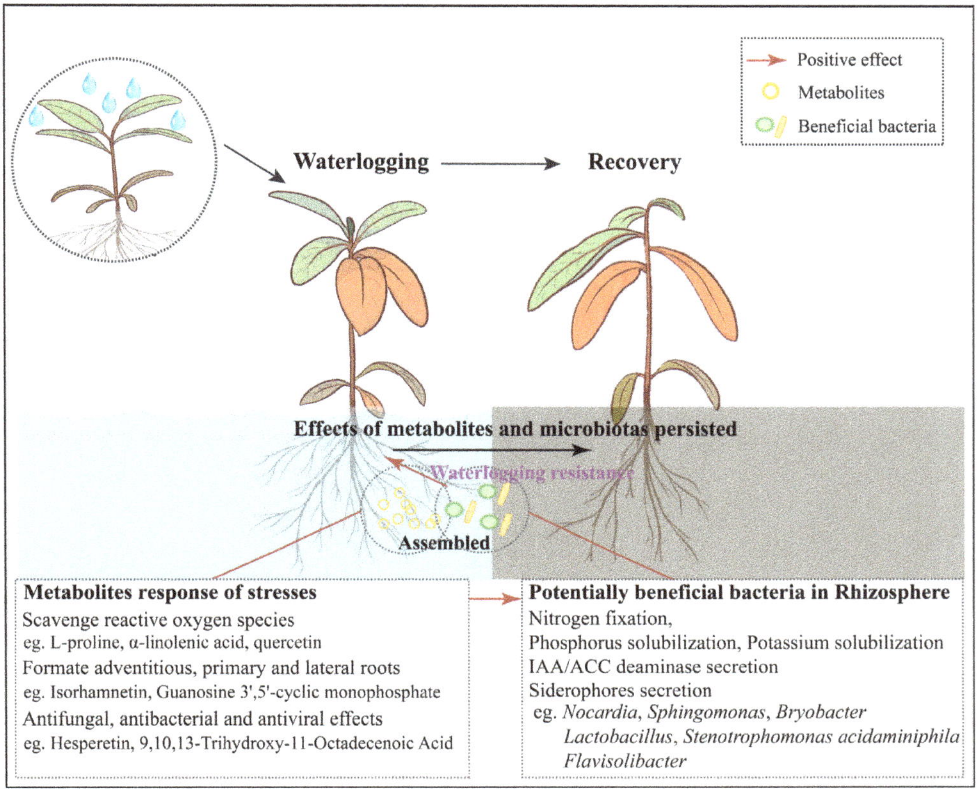

Figure 4. The response patterns of root metabolites and rhizosphere microbiota complex in *R. delavayi* under waterlogging stress and post-waterlogging recovery.

4. Materials and Methods

4.1. Plant Material and Pot Water Control Treatment

Three-year-old *R. delavayi* seedlings were selected for the pot waterlogging experiment, which included two treatments: waterlogging and recovery. These seedlings were cultured in plastic flowerpots (cylinders of diameter 8 cm and height 15 cm) with wild, healthy rhododendron soil from the Baili Rhododendron area. The soil pH was 4.8 ± 0.3, characteristic of acidic conditions. The culture conditions in a greenhouse were a photoperiod of 16 h/8 h, a temperature of 22 °C, a light intensity of 400 μmol $m^{-2}s^{-1}$, and a relative humidity of 60–70%. The plastic flowerpots were placed in trays (length × width × height, 55 cm × 30 cm × 4.5 cm); after that, the *R. delavayi* seedlings were continuously watered until the water level in the trays reached 4 cm. In the waterlogging treatment groups, daily irrigation maintained a waterlogging depth of 4 cm, with stress durations of 0 (W0), 10 (W10), 20 (W20), and 30 (W30) days. After the waterlogging stress period, water in the trays was removed, and the seedlings were allowed to recover for durations corresponding to their respective waterlogging times: 10 days of waterlogging followed by 10 days of recovery (WR10), 20 days of waterlogging followed by 20 days of recovery (WR20), and 30 days of waterlogging followed by 30 days of recovery (WR30). The soil was normally irrigated in the later part of the restoration process. Three biological replicates were included for each treatment.

4.2. Analysis of Root Metabolites by UPLC–MS/MS

Using a mixer mill (MM 400, Retsch, Haan, Germany), the freeze-dried root sample was crushed with a zirconia bead for 1.5 min at 30 Hz. The 100 mg powder sample was then weighed and dissolved in 1.2 mL of 70% methanol solution, followed by vortexing for 30 s every 30 min for 6 times in total. Subsequently, the samples were placed overnight at a temperature of 4 °C. Before UPLC–MS/MS analysis, the sample extracts were filtered (SCAA–104, 0.22 µm pore size, ANPEL, Shanghai, China) after centrifugation at a speed of 12,000× *g* rpm for a duration of 10 min. The sample extracts were analyzed using a UPLC–ESI–MS/MS system (UPLC, SHIMADZU Nexera X2, Shimadzu, Japan; MS, Applied Biosystems 4500 Q TRAP, Foster City, CA, USA). The UPLC conditions were as follows: Column, Agilent SB–C18 (1.8 µm, 2.1 mm × 100 mm). Mobile Phase, solvent A: pure water with 0.1% formic acid; solvent B: acetonitrile with 0.1% formic acid. Sample Measurements: The starting conditions were 95% A and 5% B. A linear gradient to 5% A and 95% B was applied over 9 min, and 5% A and 95% B were kept for 1 min. Subsequently, 95% A and 5.0% B were adjusted within 1.10 min and kept for 2.9 min. Flow velocity: 0.35 mL per minute. Column Oven, 40 °C. Injection Volume, 4 µL. The effluent was alternatively connected to an ESI–triple quadrupole–linear ion trap (QTRAP)–MS. The ESI source was operated in turbo spray mode with the following parameters: source temperature 550 °C, ion spray voltage ±5500 V (positive/negative modes), gas flows (GS1 50 psi, GS2 60 psi, curtain gas 25 psi), and high collision–activated dissociation. System calibration used 10 µmol/L (QQQ mode) and 100 µmol/L (LIT mode) polypropylene glycol solutions. MRM acquisitions employed nitrogen collision gas at medium pressure. Qualitative analysis of the secondary mass spectrometry data was performed using a self–built database, MWDB (v2.0), and the quantitative analysis of the metabolites was performed using the multiple reaction monitoring mode. After operation in positive– and negative–ion modes, the raw data were controlled and processed by Analyst 1.6.3 software (AB Sciex, Framingham, MA, USA) [55]. Chromatographic peak areas (Area) reflect the relative abundance of corresponding metabolites. To enable cross–sample comparison of metabolite abundance while ensuring accurate annotation and quantification, we implemented retention time–based alignment and chromatographic peak shape validation across all detected features.

4.3. Microbial DNA Extraction and 16S rRNA Gene V4 Region Sequencing

5 g root samples were first vortexed with 5 mL of phosphate–buffered saline (PBS) solution (pH 7.2) and then centrifuged (5 min at 3000× *g*) to collect all the sediment from the root surfaces, which was treated as the rhizosphere microbial samples. Total genome DNA from the samples was extracted using the MOBIO PowerSoil ® DNA Isolation Kit (12888–50, Carlsbad, CA, USA). A sample was extracted three times and mixed before being dissolved in 60 µL TE buffer. Then we quantified the DNA ND1000 and stored it at −80 °C. The V4 region of the 16S rRNA gene was amplified using the 20~30 ng DNA from each sample and then sequenced using the Illumina NovaSeq platform. The PCR primers were 515F′–'GTGCCAGCMGCCGCGGTAA,' and 806R′–'GGACTACHVGGGTWTCTAAT'. PCR reactions were carried out with 15 µL of Phusion® High–Fidelity PCR Master Mix (New England Biolabs), 2 µM of forward and reverse primers, and about 10 ng template DNA. Sequencing libraries were generated using the TruSeq® DNA PCR–Free Sample Preparation Kit (Illumina, San Diego, CA, USA) following the manufacturer's recommendations, and index codes were added.

4.4. Sequencing Data Analysis

Quality filtering on the raw tags was performed to obtain the high–quality clean tags according to QIIME. Quality filtering parameters included maximum expected errors per read (2.0), removing low–quality regions ($\leq$19), and removing the chimeric sequences. Using the VSERCH algorithm to detect chimera sequences, the Effective Tags were finally obtained. Tags with $\geq$97% similarity were assigned to the same OTUs using the UPARSE method. In total, 1,386,614 high–quality 16S rRNA V4 amplicons for 21 rhizosphere samples were obtained and analyzed (Supplementary Table S1). Then, using the Silva Database based on the USEARCH algorithm, OTUs were annotated with taxonomic information [56]. OTU abundance information was normalized using a standard of sequence number corresponding to the sample with the least sequences.

4.5. Network Construction and Characterization

The OTUs were screened for significant clusters using STEM. OTUs that could not be annotated to a genus or had an abundance of less than 10 were excluded [57]. A total of 634 OTUs were screened. Based on OTU abundance, a microbial co–occurrence network was established using Pearson co–occurrence coefficients (r) exceeding 0.7 or falling below -0.7, in conjunction with false discovery rate-adjusted p-values of less than 0.05 [58]. The co-occurrence networks were visualized with the interactive R package v4.2.2 ggClusterNet [59]. To estimate the complexity of networks, we calculated the characterization of the networks by R package igraph, including total number of nodes, number of links, network diameter, relative modularity (RM), average clustering coefficient (Average CC), and average degree [60]. Robustness is defined as the proportion of the remaining species in this network after random removal, in which 50% of the nodes were randomly removed. Cohesion can be used as an indicator to measure microbial interactions [44]. The robustness, cohesion, and negative co–occurrence ratio were calculated and visualized by the R package ggClusterNet. In this study, two different methods were used to jointly define core microbes. First, potential specialist microbiotas were identified by SPEC–OCCU plots [61]. Second, according to within–module connectivity (Zi) and among–module connectivity (Pi), hub microbiotas were identified [32]. Cytoscape 3.7.0 was utilized to visualize and clarify the network of root metabolites and microbiotas, which was established based on Spearman's co–occurrence coefficients ($r > 0.7$ or $r < -0.7$, $p < 0.05$).

4.6. Analysis of the Microbial Community Assembly Processes and the Effect of Root Metabolites on Community Assembly

By calculating the βNTI among samples, the phylogenetic pattern of the rhizosphere microbial community is assessed by employing the null model approach [62]. Sample pairs with $|\beta NTI| > 2$ were considered to result from a deterministic process, while $|\beta NTI| < 2$ indicated that the selective pressure was relatively weak, suggesting that community assembly may be under the control of a stochastic process [63]. The values obtained from the βNTI analysis reflected the driving force of the relevant factors. For the computation of RCbray, the number of simulated communities with a Bray–Curtis dissimilarity exceeding the observed value was aggregated with half of the number of simulated communities with dissimilarity equal to the observed value, and the sum was subsequently divided by the total number of simulations (999). By combining the results of βNTI and RCbray, the relative contributions of deterministic and stochastic processes to overall community assembly within each sample were determined. All analyses of microbial community assembly processes and the effect of root metabolites on community assembly were carried out within the R "stats", "minpack.lm", "Hmisc", and "picante" packages [52].

4.7. Statistical Analyses

DAMs were identified using OPLS–DA in the R package *MetaboAnalystR* (VIP > 1, $|\log_2\text{Fold change}| \geq 1$) [64], with permutation testing (200 iterations) to validate model robustness. Enriched metabolic pathways were derived via metabolite set enrichment analysis using hypergeometric tests ($p < 0.05$). Root metabolite structures were visualized by PCA of Bray–Curtis distances (vegan package). For rhizosphere microbiota, α– and β–diversity indices were computed in QIIME and visualized in R. Community structure differences were assessed via Non–Metric Multidimensional Scaling (NMDS; Bray–Curtis) and permutational multivariate analysis of variance (PERMANOVA) by the vegan package. Taxonomic differentials between groups were evaluated using likelihood ratio tests ($p < 0.05$). The co–occurrence between microbial community and root metabolites was tested via Mantel tests (*vegan*; Bray–Curtis vs. Euclidean matrices, 999 permutations). Analyses were conducted in R version v4.2.2 with vegan v2.6-6.

5. Conclusions

This study provides an analysis of root metabolites and rhizosphere microbiota in the rhizosphere soil of *R. delavayi* under waterlogging stress and post–waterlogging recovery. Our findings revealed significant alterations in the root metabolites, rhizosphere microbiota, and microbial co–occurrence network during waterlogging stress, with these changes persisting throughout the recovery period. Moreover, we found that the root metabolites are closely related to the composition and assembly process of rhizosphere microorganisms. Notably, certain root metabolites were found to enrich potentially beneficial rhizosphere microbiota, which collectively played a role in mitigating waterlogging stress. For example, the addition of L-homoserine and the beneficial bacteria *Sphingomonas* sp. may synthesize highly advantageous plant hormones to promote plant growth and resist waterlogging stress. However, further detailed time–series studies are needed to better understand the dynamic interactions between root metabolites, rhizosphere microbiota, and the duration of waterlogging stress and recovery. Such studies would provide critical insights into the intricate relationships between the root metabolome and rhizosphere microbiome. This knowledge is essential for developing future strategies that integrate "metabolites + rhizosphere microbiota" to enhance the waterlogging resistance of *R. delavayi*.

Supplementary Materials: The following supporting information can be downloaded at: https://www.mdpi.com/article/10.3390/horticulturae11070770/s1.

Author Contributions: J.T.: Writing—review and editing, Writing—original draft, Visualization, Methodology, Data curation, Conceptualization, Funding acquisition. Q.H.: Writing—original draft, Investigation, Visualization, Methodology, Data curation. Q.W.: Data curation, Investigation, Software. F.S.: Investigation. S.W.: Methodology and execution of experiments. X.Z.: Funding acquisition, Supervision. M.T.: Investigation, Project administration; Y.Y.: Writing—review and editing, Supervision, Project administration, Funding acquisition, Conceptualization. All authors have read and agreed to the published version of the manuscript.

Funding: This research was funded by Guizhou Provincial Natural Science Foundation Project (QKHJC–ZD [2025]046), Science and Technology Support Project of Guizhou Province (QKHZC [2021]YB459), the Natural Science Foundation of China(NSFC) (32260393); Key Laboratory of Environment Friendly Management on Alpine Rhododendron Diseases and Pests of Institutions of Higher Learning in Guizhou Province ([2022]044) and Karst Mountain Ecological Security Engineering Re–search Center, grant number [2021]007.

Data Availability Statement: The datasets presented in this study can be found in NCBI online repositories. The names of the repositories and accession number(s) can be found below: https://www.ncbi.nlm.nih.gov/sra, (12-Nov-2024) PRJNA1184993.

Conflicts of Interest: The authors declare that the research was conducted in the absence of any commercial or financial relationships that could be construed as a potential conflicts of interest.

References

1. Ma, H.; Liu, Y.B.; Liu, D.T.; Sun, W.B.; Liu, X.F.; Wan, Y.M.; Zhang, X.J.; Zhang, R.G.; Yun, Q.Z.; Wang, J.H.; et al. Chromosomelevel genome assembly and population genetic analysis of a critically endangered rhododendron provide insights into its conservation. *Plant J.* **2021**, *107*, 1533–1545. [CrossRef] [PubMed]
2. Zhang, X.M.; Duan, S.G.; Xia, Y.; Li, J.T.; Liu, L.X.; Tang, M.; Tang, J.; Sun, W.; Yi, Y. Transcriptomic, Physiological, and Metabolomic Response of an Alpine Plant, Rhododendron delavayi, to Waterlogging Stress and PostWaterlogging Recovery. *Int. J. Mol. Sci.* **2023**, *24*, 10509. [CrossRef]
3. Barbara, R.H.; Wiebke, B.; Claudia, S.B.; Mugdha, S.; Thomas, H. Roots Shaping Their Microbiome: Global Hotspots for Microbial Activity. *Annu. Rev. Phytopathol.* **2015**, *53*, 403–424. [CrossRef]
4. Zhou, S.Y.D.; Lie, Z.Y.; Liu, X.J.; Zhu, Y.G.; Peñuelas, J.; Neilson, R.; Su, X.X.; Liu, Z.F.; Chu, G.W.; Meng, Z.; et al. Distinct patterns of soil bacterial and fungal community assemblages in subtropical forest ecosystems under warming. *Glob. Change Biol.* **2023**, *29*, 1501–1513. [CrossRef]
5. Compant, S.; Samad, A.; Faist, H.; Sessitsch, A. A review on the plant microbiome: Ecology, functions, and emerging trends in microbial application. *J. Adv. Res.* **2019**, *19*, 29–37. [CrossRef] [PubMed]
6. Li, T.; Wang, M.H.; Cui, R.F.; Li, B.C.; Wu, T.; Liu, Y.L.; Geng, G.; Xu, Y.; Wang, Y.G. Waterlogging stress alters the structure of sugar beet rhizosphere microbial community structure and recruiting potentially beneficial bacterial. *Ecotoxicol. Environ. Saf.* **2023**, *262*, 115172. [CrossRef]
7. Francioli, D.; Cid, G.; Kanukollu, S.; Ulrich, A.; Hajirezaei, M.R.; Kolb, S. Waterlogging Causes Dramatic Compositional Shifts and Depletion of Putative Beneficial Bacteria on the Spring Wheat Microbiota. *Front. Microbiol.* **2021**, *12*, 773116. [CrossRef]
8. Yu, T.; Cheng, L.; Liu, Q.; Wang, S.; Zhou, Y.; Zhong, H.; Tang, M.; Nian, H.; Lian, T. Effects of Waterlogging on Soybean Rhizosphere Bacterial Community Using V4, LoopSeq, and PacBio 16S rRNA Sequence. *Microbiol. Spectr.* **2022**, *10*, e0201121. [CrossRef]
9. Martínez-Arias, C.; Witzell, J.; Solla, A.; Martin, J.A.; Rodríguez-Calcerrada, J. Beneficial and pathogenic plant–microbe interactions during waterlogging stress. *Plant Cell Environ.* **2022**, *45*, 2875–2897. [CrossRef]
10. Moche, M.; Gutknecht, J.; Schulz, E.; Langer, U.; Rinklebe, J. Monthly dynamics of microbial community structure and their controlling factors in three floodplain soils. *Soil Biol. Biochem.* **2015**, *90*, 169–178. [CrossRef]
11. Bhaskar, J.P.; Indrani, S.; Niraj, A. Root exudation drives abiotic stress tolerance in plants by recruiting beneficial microbes. *Appl. Soil Ecol.* **2024**, *198*, 105351. [CrossRef]
12. Bouwmeester, H.J.; Fonne-Pfister, R.; Screpanti, C.; De Mesmaeker, A. Strigolactones: Plant Hormones with Promising Features. *Angew. Chem. Int. Ed.* **2019**, *58*, 12778–12786. [CrossRef] [PubMed]
13. Zhang, P.; Lyu, D.G.; Jia, L.T.; He, J.L.; Qin, S.J. Physiological and de novo transcriptome analysis of the fermentation mechanism of Cerasus sachalinensis roots in response to short-term waterlogging. *BMC Genom.* **2017**, *18*, 649. [CrossRef]
14. Song, Y.; Wilson, A.J.; Zhang, X.C.; Thoms, D.; Sohrabi, R.; Song, S.Y.; Geissmann, Q.; Liu, Y.; Walgren, L.; He, S.Y.; et al. FERONIA restricts Pseudomonas in the rhizosphere microbiome via regulation of reactive oxygen species. *Nat. Plants* **2021**, *7*, 644–654. [CrossRef] [PubMed]
15. Bulgarelli, D.; Rott, M.; Schlaeppi, K.; Ver Loren van Themaat, E.; Ahmadinejad, N.; Assenza, F.; Rauf, P.; Huettel, B.; Reinhardt, R.; Schmelzer, E.; et al. Revealing structure and assembly cues for Arabidopsis root-inhabiting bacterial microbiota. *Nature* **2012**, *488*, 91–95. [CrossRef] [PubMed]
16. Vives-Peris, V.; de Ollas, C.; Gómez-Cadenas, A.; Pérez-Clemente, R.M. Root exudates: From plant to rhizosphere and beyond. *Plant Cell Rep.* **2020**, *39*, 3–17. [CrossRef]
17. Savchenko, T.; Rolletschek, H.; Heinzel, N.; Tikhonov, K.; Dehesh, K. Waterlogging tolerance rendered by oxylipin-mediated metabolic reprogramming in Arabidopsis. *J. Exp. Bot.* **2019**, *70*, 2919–2932. [CrossRef]
18. Yeung, E.; Bailey-Serres, J.L.; Sasidharan, R. After The Deluge: Plant Revival Post-Waterlogging. *Trends Plant Sci.* **2019**, *24*, 443–454. [CrossRef]
19. Herren, C.M.; McMahon, K.D. Cohesion: A method for quantifying the connectivity of microbial communities. *ISME J.* **2017**, *11*, 2426–2438. [CrossRef]
20. Coyte, K.Z.; Schluter, J.; Foster, K.R. The ecology of the microbiome: Networks, competition, and stability. *Science* **2015**, *350*, 663–666. [CrossRef]
21. Zhou, Y.Y.; Liu, D.H.; Li, F.Q.; Dong, Y.H.; Jin, Z.L.; Liao, Y.W.K.; Li, X.H.; Peng, S.G.; Delgado-Baquerizo, M.; Li, X.G. Superiority of native soil core microbiomes in supporting plant growth. *Nat. Commun* **2024**, *15*, 6599. [CrossRef] [PubMed]
22. Huang, A.C.; Jiang, T.; Liu, Y.X.; Bai, Y.C.; Reed, J.; Qu, B.Y.; Goossens, A.; Nützmann, H.W.; Bai, Y.; Osbourn, A. A specialized metabolic network selectively modulates Arabidopsis root microbiota. *Science* **2019**, *364*, eaau6389. [CrossRef]
23. Wang, L.X.; Chen, M.X.; Lam, P.Y.; Dini-Andreote, F.; Dai, L.; Wei, Z. Multifaceted roles of flavonoids mediating plant-microbe interactions. *Microbiome* **2022**, *10*, 233. [CrossRef]

24. Yuan, L.B.; Chen, M.X.; Wang, L.N.; Sasidharan, R.; Voesenek, L.A.C.J.; Xiao, S. Multi-stress resilience in plants recovering from submergence. *Plant Biotechnol. J.* **2023**, *21*, 466–481. [CrossRef] [PubMed]
25. Yeung, E.; van Veen, H.; Vashisht, D.; Sobral Paiva, A.L.; Hummel, M.; Rankenberg, T.; Steffens, B.; Steffen-Heins, A.; Sauter, M.; de Vries, M.; et al. A stress recovery signaling network for enhanced waterlogging tolerance in Arabidopsis thaliana. *Proc. Natl. Acad. Sci. USA* **2018**, *115*, E6085–E6094. [CrossRef] [PubMed]
26. Shikov, A.E.; Chirkova, T.V.; Yemelyanov, V.V. Post-Anoxia in Plants: Reasons, Consequences, and Possible Mechanisms. *Russ. J. Plant Physiol.* **2020**, *67*, 45–59. [CrossRef]
27. Bai, H.; He, S.; Qin, T.L.; Yan, D.H.; Weng, B.S.; Zhao, X.X.; Li, X.N.; Bai, Y.J.; Ma, J. Influences of irrigation amount on the rhizospheric microorganism composition and carbon dioxide flux of maize crops. *Geoderma* **2019**, *343*, 1–9. [CrossRef]
28. Etesami, H.; Maheshwari, D.K. Use of plant growth promoting rhizobacteria (PGPRs) with multiple plant growth promoting traits in stress agriculture: Action mechanisms and future prospects. *Ecotoxicol. Environ. Saf.* **2018**, *156*, 225–246. [CrossRef]
29. Yin, J.; Guo, H.; Fry, E.L.; De Long, J.R.; Tang, S.; Yuan, T.; Ren, W. Plant roots send metabolic signals to microbes in response to long-term overgrazing. *Sci. Total Environ.* **2022**, *842*, 156241. [CrossRef]
30. Li, P.F.; Tedersoo, L.; Crowther, T.W.; Dumbrell, A.J.; Dini-Andreote, F.; Bahram, M.; Kuang, L.; Li, T.; Wu, M.; Jiang, Y.J.; et al. Fossil-fuel-dependent scenarios could lead to a significant decline of global plant-beneficial bacteria abundance in soils by 2100. *Nat. Food* **2023**, *4*, 996–1006. [CrossRef]
31. Compant, S.; Cassan, F.; Kostić, T.; Johnson, L.; Brader, G.; Trognitz, F.; Sessitsch, A. Harnessing the plant microbiome for sustainable crop production. *Nat. Rev. Microbiol.* **2024**, *23*, 9–23. [CrossRef] [PubMed]
32. Qiao, Y.Z.; Wang, T.T.; Huang, Q.W.; Guo, H.Y.; Zhang, H.; Xu, Q.C.; Shen, Q.R.; Ling, N. Core species impact plant health by enhancing soil microbial cooperation and network complexity during community coalescence. *Soil Biol. Biochem.* **2024**, *188*, 109231. [CrossRef]
33. Trivedi, K.; Kumar, R.; Vijay Anand, K.G.; Bhojani, G.; Kubavat, D.; Ghosh, A. Structural and functional changes in soil bacterial communities by drifting spray application of a commercial red seaweed extract as revealed by metagenomics. *Arch. Microbiol.* **2021**, *204*, 72. [CrossRef] [PubMed]
34. Ajilogba, C.F.; Olanrewaju, O.S.; Babalola, O.O. Plant Growth Stage Drives the Temporal and Spatial Dynamics of the Bacterial Microbiome in the Rhizosphere of Vigna subterranea. *Front. Microbiol.* **2022**, *13*, 825377. [CrossRef]
35. Das, V.A.; Gautam, B.; Yadav, P.K.; Singh, S. Identification of Conserved Pathways in Bacillus Strains Known for Plant Growth-Promoting Behavior Using a Multifaceted Computational Approach. *Agriculture* **2024**, *14*, 838. [CrossRef]
36. Zarraonaindia, I.; Martínez-Goñi, X.S.; Liñero, O.; Muñoz-Colmenero, M.; Aguirre, M.; Abad, D.; Baroja-Careaga, I.; de Diego, A.; Gilbert, J.A.; Estonba, A. Response of Horticultural Soil Microbiota to Different Fertilization Practices. *Plants* **2020**, *9*, 1051. [CrossRef]
37. Li, D.; Zhou, C.R.; Wu, Y.L.; An, Q.S.; Zhang, J.B.; Fang, Y.; Li, J.Q.; Pan, C.P. Nanoselenium integrates soil-pepper plant homeostasis by recruiting rhizosphere-beneficial microbiomes and allocating signaling molecule levels under Cd stress. *J. Hazard. Mater.* **2022**, *432*, 128763. [CrossRef]
38. Chen, J.Y.; Liu, S.; Deng, W.K.; Niu, S.H.; Liao, X.D.; Xiang, L.; Xing, S.C. The effect of manure-borne doxycycline combined with different types of oversized microplastic contamination layers on carbon and nitrogen metabolism in sandy loam. *J. Hazard. Mater.* **2023**, *456*, 131612. [CrossRef]
39. Asaf, S.; Numan, M.; Khan, A.L.; Al-Harrasi, A. Sphingomonas: From diversity and genomics to functional role in environmental remediation and plant growth. *Crit. Rev. Biotechnol.* **2020**, *40*, 138–152. [CrossRef]
40. Zheng, B.X.; Bi, Q.F.; Hao, X.L.; Zhou, G.W.; Yang, X.R. Massilia phosphatilytica sp. nov., a phosphate solubilizing bacteria isolated from a long-term fertilized soil. *Int. J. Syst. Evol. Microbiol.* **2017**, *67*, 2514–2519. [CrossRef]
41. Wang, D.N.; He, X.M.; Baer, M.; Lami, K.; Yu, B.G.; Tassinari, A.; Salvi, S.; Schaaf, G.; Hochholdinger, F.; Yu, P. Lateral root enriched Massilia associated with plant flowering in maize. *Microbiome* **2024**, *12*, 124. [CrossRef]
42. Steenhoudt, O.; Vanderleyden, J. Azospirillum, a free-living nitrogen-fixing bacterium closely associated with grasses: Genetic, biochemical and ecological aspects. *FEMS Microbiol. Rev.* **2000**, *24*, 487–506. [CrossRef] [PubMed]
43. Pereyra, M.A.; Zalazar, C.A.; Barassi, C.A. Root phospholipids in Azospirillum-inoculated wheat seedlings exposed to water stress. *Plant Physiol. Biochem.* **2006**, *44*, 873–879. [CrossRef]
44. Wu, H.; Yang, J.J.; Fu, W.; Rillig, M.C.; Cao, Z.J.; Zhao, A.; Hao, Z.P.; Zhang, X.; Chen, B.D.; Han, X.G. Identifying thresholds of nitrogen enrichment for substantial shifts in arbuscular mycorrhizal fungal community metrics in a temperate grassland of northern China. *New Phytol.* **2023**, *237*, 279–294. [CrossRef]
45. Aurilia, V.; Martin, J.C.; McCrae, S.I.; Scott, K.P.; Rincon, M.T.; Flint, H.J. Three multidomain esterases from the cellulolytic rumen anaerobe Ruminococcus flavefaciens 17 that carry divergent dockerin sequences. *Microbiology* **2000**, *146*, 1391–1397. [CrossRef]
46. Tanaka, K.; Stackebrandt, E.; Tohyama, S.; Eguchi, T. Desulfovirga adipica gen. nov., sp. nov., an adipate-degrading, gram-negative, sulfate-reducing bacterium. *Int. J. Syst. Evol. Microbiol.* **2000**, *50*, 639–644. [CrossRef] [PubMed]

47. Morotomi, M.; Nagai, F.; Watanabe, Y. *Parasutterella secunda* sp. nov., isolated from human faeces and proposal of *Sutterellaceae* fam. nov. in the order Burkholderiales. *Int. J. Syst. Evol. Microbiol.* **2011**, *61*, 637–643. [CrossRef] [PubMed]
48. Li, M.; Lee, K.; Hsu, M.; Nau, G.; Mylonakis, E.; Ramratnam, B. Lactobacillus-derived extracellular vesicles enhance host immune responses against vancomycin-resistant enterococci. *BMC Microbiol.* **2017**, *17*, 66. [CrossRef] [PubMed]
49. de la Maza, L.M.; Pezzlo, M.T.; Bittencourt, C.E.; Peterson, E.M. Bacteroides, Porphyromonas, Prevotella, Fusobacterium, and Other Anaerobic Gram-Negative Bacteria. In *Color Atlas of Medical Bacteriology*; American Society for Microbiology: Washington, DC, USA, 2020; pp. 252–260. [CrossRef]
50. Northen, T.R.; Kleiner, M.; Torres, M.; Kovács, Á.T.; Nicolaisen, M.H.; Krzyżanowska, D.M.; Sharma, S.; Lund, G.; Jelsbak, L.; Baars, O.; et al. Community standards and future opportunities for synthetic communities in plant-microbiota research. *Nat. Microbiol.* **2024**, *9*, 2774–2784. [CrossRef]
51. Vorholt, J.A.; Vogel, C.; Carlström, C.I.; Müller, D.B. Establishing Causality: Opportunities of Synthetic Communities for Plant Microbiome Research. *Cell Host Microbe* **2017**, *22*, 142–155. [CrossRef]
52. Wen, T.; Xie, P.; Penton, C.R.; Hale, L.; Thomashow, L.S.; Yang, S.; Ding, Z.; Su, Y.; Shen, Q.R.; Yuan, J. Specific metabolites drive the deterministic assembly of diseased rhizosphere microbiome through weakening microbial degradation of autotoxin. *Microbiome* **2022**, *10*, 177. [CrossRef] [PubMed]
53. Baker, N.R.; Zhalnina, K.; Yuan, M.T.; Herman, D.; Ceja-Navarro, J.A.; Sasse, J.; Jordan, J.S.; Bowen, B.P.; Wu, L.Y.; Fossum, C.; et al. Nutrient and moisture limitations reveal keystone metabolites linking rhizosphere metabolomes and microbiomes. *Proc. Natl. Acad. Sci. USA* **2024**, *121*, e2303439121. [CrossRef] [PubMed]
54. Rizaludin, M.S.; Stopnisek, N.; Raaijmakers, J.M.; Garbeva, P. The Chemistry of Stress: Understanding the 'Cry for Help' of Plant Roots. *Metabolites* **2021**, *11*, 357. [CrossRef]
55. Chen, W.; Gong, L.; Guo, Z.L.; Wang, W.S.; Zhang, H.Y.; Liu, X.Q.; Yu, S.B.; Xiong, L.Z.; Luo, J. A novel integrated method for large-scale detection, identification, and quantification of widely targeted metabolites: Application in the study of rice metabolomics. *Mol. Plant* **2013**, *6*, 1769–1780. [CrossRef] [PubMed]
56. Liu, Y.X.; Chen, L.; Ma, T.F.; Li, X.F.; Zheng, M.S.; Zhou, X.; Chen, L.; Qian, X.B.; Xi, J.; Lu, H.Y.; et al. EasyAmplicon: An easy-to-use, open-source, reproducible, and community-based pipeline for amplicon data analysis in microbiome research. *IMeta* **2023**, *2*, e83. [CrossRef]
57. Ernst, J.; Bar-Joseph, Z. STEM: A tool for the analysis of short time series gene expression data. *BMC Bioinform.* **2006**, *7*, 191. [CrossRef]
58. Jiao, S.; Qi, J.J.; Jin, C.J.; Liu, Y.; Wang, Y.; Pan, H.B.; Chen, S.; Liang, C.L.; Peng, Z.H.; Chen, B.B.; et al. Core phylotypes enhance the resistance of soil microbiome to environmental changes to maintain multifunctionality in agricultural ecosystems. *Glob. Change Biol.* **2022**, *28*, 6653–6664. [CrossRef]
59. Wen, T.; Xie, P.H.; Yang, S.D.; Niu, G.Q.; Liu, X.Y.; Ding, Z.X.; Xue, C.; Liu, Y.X.; Shen, Q.; Yuan, J. ggClusterNet: An R package for microbiome network analysis and modularity-based multiple network layouts. *IMeta* **2022**, *1*, e32. [CrossRef]
60. Yuan, M.M.; Guo, X.; Wu, L.W.; Zhang, Y.; Xiao, N.J.; Ning, D.L.; Shi, Z.; Zhou, X.S.; Wu, L.Y.; Yang, Y.F.; et al. Climate warming enhances microbial network complexity and stability. *Nat. Clim. Change* **2021**, *11*, 343–348. [CrossRef]
61. Gweon, H.S.; Bowes, M.J.; Moorhouse, H.L.; Oliver, A.E.; Bailey, M.J.; Acreman, M.C.; Read, D.S. Contrasting community assembly processes structure lotic bacteria metacommunities along the river continuum. *Environ. Microbiol.* **2021**, *23*, 484–498. [CrossRef]
62. Stegen, J.C.; Lin, X.J.; Konopka, A.E.; Fredrickson, J.K. Stochastic and deterministic assembly processes in subsurface microbial communities. *ISME J.* **2012**, *6*, 1653–1664. [CrossRef] [PubMed]
63. Kembel, S.W.; Cowan, P.D.; Helmus, M.R.; Cornwell, W.K.; Morlon, H.; Ackerly, D.D.; Blomberg, S.P.; Webb, C.O. Picante: R tools for integrating phylogenies and ecology. *Bioinformatics* **2010**, *26*, 1463–1464. [CrossRef] [PubMed]
64. Chong, J.; Xia, J.G. MetaboAnalystR: An R package for flexible and reproducible analysis of metabolomics data. *Bioinformatics* **2018**, *34*, 4313–4314. [CrossRef] [PubMed]

horticulturae

MDPI

Article

Comprehensive Analysis of Microbiomes and Metabolomics Reveals the Mechanism of Adaptation to Cadmium Stress in Rhizosphere Soil of *Rhododendron decorum* subsp. *Diaprepes*

Ming Tang [1,2,3,†], Lanlan Chen [1,2,3,†], Li Wang [1,2,3,†], Yin Yi [1,2,3], Jianfeng Wang [4,5,6,7,8], Chao Wang [4,5,6,7,8], Xianlei Chen [1,2,3], Jie Liu [1,2,3], Yongsong Yang [9], Kamran Malik [4,5,6,7,8] and Jiyi Gong [1,2,3,*]

[1] Key Laboratory of National Forestry and Grassland Administration on Biodiversity Conservation in Karst Mountainous Areas of Southwestern China, Guizhou Normal University, Guiyang 550025, China; mingtang@gznu.edu.cn (M.T.); lanlanchen222@163.com (L.C.); gznu_wangli0521@163.com (L.W.); yiyin@gznu.edu.cn (Y.Y.); xianleichen321@163.com (X.C.); liujie791204@126.com (J.L.)

[2] College of Life Sciences, Guizhou Normal University, Guiyang 550025, China

[3] Engineering Research Center of Carbon Neutrality in Karst Areas, Ministry of Education, Guizhou Normal University, Guiyang 550025, China

[4] State Key Laboratory of Herbage Improvement and Grassland Agro-Ecosystems, Lanzhou University, Lanzhou 730000, China; wangjf12@lzu.edu.cn (J.W.); cwang@lzu.edu.cn (C.W.); malik@lzu.edu.cn (K.M.)

[5] Center for Grassland Microbiome, Lanzhou University, Lanzhou 730000, China

[6] Key Laboratory of Grassland Livestock Industry Innovation, Ministry of Agriculture and Rural Affairs, Lanzhou University, Lanzhou 730000, China

[7] Engineering Research Center of Grassland Industry, Ministry of Education, Lanzhou University, Lanzhou 730000, China

[8] College of Pastoral Agriculture Science and Technology, Lanzhou University, Lanzhou 730000, China

[9] Guizhou Baili Dujuan Management Area Forestry Bureau, Bijie 551614, China; yangyongsong2024@163.com

* Correspondence: 201307048@gznu.edu.cn

† These authors contributed equally to this work.

Abstract: The toxicity of cadmium (Cd) not only affects the growth and development of plants but also has an impact on human health. In this study, high-throughput sequencing and LC-MS were conducted to analyze the effect of $CdCl_2$ treatment on the microbial community and soil metabolomics of rhizosphere soil in *Rhododendron decorum* subsp. *diaprepes*. The results showed that $CdCl_2$ treatment reduced the quality of the rhizosphere soil by significantly decreasing the soil organic carbon (SOC) content, urease, and invertase activities, increasing the percentage of the exchangeable Cd fraction. $CdCl_2$ treatment did not significantly change the Chao1 and Shannon indices of bacterial and fungal communities in the rhizosphere soil. *R. decorum* was more likely to recruit Cd-resistant bacteria (e.g., Proteobacteria, Chloroflexi) and increase the abundance of Cd-resistant fungi (e.g., Basidiomycota, Rozellomycota). Moreover, $CdCl_2$ treatment decreased the content of secondary metabolites associated with plants' resistance to Cd. Rhizosphere soil urease, invertase activities, alkaline phosphatase (ALP), SOC, total potassium (TK), Cd, and nitrate nitrogen (NN) were the main drivers of the composition of rhizosphere bacterial and fungal communities. $CdCl_2$ treatment weakened the relationships among bacterial/fungi, differential metabolites, and physicochemical properties in rhizosphere soil.

Keywords: cadmium; soil metal contamination; soil microbial community; rhizosphere; soil metabolome

1. Introduction

Soil contamination caused by cadmium (Cd) is a serious problem worldwide, representing a broadly distributed contaminant in the soil ecosystem [1]. It not only causes nutritional imbalances and deficiencies in plants growing in contaminated soil but also passes through those same plants into the food chain, posing a threat to environmental and food security [2,3].

Microorganisms are essential for the soil ecosystem to maintain biodiversity, control nutrient cycling, and regulate soil fertility [4]. Cd has been reported to inhibit soil microbial diversity and plant growth even at low doses [5–7]. According to Jin et al. (2023), the composition of the rhizosphere's microbial community was significantly negatively affected by Cd [8]. Since the adaptation of microbes to the soil environment influences the accumulation of metals in plant species in contaminated environments, the characteristics of the microbial community structure have been used as markers of soil quality [9,10]. Rhizosphere microorganisms show different responses to Cd-contaminated soil. For example, Actinobacteria tolerate high Cd concentrations found in heavy-metal-contaminated soil [11,12]. In addition, rhizosphere microorganisms stimulate plant growth and enhance resistance to heavy metals through releasing amino acids and plant hormones [13]. In heavy-metal-contaminated soils, the metabolic capacity of microbial communities has been constantly altered to adapt to different environments [14]. Rhizosphere soil metabolites reflect changes in the soil microbial community, as changes experienced at the organism and enzyme levels will manifest as a modified metabolite profile. In addition, under stress, plants play a driving role in regulating the distribution of micro-organisms to ensure metabolite activity [15]. Therefore, rhizosphere soil metabolite communication is a key mechanism of plant–microbe interactions [16].

Rhododendron decorum subsp. *diaprepes* belongs to the Ericaceae and is a variety of *Rhododendron decorum*. *R. decorum* has certain ornamental and ecological restoration value. *R. decorum* is mainly distributed in Yunnan, Guizhou, and Sichuan, China, and usually grows in shrubland or forests at an altitude of 1000–4000 m. More and more researchers have been paying attention to the growth of rhododendrons under alkaline soil conditions, and they select alkali-resistant species to adapt to the growth of an alkaline soil environment. Research has reported that *R. decorum* can survive in heavy-metal-contaminated soil [17,18]. Furthermore, a study showed that Cd stress altered the composition of endophytic microbial communities in *R. decorum* leaves [19]. However, there is limited information on rhizosphere soil microbial communities, metabolites, and physicochemical properties, as well as responses to Cd stress in *R. decorum*.

Therefore, this study aimed to (1) explore the effects of different $CdCl_2$ treatments on rhizosphere soil physicochemical properties in alkaline soil, (2) evaluate the effects of different $CdCl_2$ treatments on the composition of rhizosphere bacterial and fungal communities in alkaline soil and rhizosphere soil metabolomic characteristics in alkaline soil, (3) determine the key factors affecting the changes in the rhizosphere microbial communities under $CdCl_2$ treatment, and (4) investigate the relationships among the rhizosphere bacterial/fungal community, differential metabolites, and physicochemical properties of rhizosphere soil under $CdCl_2$ treatment.

2. Materials and Methods

2.1. Experiments on Plant Growth and Cd Addition

The seeds of *Rhododendron decorum* subsp. *diaprepes* were provided by associate professor Jie Liu from the Key Laboratory of Plant Physiology and Development Regulation, School of Life Sciences, Guizhou Normal University. Seeds with full and evenly sized particles were selected and disinfected by soaking in a 10% NaClO solution for 10 min, followed by rinsing 5 times with sterile water. Then, the seeds were disinfected again with 75% ethanol for 30 s and rinsed another 5 times with sterile water. Later, the seeds were soaked in 600 mg/L gibberellin (GA_3) for 24 h and evenly dispersed in a 90 mm culture dish containing sterile water filter paper with sufficient absorption. The seeds were allowed to germinate in an incubator at a constant temperature of 22 ± 2 °C, with 50% relative humidity, a light cycle of 16 h/8 h (light/dark), and light intensity of 2500 lx. After growing 3 leaves, the seedlings were transplanted into pinnacle peat soil/perlite = (3:1 ratio) and allowed to grow for one and a half years in a culture chamber under the same growth parameters as mentioned above.

The 1.5-year-old robust seedlings with consistent growth status were selected and transplanted to flower pots (top diameter: 92 mm, bottom diameter: 81 mm, height: 80 mm), with 50 g of soil for Cd stress experiments (the soil was obtained from Fengxiang Mountain of Guizhou Normal University, pH 8.05 ± 0.04). There were five biological replicates for each treatment. For $CdCl_2$ treatment, 1.5-year-old robust seedlings were treated with 50 mL of distilled water and $CdCl_2$ once every five days. The soil samples were air-dried and sieved to a 2 mm size. The three treatments were as follows: (1) CK (soil + *R. decorum*); (2) 2.5 mM $CdCl_2$ (1690 mg kg^{-1} Cd^{2+} contaminated soil + *R. decorum*); and (3) 5 mM $CdCl_2$ (3370 mg kg^{-1} Cd^{2+} contaminated soil + *R. decorum*). After 6 Cd treatments, the treatments were stopped, and the seedlings were allowed to grow for another two weeks. After the experiment, the loose soil on the surface of the roots was gently removed, and the soil firmly attached to the roots was collected as rhizosphere soil. After gently shaking away the loosely attached soil, the soil attached to the root system (0–0.5 cm from the root) was carefully brushed to remove the mixed roots in the rhizosphere soil. The rhizosphere soil samples were collected and divided into two parts. A portion of the rhizosphere soil was stored at −80 °C to analyze soil enzyme activities. Another portion was air-dried to measure nutrient contents, pH, and chemical forms of Cd. The separation for roots and rhizosphere soil was carried out as described in our previously published paper [20]. Briefly, the excised roots were placed in a 50 mL sterile centrifuge tube, and 35 mL of sterile phosphate buffer (PBS: 3 mM NaH_2PO_4, 7 mM Na_2HPO_4, 130 mM NaCl, pH 7.4) was added. Then, the sterile centrifuge tubes were shaken for 3 min and centrifuged at 3000× *g* for 5 min to obtain the rhizosphere soil. After the supernatant was discarded, the rhizosphere soil was placed in a new 2 mL sterile centrifuge tube and stored at −80 °C until we extracted DNA.

2.2. Analysis of Rhizosphere Soil's Physical Physicochemical Properties

The rhizosphere soil's available phosphorus (AP) was extracted with 0.5 M $NaHCO_3$ and measured using the molybdenum blue method at 710 nm. The rhizosphere soil's total phosphorus (TP) was extracted HCl and then with 0.5 M $NaHCO_3$. After mixing, the extracts were mineralized with a triple acid mixture of HNO_3-$HClO_4$-H_2SO_4 (10:1:4, *v*/*v*) and determined at 660 nm using a spectrophotometer (made by Shimadzu Co., Ltd., Kyoto, Japan). The rhizosphere soil's total nitrogen (TN) was determined using the Kjeldahl method. Soil ammonium nitrogen (AN) and nitrate nitrogen (NN) were extracted with 1 M KCl and measured by a continuous-flow analyzer (Skalar san++, Skalar, Breda, The Netherlands). Soil organic carbon (SOC) was determined by the dichromate oxidation method. Available potassium (AK) was extracted with 1 M NH_4OAc and determined by inductively coupled plasma mass spectrometry (ICP-MS) (SHIMADZU, Kyoto, Japan, 2030). The rhizosphere soil's total potassium (TK) was digested with HF-$HClO_4$, and TK in the digest was determined by ICP-MS. A pH meter (Mettler Toledo Delta 320, Greifensee, Switzerland) was used to measure the rhizosphere soil's pH (soil/distilled water = 1:2.5). A 5 g rhizosphere soil was used to analyze the activity of invertase, and alkaline phosphatase was measured using 3,5-dinitrosalicylic acid colorimetry and disodium phenyl phosphate was measured using the colorimetric method [21]. The urease activity was measured with 5 g of rhizosphere soil, 1 mL of toluene, 10 mL of 10% urea solution, and 20 mL of citrate buffer (pH 6.7) added to a 50 mL volumetric flask, which was then incubated at 37 °C for 24 h. The reaction mixture was filtered, and 1 mL of the filtered solution was reacted with 3 mL of 0.9% NaClO solution and 4 mL of sodium phenol solution. After 20 min, the mixture was measured with an ultraviolet spectrophotometer (made by Shimadzu Co., Ltd.) at 578 nm. Catalase activity was determined with 2 g of rhizosphere soil, 40 mL of water, and 5 mL of 0.3% H_2O_2, which were thoroughly mixed for 20 min at 150 rpm. The mixture was added to 5 mL of 1.5 mol/L H_2SO_4 and then filtered. Finally, the filtrate was titrated with 0.1 mol/L $KMnO_4$. The activity of catalase was analyzed in soil according to the method of Hackenberger et al. (2018) [22]. Total Cd was extracted with a triple acid mixture of HF-HNO_3:$HClO_4$ (5:5:1, *v*/*v*) and determined by ICP-MS. The Cd fractions were determined according to the method of He et al. (2018) [23]. Five fractions were determined as follows: extractable with 1 M $MgCl_2$ (pH 7) (exchangeable, Ex-Cd), extractable with 1 M NaOAc (pH 5) (carbonate-bound, CB-Cd), extractable with 0.04 M $NH_2OH{\cdot}HCl$ in 25% (*v*/*v*%) CH_3COOH solution (Fe-Mn oxide-bound, OX-Cd), extractable with 0.02 M HNO_3 in 30% H_2O_2 (organic matter-bound, OM-Cd), and HNO_3-HF-$HClO_4$ digested (residual, Res-Cd). The extracted Cd solutions were filtered into 25 mL tubes, diluted with distilled water to volume, and stored at 4 °C. The extracted Cd concentrations were determined by ICP-MS. The validation of ICP-MS measurement results was carried out using working calibration solutions of all investigated Cd ions. These solutions were prepared using appropriate stepwise dilutions of 100 mg L^{-1} certified standard stock solutions. The ICP-MS instrument (SHIMADZU, Kyoto, Japan, 2030) was regulated to measure the samples and the correlation coefficient; the RSE and BEC were 1.00, 3.03% and 0.003, respectively. The recovery percentages between the four steps used to obtain Cd fractions and total Cd (BCR/total) are shown in Table S1. Blanks and standard soil material (GBW07564) from the Chinese Academy of Geological Science were used for quality control.

2.3. DNA Extraction, Amplicon Sequencing, and Bioinformatics Analysis

The total genomic DNA was extracted from 350 mg of rhizosphere soil using the TGuide S96 Magnetic Soil/Stool DNA Kit (Tiangen Biotech (Beijing) Co., Ltd., Beijing, China), according to the manufacturer's instructions. A NanoDrop 2000 UV-Vis spectrophotometer (Thermo Scientific, Wilmington, NC, USA) was used to measure the concentration and purity of the extracted DNA, and electrophoresis on a 1.8% agarose gel was used to evaluate the quality and amount of the material. The V3V4 region of the 16S rRNA and the ITS1 region of the rRNA gene was used with 338F (5′-ACTCCTACGGGAGGCAGCA-3′) and 806R (5′-GGACTACHVGGGTWTCTAAT-3′) primers and ITS1F (5′-CTTGGTCATTTAGAG GAAGTAA-3′) and ITS2 (5′-GCTGCGTTCTTCATCGATGC-3′) primers, respectively. The PCR was carried out in a reaction volume of 10 μL: 4 ng of DNA template, 0.3 μL of forward primer (10 μM), 0.3 μL of reverse primer (10 μM), 5 μL of KOD FX Neo Buffer, 2 μL of dNTP (2 mM each), and 0.2 μL of KOD FX Neo. The remaining volume was adjusted to 20 μL with ddH_2O. The amplified products were quantified using the Qsep-400 (BiOptic, Inc., New Taipei City, Taiwan) and purified with the Omega DNA purification kit (Omega Inc., Norcross, GA, USA). Utilizing an Illumina novaseq6000 (Beijing Biomarker Technologies Co., Ltd., Beijing, China), the amplicon library underwent paired-end sequencing (2 × 250). The raw data processing followed the method described by Jin et al. [8].

2.4. Rhizosphere Soil Metabolomics and Data Analysis

A 50 mg sample was added to 1000 μL of extraction solution containing the inner target. The extraction solution was prepared with methanol, acetonitrile, and water at a ratio of 2:2:1 and an internal standard concentration of 20 mg/L. The solution was thoroughly mixed for 30 s by swirling and vortexing, followed by the addition of steel balls, a 45 Hz grinding machine for 10 min, and ultrasonication for 10 min (ice-water bath). After being stored at −20 °C for an hour, the sample was centrifuged for 15 min at 12,000 rpm and 4 °C. Then, 500 μL of supernatant was carefully removed, transferred to an EP tube, and dried in a vacuum concentrator. The dried metabolites were then redissolved by adding 160 μL of extraction solution (acetonitrile-to-water volume ratio = 1:1), vortexing for 30 s, and ice-water bath sonication for 10 min. The samples were centrifuged at 12,000 rpm for 15 min at 4 °C. Then, we carefully removed 120 μL of supernatant into a 2 mL injection vial; 10 μL of each sample was taken and mixed into the QC sample for automated testing. The chromatographic column used for LC-MS analysis was the Acquire UPLC HSS T3 (1.8 μm, 2.1 × 100 mm), (mobile phase A: 0.1% aqueous formic acid solution; mobile phase B: 0.1% formic acid acetonitrile). ESI ion source parameters and raw data were obtained using the method of Jin et al. [8]. Metabolite data analysis was performed on the BMKCloud platform (www.biocloud.net).

2.5. Statistical Analysis

Statistical analyses of the rhizosphere soil's physicochemical properties were performed using the SPSS software (v27.0.1). The significant differences in the rhizosphere soil's physicochemical properties were determined using Duncan's test in a one-way analysis of variance (ANOVA). Statistical significance was defined as the 95% confidence level, and the data results are expressed as the mean ± standard deviation (mean ± SD) and were plotted using GraphPad Prism software version 8 (GraphPad Software Inc.; San Diego, CA, USA). The numbers of up- and down-regulated differential metabolites in the rhizosphere soil were analyzed by Origin (2021). Based on the Bray–Curtis dissimilarity, PCoA analysis was performed using the R "vegan" package (v3.1.1). The Chao1 and Shannon indices, the relative abundance of rhizosphere bacterial and fungal communities at the phylum and genus levels, and a random forest classification were analyzed using R (v3.1.1). We applied linear discriminant analysis (LDA) of effect size (LEfSe) to identify the differential abundance of rhizosphere bacteria and fungi, with a threshold set to LDA $\geq$ 3.0, $p < 0.05$. Principal component (PCA) analysis and heat map analysis of rhizosphere soil metabolites were carried out in R (v3.1.1). The rhizosphere soil's differential metabolite abundance score map clustering and supervised orthogonal partial least squares analysis (OPLS-DA) were performed using the BMKCloud platform (www.biocloud.net). The co-occurrence network analysis was conducted using the "igraph" package in R and illustrated using Gephi software (v0.9.2). Redundancy analysis (RDA) was conducted using R (v3.1.1) to determine the impact of environmental factors on the composition of rhizosphere bacterial and fungal communities.

3. Results

3.1. The Physicochemical Properties of Rhizosphere Soil at Different $CdCl_2$ Concentrations

Compared with CK, 2.5 and 5 mM $CdCl_2$ had significant toxic effects on *R. decorum* and inhibited the growth of *R. decorum* (Figure 1). After one month of treatment with different concentrations of $CdCl_2$ (CK, 2.5 mM, and 5 mM), there was a noticeable difference in the physicochemical properties of the rhizosphere soil in *R. decorum*. The concentration of $CdCl_2$ had a significant impact on the total Cd content of the rhizosphere soil. The maximum value was reached under 5 mM treatment (Figure 2A). The Cd content of the different chemical forms was the highest in EX-Cd (12.8–39.5%), followed by CB-Cd (25.4–36.6%), OX-Cd (29.6–30.4%), OM-Cd (2.4–16.0%), and Res-Cd (0.5–4.3%) (Figure 2G). The results indicated that the chemical forms of EX-Cd and CB-Cd were dominant in the rhizosphere soil. Under the CK, 2.5 mM and 5 mM $CdCl_2$ treatments, EX-Cd occupied 12.8%, 39.8%, and 39.5%; CB-Cd occupied 36.6%, 25.4%, and 29.6%; OX-Cd occupied 30.4%, 31.5%, and 29.6%; OM-Cd occupied 16.0%, 2.8%, and 2.4%; and Res-Cd occupied 4.3%, 0.5%, and 0.5%, respectively.

Compared with the CK, the addition of $CdCl_2$ did not significantly alter the catalase activity and the contents of TN, AN, NN, TP, and AK in the rhizosphere soil of *R. decorum*, but the TK content was significantly altered by $CdCl_2$ treatment (Figure 3). Compared with CK, the $CdCl_2$ treatments significantly decreased the SOC content and the urease and invertase activities in rhizosphere soil (Figure 3). The AP content only significantly increased under the 2.5 mM treatment compared to CK (Figure 3E). Compared with CK, the 2.5 mM $CdCl_2$ treatment did not significantly change the ALP activity and pH, but they were decreased by the 5 mM $CdCl_2$ treatment (Figure 3J).

Figure 1. The growth of *R. decorum* seedings under CK, 2.5 and 5 mM $CdCl_2$ treatments, respectively.

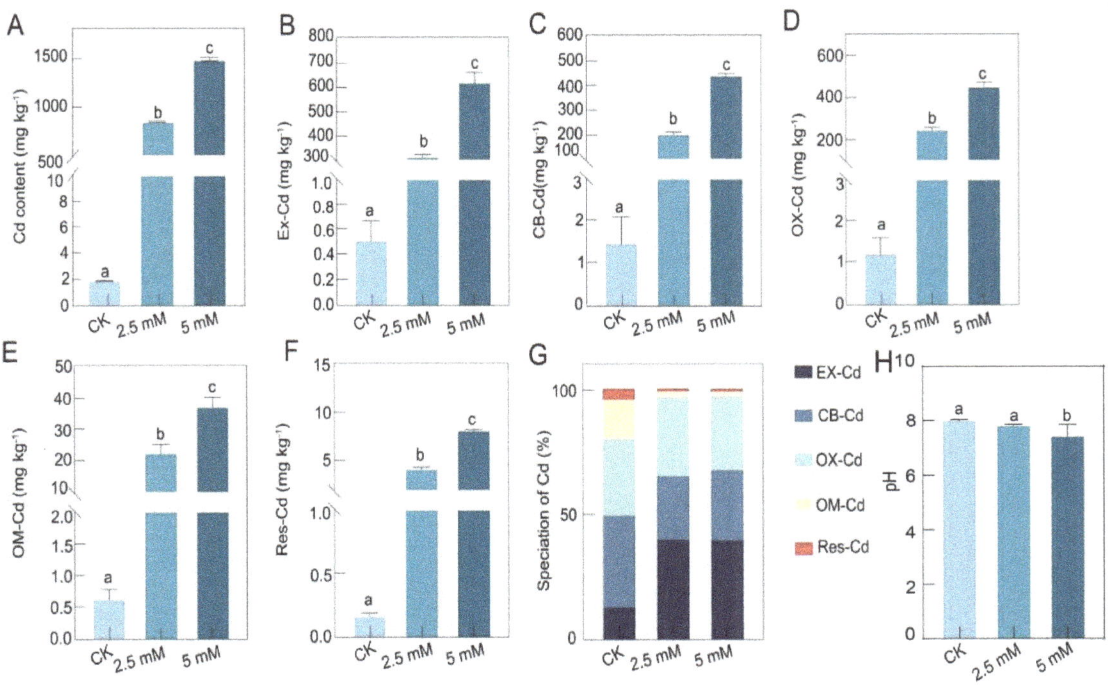

Figure 2. The effects of the different concentrations of $CdCl_2$ on the physicochemical properties of *R. decorum* rhizosphere soil. (**A**) Cd content; (**B**) EX-Cd: exchangeable Cd; (**C**) OB-Cd: carbonate-bound Cd; (**D**) OX-Cd: Fe-Mn oxide-bound Cd; (**E**) OM-Cd: organic matter-bound Cd; (**F**) Res-Cd: residual Cd in rhizosphere soil; (**G**) percentage diagram of Cd forms in rhizosphere soil. (**H**) pH of rhizosphere soil. The data were expressed as the mean ± standard deviation (mean ± SD); the different letters indicate significantly differences ($p < 0.05$ by Duncan's test).

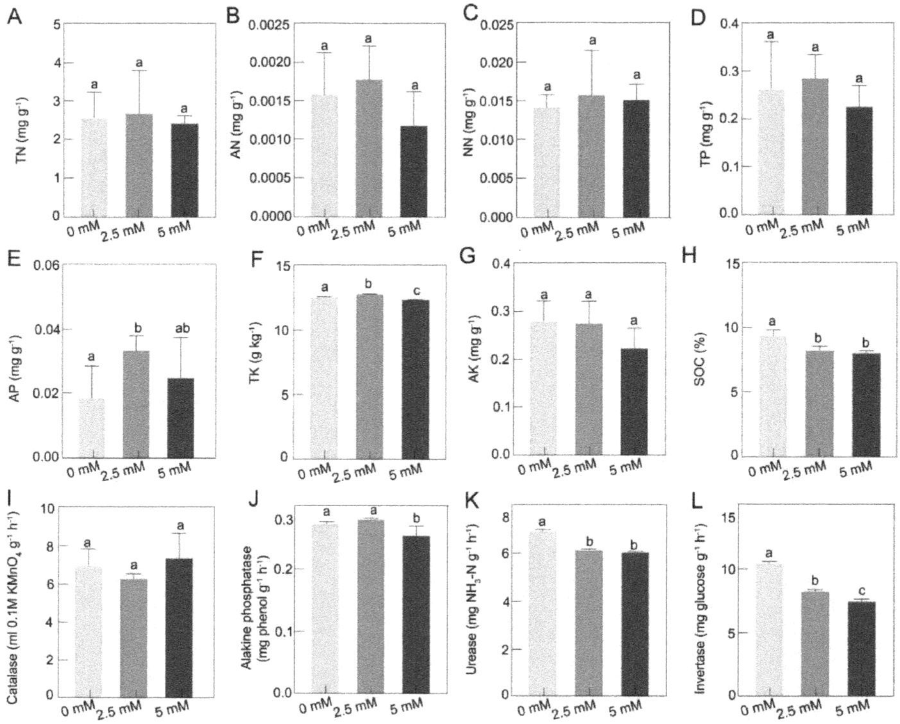

Figure 3. The effects of the different concentrations of $CdCl_2$ on rhizosphere soil properties of *R. decorum*. (**A**) TN: soil's total N, (**B**) AN: ammonium N, (**C**) NN: nitrate N, (**D**) TP: soil's total P, (**E**) AP: plant-available P, (**F**) TK: soil's total potassium, (**G**) AK: ammonium potassium, (**H**) SOC: soil's organic carbon, (**I**) catalase, (**J**) alkaline phosphatase, (**K**) urease, and (**L**) invertase enzymatic activities in rhizosphere soils of *R. decorum* under control and $CdCl_2$ addition treatments. The different letters indicate significantly differences ($p < 0.05$ by Duncan's test).

3.2. Effects of $CdCl_2$ on Diversity and Composition of Rhizosphere Soil Bacterial and Fungal Communities

Our results suggested that the Chao1 and Shannon indices of bacterial and fungal communities in the rhizosphere soil showed no significant difference among all groups (Figure 4A,B,D,E). Furthermore, principal coordinates analysis (PCoA) revealed that the rhizosphere's bacterial and fungal community composition showed differences at different soil Cd content. Moreover, the first and second principal coordinates accounted for 48.9% and 53.9% of the total variation in the rhizosphere bacterial and fungal communities under different treatments, respectively (Figure 4C,F).

The most abundant phyla in the bacterial community of the rhizosphere soil, across all three treatments, were Proteobacteria, Acidobacteriota, and Actinobacteriota, which accounted for over 70% of the total sequences (Figure 5A). We observed that the increase in $CdCl_2$ concentration could significantly affect the relative abundance of Proteobacteria (45.4%, 45.9%, 45.5%), Acidobacteriota (12.1%, 11.8%, 13.0%), Myxococcota (9.5%, 9.9%, 8.8%), Bacteroidota (7.9%, 9.4%, 7.7%), and Chloroflexi (3.8%, 5.3%, 4.0%) in rhizosphere soil (Figure 5A). The relative abundance of the genera *SWB02*, *Haliangium*, *Bauldia*, and *Acidibacter* was the highest under the 2.5 mM $CdCl_2$ treatment (Figure 5B). The relative abundance of genera *Pseudolabrys* and *Streptomyces* was the highest at 5 mM $CdCl_2$ treatment (Figure 5B). As shown by the cladogram, the bacterial genera *Terrimonas*, *Actinospica*, *Dokdonella*, *Pseudolabrys*, and *Streptomyces* were significantly more abundant in rhizosphere soil with $CdCl_2$ treatment (Figure S1A). In addition, the random forest analysis showed that the bacteria Chloroflexi, Bacteroidota, and Chytridiomycota were the key phyla in the CK vs 2.5 mM group, whereas Firmicutes, Fibrobacterota, and Chytridiomycota were the key phyla in the CK vs. 5 mM group (Figure S2).

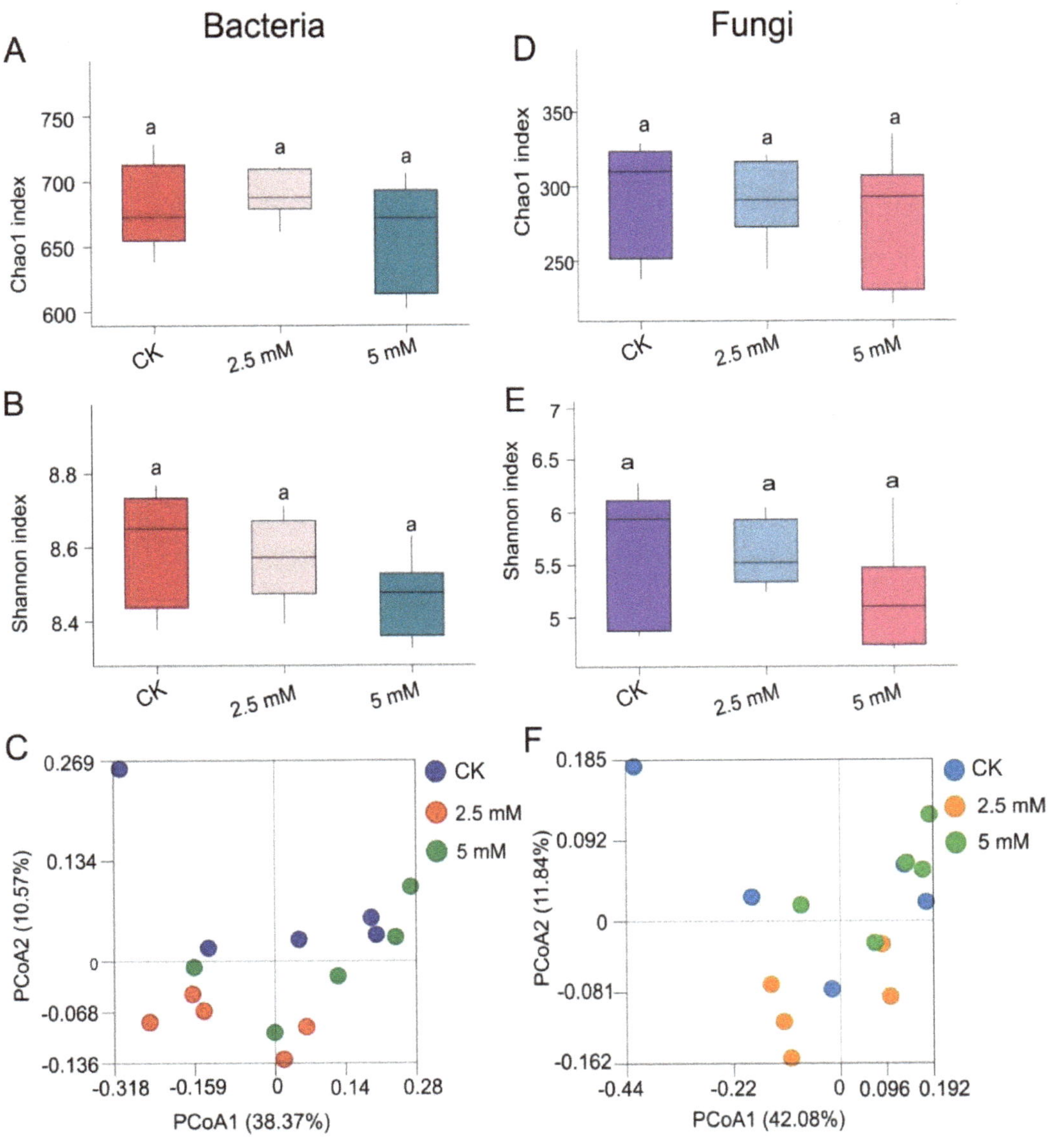

Figure 4. Bacterial and fungal diversity of rhizosphere soil in *R. decorum* under the different concentrations of $CdCl_2$. The α diversity of bacterial community showed by the (**A**) Chao1 index and (**B**) Shannon index. The α diversity of fungal community showed by the (**D**) Chao1 index and (**E**) Shannon index. Principal coordinates analysis (PCoA) of the (**C**) bacterial and (**F**) fungal communities. The different letters indicate significantly differences ($p < 0.05$ by Duncan's test).

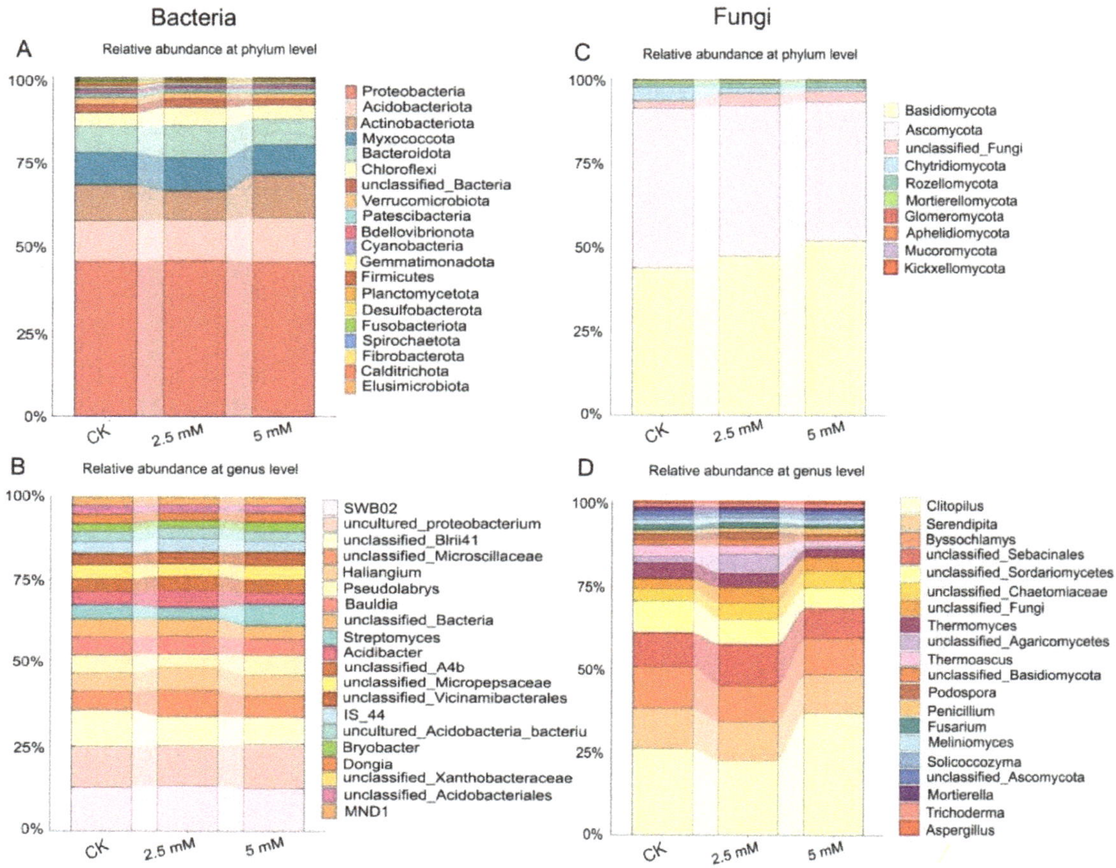

Figure 5. Effects of $CdCl_2$ on the rhizosphere's microbial community. Relative abundance of (**A**) bacterial and (**C**) fungal community composition components at the phylum level. Relative abundance of (**B**) bacterial and (**D**) fungal community composition components at the genus level.

The most abundant phyla in the fungal community of rhizosphere soil, across all three treatments were Basidiomycota, Ascomycota, unclassified_Fungi, and Chytridiomycota, which accounted for over 95% of the total sequences (Figure 5C). $CdCl_2$ treatment increased the relative abundance of Basidiomycota and Rozellomycota, while it decreased the relative abundance of Ascomycota, Chytridiomycota, Mortierellomycota, and Glomeromycota in rhizosphere soil. Compared with CK, the 2.5 mM $CdCl_2$ treatment decreased the relative abundance of *Clitopilus*, *Byssochlamys*, and *Serendipita* in rhizosphere soil, but the 5 mM $CdCl_2$ treatment increased the relative abundance of *Clitopilus* (Figure 5D). As shown by the cladogram, the fungal genera *unclassified_Agaricomycetes*, *Simplicillium*, *Leptodontidium*, *Phialemonium*, and *Penicillium* were significantly enriched in rhizosphere soil under Cd stress (Figure S1B).

For RDA, the rhizosphere soil's physicochemical characteristics and the enzyme activities were selected. Our results showed that the three rhizosphere soil samples could be distinguished from one another (Figure 6), implying that the $CdCl_2$ treatment altered the rhizosphere soil's environment. At the phylum level, the first and second coordinate axes of the RDA explained 35.8% and 22.2% of the total variance in the rhizosphere bacterial and fungal communities, respectively. Rhizosphere soil urease, invertase, and ALP enzymes, along with SOC, TK, Cd, and NN were the main factors in the composition of the rhizosphere bacterial and fungal communities. Most of the phyla, including Chytridiomycota, Verrucomicrobiota, Ascomycota, and Actinobacteriota, had positive correlations with invertase, urease, and catalase enzymes, as well as SOC and pH, while they had negative correlations with Cd, NN, and AP. Myxococcota, Chloroflexi, and Bacteroidota were positively correlated with ALP, TP, AK, TK, AN, and NN, while Myxococcota were negatively correlated with NN and Cd. Basidiomycota, Proteobacteria, and Rozellomycota were positively correlated with NN, Cd, AP, and AN, while they were negatively correlated with pH, SOC, catalase, invertase, and urease.

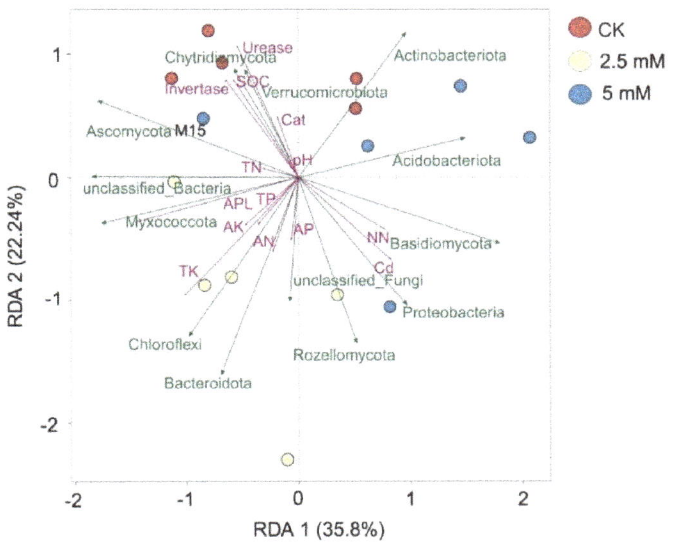

Figure 6. The results of redundancy analysis (RDA) showed the correlation between soil physicochemical properties and enzyme activities and bacteria (top 8) and fungi (top 5) at the phylum level, respectively, in rhizosphere soils of *R. decorum* under different $CdCl_2$ treatment concentrations.

3.3. Measurement of Metabolites in Rhizosphere Soil Treated with $CdCl_2$

To determine the response of rhizosphere soil's metabolic profiles to the different $CdCl_2$ treatments, we used LC-MS to identify the main metabolites in rhizosphere soil. PCA demonstrated that the different $CdCl_2$ treatment concentrations significantly altered the rhizosphere soil's metabolic profiles in *R. decorum*, where PC1 and PC2 explained 49.6% and 17.9% of the total variance, respectively (Figure 7A). We also used the supervised discriminant analysis OPLS-DA model to evaluate the effect of $CdCl_2$ stress on rhizosphere soil metabolic profiles, and the results showed a significant separation between CK and 2.5 mM (R2Y = 1, Q2Y = 1.0) (Figure S3A), as well as between CK and 5 mM (R2Y = 1, Q2Y = 1.0) (Figure S3B). Compared with CK, the 2.5 mM $CdCl_2$ up-regulated 418 differential metabolites (DMs) and down-regulated 699 DMs, while the 5 mM $CdCl_2$ up-regulated 441 DMs and down-regulated 791 DMs (Figure 7B and Figure S5A,B).

Compared with CK, the DMs up-regulated by the 2.5 mM $CdCl_2$ treatment were mainly enriched in the "Aminobenzoate degradation pathway", while down-regulated DMs were mainly enriched in the "Cutin, suberine", "wax biosynthesis" and "Tryptophan metabolism pathway" (Figure S3B). Compared with CK, the DMs up-regulated by the 5 mM $CdCl_2$ treatment were mainly enriched in the "Aminobenzoate degradation pathway", while down-regulated DMs were mainly enriched in the "Purine metabolism" and "Tryptophan metabolism" pathways (Figure S3D). Compared with CK, the 2.5 mM $CdCl_2$ treatment increased the content of taurine, but 2.5 mM and 5 mM $CdCl_2$ treatment decreased the contents of hydroxyspirilloxanthin, 8 (R)-HPETE, trioxilin A3, and farnesylcysteine in rhizosphere soil (Figure S4).

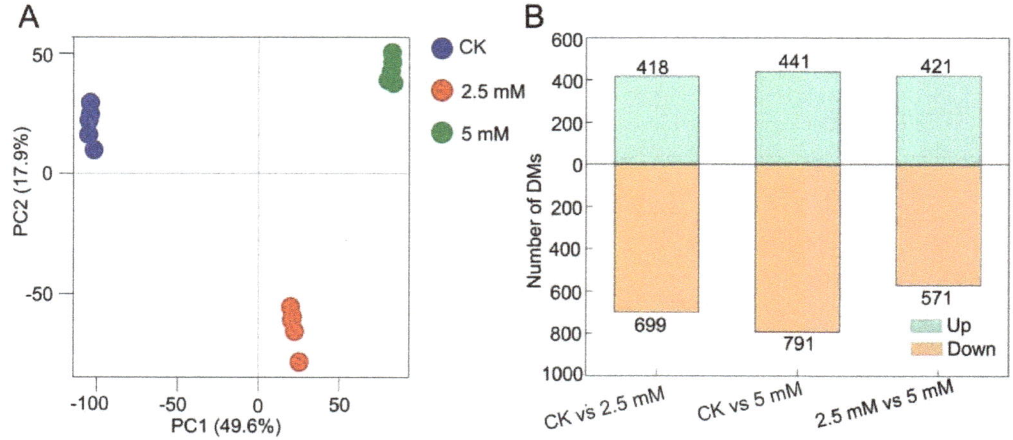

Figure 7. Metabolite profiles in rhizosphere soil under different $CdCl_2$ treatments. (**A**) Principal component analysis of soil metabolites in different treatment groups. (**B**) The numbers of up- and down-regulated differential metabolites (DMs) in comparative groups of CK vs. 2.5 mM, and CK vs. 5 mM.

3.4. Correlations among Microbial Communities, Physicochemical Properties, and Metabolites in Rhizosphere Soil

In the rhizosphere's microbial networks, angelicin, dopaxanthin, epsilon-Caprolactam, dihydroxyflavone, nicotinamide D-ribonucleotide, and L-1-pyrroline-3-hydroxy-5-carboxylate were identified as key differential metabolites that were highly connected to other nodes in CK vs. 2.5 mM $CdCl_2$ treatment (Figure 8A,C), while scolymoside, farnesylcysteine, triethanolamine, L-threonine emerged in CK vs. 5 mM $CdCl_2$ (Figure 8B,D). Moreover, invertase, urease, and ALP activity, along with CB-Cd, AP, OX-Cd, TK, OM-Cd, Res-Cd, pH, EX-Cd, and SOC were the key rhizosphere soil properties to connect other nodes in CK vs. 2.5 mM $CdCl_2$ treatment (Figure 8A,C), while urease and invertase activity, EX-Cd, pH, CB-Cd, OM-Cd, Cd, and TK were identified in CK vs. 5 mM $CdCl_2$ treatment (Figure 8B,D). Additionally, bacterial species hubs affiliated with Actinobacteriota, Chloroflexi, and Bacteroidota were identified in CK vs. 2.5 mM $CdCl_2$ treatment (Figure 8A), while bacterial hubs at the phylum level affiliated with Firmicutes, Actinobacteriota, and Planctomycetota were identified in CK vs. 5 mM $CdCl_2$ treatment (Figure 8B). Meanwhile, the fungal phylum Chytridiomycota was identified as the keystone taxon in samples for CK vs. 2.5 mM $CdCl_2$ treatment and CK vs. 5 mM $CdCl_2$ treatment (Figure 8C,D). The fungal phylum Basidiomycota was the keystone taxon for CK vs. 5 mM $CdCl_2$ treatment (Figure 8D).

As shown in Table 1, in the rhizosphere bacterial and fungal networks, $CdCl_2$ treatments decreased the numbers of total edges, negative edges, connectance, average degree, and diameter, indicating that $CdCl_2$ treatments decreased the complexity of the networks.

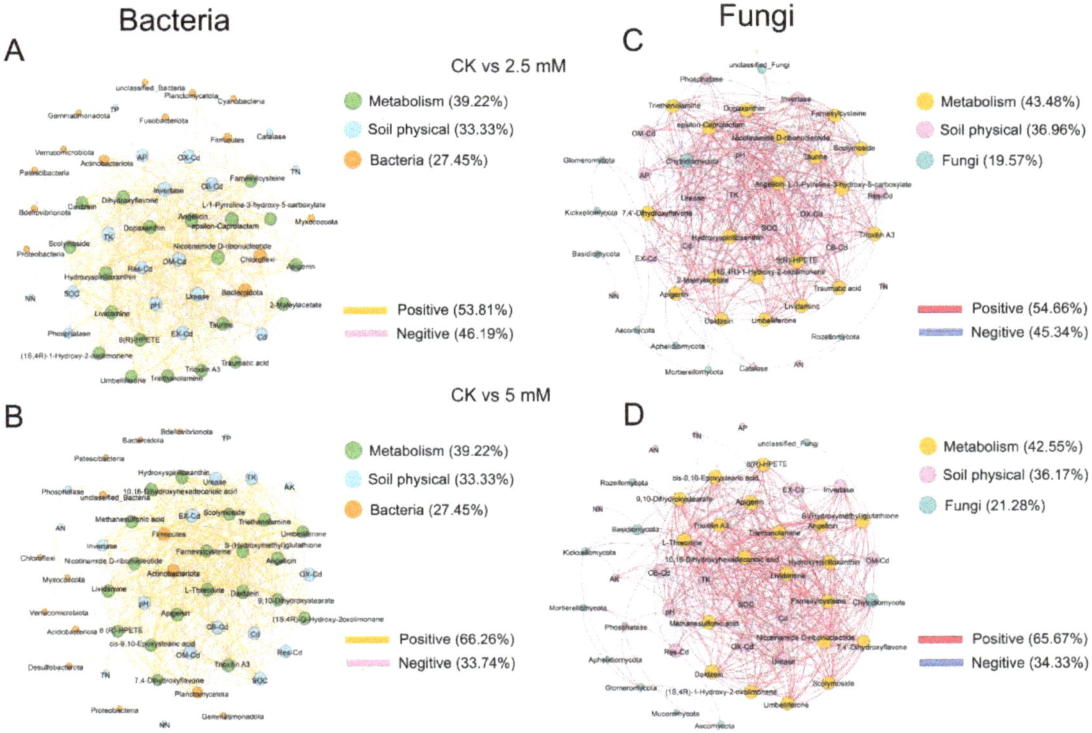

Figure 8. Bacterial and fungal co-occurrence network analysis. Co-occurrence networks were constructed using the rhizosphere soil differential metabolites (top 20), physicochemical properties, and bacteria (top 15 phyla) under (**A**) CK vs. 2.5 mM treatment and (**B**) CK vs. 5 mM treatment. Co-occurrence networks were constructed using the rhizosphere soil differential metabolites (top 20), physicochemical properties, and fungi (top 10 phyla) under (**C**) CK vs. 2.5 mM treatment and (**D**) CK vs. 5 mM treatment. The size of the nodes is proportional to the abundance of these variables. The yellow and red edges represent positive Pearson correlations, while the pink and blue edges represent negative ones. The top 20 metabolites were selected based on their VIP values.

Table 1. Topological features of the co-occurrence networks depicting the interactions between rhizosphere bacterial/fungal taxa, rhizosphere soil physicochemical properties, and metabolites under different Cd stress. Bac: bacteria, Fun: fungi.

		Total Edges	Positive Edges	Negative Edges	Connectance	Average Degree	Average Pathway Length	Diameter	Clustering Coefficient
Bac	CK vs. 2.5 mM	630	339	291	0.49	24.7	1.31	4.79	0.87
	CK vs. 5 mM	569	377	192	0.45	22.3	1.33	4.20	0.90
Fun	CK vs. 2.5 mM	558	305	253	0.54	24.3	0.96	2.88	0.90
	CK vs. 5 mM	533	350	183	0.49	22.7	1.15	2.87	0.89

4. Discussion

Microorganisms are recognized as perturbation sensors, being much more sensitive to environmental stress than macro-organisms [24,25]. They make crucial contributions to environmental adaptation, but there has been little research on the response of the rhizosphere environment and metabolites to Cd stress, especially the bacterial and fungal communities associated with rhizosphere soil. In this study, the prime objective was to explore the effects of Cd stress on the composition of rhizosphere microbial communities and rhizosphere soil metabolites in *R. decorum*.

Cd leads to a change in rhizosphere soil's plant environment, which in turn affects plant growth. Boampong reported that Cd treatment (1000 mg kg^{-1}) decreased the dry shoot weight by 39%, 45.8%, and 49.3% in *Acacia saligna*, *Eucalyptus rostrate*, and *Conocarpus erectu*, respectively [26]. A study reported that a 500 μM Cd nutrient solution clearly inhibited the physiological and biochemical process in *K. paniculata*, caused membrane lipid peroxidation and severe membrane damages, and increased MDA and H_2O_2 contents [27]. Previous studies have reported that $CdCl_2$ treatment resulted in significant enrichment of Cd^{2+} in the soil [8,28]. In this study, 2.5 and 5 mM $CdCl_2$ treatments had significant toxic effects on *R. decorum* and inhibited the growth of *R. decorum* (Figure 1), suggesting that $CdCl_2$ treatment might inhibit the physiological, biochemical, and molecular process in *R. decorum*. In addition, the biological toxicity of heavy metals in soil is not only determined by their total content but is also closely related to their chemical forms [29]. Soil pH significantly affects the availability of heavy metals in soil [30]. Under alkaline conditions, Cd is present as $CdHCO^{3+}$ or $CdCO_3$, which is conducive to the adsorption of heavy metal ions in soil and reduces the availability of Cd in soil [31]. Jin et al. (2023) reported that $CdCl_2$ decreased the pH in *A. inebrians* rhizosphere soil [8], which was consistent with our results. The decrease in pH results in more H^+ to exchange with the Cd absorbed by soil colloids, increasing the Cd^{2+} content in the root–soil system, thereby increasing Cd's mobility [32]. A decrease in soil pH will increase the transformation of Cd from a stable form (e.g., carbonates, Fe, and Mn oxide-bound forms) to an effective bioavailable (e.g., exchangeable form) form [33]. In addition, the possible risks of Cd transfer from rhizosphere soil to plants can be divided into three categories: high-risk, including EX-Cd (exchangeable fraction) and CB-Cd (carbonate-bound fraction); medium-risk, including OX-Cd (Fe/Mn oxides-bound fraction) and OM-Cd (organics fraction); and low-risk, including Res-Cd (residual fraction) [34]. In this study, $CdCl_2$ increased the percentage of the exchangeable Cd fraction but decreased the percentages of the organic matter-bound and residual fractions in the rhizosphere soil, indicating that $CdCl_2$ increased the effective bioavailability in *R. decorum* rhizosphere soil. These results are consistent with the findings of Dong et al. (2016) on rice under Cd stress [35]. In addition, the recovery percentages of Cd were 94% and 105% under 2.5 and 5 mM CdCl2 treatments, respectively. The results are in accordance with the specification, and the test results are reliable. However, in the CK treatment, the recovery percentage of Cd was 197%, which might be due to contamination of the container used during the operation, and the error could be reduced by increasing the number of replicates of experiments.

Soil enzyme activities are very sensitive to heavy metal pollution and are involved in functions related to soil C, N, and P cycling [36]. In rhizosphere soil, $CdCl_2$ treatment significantly decreased the SOC content, urease, and invertase activity. SOC plays an important role in providing carbon sources, promoting soil fertility and contributing to microorganism energy [37]. Therefore, the reduction in SOC affected soil quality. A study reported that Cd pollution reduces the distribution of carbon in roots and leads to a decrease in root exudation, resulting in a decrease in soil organic carbon content [38]. Moreover, the increases in Basidiomycota abundance under Cd stress were associated with increased lignin degradation, releasing SOC [39]. Xiao et al. (2023) reported that SOC loss is caused by increasing the abundance of Basidiomycota that promotes carbon degradation [40], which is consistent with our findings. Soil invertase and urease are key enzymes in the soil–plant C and N cycles [41,42]. Previous studies have shown that heavy metals decrease the activities and community composition of soil microorganisms, thereby affecting the synthesis of soil enzymes when the soil is contaminated by heavy metals [43]. In addition, the decrease in enzyme activity may be due to the interaction of heavy metals with enzyme–substrate complexes, denaturing the enzyme proteins or interacting with the protein active moiety [44].

As a hazardous heavy metals, Cd seriously jeopardizes the community and activity of soil microorganisms [45]. In addition, the effects of different Cd fractions on microorganisms vary significantly [46]. The variety and quantity of sensitive bacteria are adversely affected in heavy-metal-contaminated soils, whereas bacteria that are resistant to the metals are able to adapt and proliferate, increasing their abundance [14,47,48]. Microbial diversity is a crucial indicator for assessing the functioning of ecosystem [49]. In this study, Cd pollution had a negative impact on fungal and bacterial diversity (according to the Shannon and Chao1 indices) (Figure 4). The PCoA results indicated that $CdCl_2$ changed the composition of rhizosphere microbial communities (Figure 4). Microorganisms possess robust metabolic and physiological adaptations to changing environmental conditions, enabling microbial communities to withstand adverse environments [50]. Studies have shown that the structure of the microbial community is extremely sensitive to changes in the toxicity of heavy metals in the soil [16,51]. In this study, $CdCl_2$ increased the relative abundance of Proteobacteria and Chloroflexi, which might be a way for bacteria to resist Cd in *R. decorum* rhizosphere soil. Hou et al. (2018) reported that Proteobacteria were significantly enriched in *Sedum alfredii* rhizosphere soil under high Cd stress [52]. It has been reported that Proteobacteria are the dominant bacterial phylum in areas with severe heavy metal contamination [53–55]. Proteobacteria are important for the carbon and nitrogen cycles [55]. The present findings are consistent with those of Zhao et al. (2019), who discovered that the relative abundance of Proteobacteria in the microbial communities of soil with long-term heavy metal pollution, was inversely linked with soil AK [14]. Previous studies have demonstrated that Chloroflexi is the dominant phylum in heavy-metal-contaminated soils. This is due to the fact that bacteria belonging to this phylum are tolerant and resistant to heavy metal stress [56–58]. Meanwhile, members of Chloroflexi are oligotrophic bacteria that prefer to live in nutrient-poor soil habitats [59–62]. Therefore, Chloroflexi could survive and thrive better than other bacteria in heavy-metal-contaminated areas characterized by low soil nutrient levels [63]. For fungal communities, with the increase in $CdCl_2$ concentration, the relative abundance of Basidiomycota and Rozellomycota increased in the rhizosphere soil. Basidiomycota play a key role in preventing the migration of Cd from the soil to the plants [64]. Basidiomycota are filamentous fungi that have been used to remove heavy metals [65]. In addition, Rozellomycota exhibits a high degree of tolerance to heavy metals [66]. Rozellomycota is able to directly obtain nutrients from the organic matter in the environment through phagocytosis [67]. Rhizosphere soil urease, invertase activities, ALP, SOC, TK, Cd, and NN were the main drivers of the composition of bacterial and fungal communities. In addition, the nutrients (TN, AN, NN, TP, AP, and TK) had a positive correlation with Myxococcota and Bacteroidota and Chloroflexi but had a negative correlation with Acidobacteriota and Actinobacteriota. This may be attributed to the observation that rhizosphere soil nutrients could influence microbial abundance [68]. The report suggests that competition among microorganisms for nutrients may result in a decreased abundance of some microorganisms [69].

Plant and rhizosphere soil microbes respond to environmental stress by regulating their metabolic processes [43]. In this study, Cd exposure reduced the content of secondary metabolites (SMs) in rhizosphere soil, such as hydroxyspirilloxanthin, 8(R)-HPETE, farnesylcysteine, trioxilin A3, and lividamine. The accumulation of SMs is thought to be how plants adapt and protect against environmental stress. The accumulation of SMs is considered to be a way for plants to adapt and protect themselves against heavy metal stress [70]. One study has shown a reduction in SMs with increasing stress time or intensity in rhizosphere soil [71]. Under this situation, plants expend all of their energy to sustain life instead of synthesizing SMs [72]. Additionally, rhizosphere metabolism may shape specific rhizosphere microbial communities through changes in the rhizosphere soil's physicochemical properties [73–75]. In addition to changes in rhizosphere soil's microbial communities, physicochemical properties, and metabolites, *R. decorum*-mediated co-occurrence networks among these variables were clearly influenced by Cd stress. In this study, $CdCl_2$ treatment decreased the number of total edges, negative edges, connectance, average degree, and diameter of bacterial and fungal co-occurrence networks, indicating that Cd decreased the complexity of the co-occurrence networks (Figure 8). It has been reported that rhizosphere microbial communities with simple networks exhibit less resistance to abiotic stress than those with complex ones [76,77]. Reduced network complexity leads to reduced ecological community stability and impaired ecosystem functioning [78,79]. Different microbial taxa may complement one another through metabolites under Cd stress, thereby improving their tolerance to adverse environmental conditions. Farnesylcysteine is one of the key rhizosphere soil metabolites nodes in bacterial and fungal networks belonging to SMs. Plants exposed to Cd stress exhibit higher levels of SMs, which help them against Cd stress [43]. As the key bacterial node, Actinobacteriota is beneficial to plant growth [80]. Meanwhile, the fungal phylum Chytridiomycota contributes to nutrient cycling and the flow of energy [81].

5. Conclusions

In this study, we found that $CdCl_2$ treatment significantly increased the total Cd content and the percentage of exchangeable Cd fraction in rhizosphere soil, while it decreased the SOC content and the urease and invertase activities, reducing the quality of the rhizosphere soil. $CdCl_2$ treatment did not significantly change the Chao1 and Shannon indices of the rhizosphere bacterial and fungal communities. *R. decorum* was more likely to recruit beneficial bacteria (Proteobacteria and Chloroflexi) to resist Cd stress and increased the relative abundance of Cd-resistant fungi (Basidiomycota and Rozellomycota) in rhizosphere soil. Rhizosphere soil urease and invertase activities, ALP, SOC, TK, Cd, and NN, were the main drivers of the composition of the bacterial and fungal communities. The random forest analysis showed that the bacterial Firmicutes and the fungal Chytridiomycota contributed the most. Moreover, $CdCl_2$ treatment decreased the contents of SMs in rhizosphere soil, such as hydroxyspirilloxanthin, farnesylcysteinen, and lividamine. Furthermore, $CdCl_2$ treatment decreased the complexity of the co-occurrence network of rhizosphere soil's differential metabolites, physicochemical properties, and bacteria/fungi. This study could enhance the understanding of the rhizosphere's microbiological mechanisms of *R. decorum*'s adaptation to Cd stress.

Supplementary Materials: The following supporting information can be downloaded at https://www.mdpi.com/article/10.3390/horticulturae10080884/s1, Figure S1: Analysis of LDA effect size (LEfSe) of soil bacteria in three treatment groups; LDA > 3.0, and the classification label is genus. Branch diagram of (A) bacterial and (B) fungal species evolution under different $CdCl_2$ concentrations, and histogram of LDA value distribution for different species. Figure S2: A random forest was used to analyze characteristic species with significant effects on phylum level bacteria in (A) CK vs. 2.5 mM and (B) CK vs. 5 mM, and fungi in (C) CK vs 2.5 mM and (D) CK vs. 5 mM. Figure S3: OPLS-DA score map showing the separation of metabolites in rhizosphere soil treated with Cd in comparison groups of (A) CK vs. 2.5 mM, and (C) CK vs 5 mM. A differential metabolite abundance score map revealed the differential metabolite pathways in the (B) CK vs. 2.5 mM and (D) CK vs. 5 mM $CdCl_2$ comparative treatment groups. Figure S4. Effects of CdCl2 on rhizosphere soil metabolites: The different rhizosphere soil metabolites (top 20) were identified (A) between CK and the 2.5 mM treatment and (B) between CK and the 5 mM treatment. Figure S5: The volcano map shows differences in metabolites for (A) CK vs. 2.5 mM and (B) CK vs. 5 mM. All identified metabolite data were plotted in a $\log_2$ (FC) versus-$\log_{10}$ (p-value) relationship. Table S1: Sequential extraction fractions of Cd.

Author Contributions: Conceptualization, J.G. and M.T.; methodology, J.G., M.T. and L.C.; validation, J.G., M.T., L.W. and C.W.; formal analysis, J.G., Y.Y. (Yin Yi) and J.W.; investigation, J.G., M.T. and X.C.; resources, J.G., J.L. and Y.Y. (Yongsong Yang); data curation, M.T., L.C. and L.W.; writing—original draft preparation, M.T. and J.G.; writing—review and editing, M.T. and K.M.; project administration, J.G.; funding acquisition, J.G. and Y.Y. (Yin Yi). All authors have read and agreed to the published version of the manuscript.

Funding: This research was supported by the Joint Fund of the Natural Science Foundation of China and the Karst Science Research Center of Guizhou Province (Grant No. U1812401), Guizhou

Provincial Science and Technology Plan Project (Qian Ke He Cheng Guo [2022] Zhong Dian010), Guizhou forestry scientific research project (Qianlinkehe [2022] No. 28), Karst Rocky Desertification Water-Fertilizer Coupling and Biodiversity Restoration (Guizhou Education Technology [2023]004), Changjiang Scholars and Innovative Research Team in University (IRT_17R50), the Natural Science Foundation of Gansu Province (22JR5RA451, 23JRRA1093), the Fundamental Research Funds for the Central Universities (lzujbky-2021-ey01) in Lanzhou University, and the China Postdoctoral Science Foundation (Grant No. 2023M731470).

Data Availability Statement: Data are contained within the article and Supplementary Materials.

Conflicts of Interest: The authors declare that they have no competing interests.

References

1. Mahar, A.; Ali, A.; Lahori, A.H.; Wahid, F.; Li, R.; Azeem, M.; Zhang, Z. Promising technologies for Cd-contaminated soils: Drawbacks and possibilities. In *Environment, Climate, Plant and Vegetation Growth*; Springer: Cham, Switzerland, 2020; pp. 63–91. [CrossRef]
2. Adrees, M.; Ali, S.; Rizwan, M. The effect of excess copper on growth and physiology of important food crops: A review. *Environ. Sci. Pollut. Res. Int.* **2015**, *22*, 8148–8162. [CrossRef] [PubMed]
3. Genchi, G.; Sinicropi, M.S.; Lauria, G.; Carocci, A.; Catalano, A. The Effects of Cadmium Toxicity. *Int. J. Environ. Res. Public Health* **2020**, *17*, 3782. [CrossRef]
4. Ramakrishna, W.; Yadav, R.; Li, K. Plant growth promoting bacteria in agriculture: Two sides of a coin. *Appl. Soil. Ecol.* **2019**, *138*, 10–18. [CrossRef]
5. Bertin, G.; Averbeck, D. Cadmium: Cellular effects, modifications of biomolecules, modulation of DNA repair and genotoxic consequences (a review). *Biochimie* **2006**, *88*, 1549–1559. [CrossRef]
6. Rizwan, M.; Ali, S.; Abbas, T. Cadmium minimization in wheat: A critical review. *Ecotoxicol. Environ. Saf.* **2016**, *130*, 43–53. [CrossRef] [PubMed]
7. Rizwan, M.; Ali, S.; Adrees, M. A critical review on effects, tolerance mechanisms and management of cadmium in vegetables. *Chemosphere* **2017**, *182*, 90–105. [CrossRef] [PubMed]
8. Jin, J.; Wang, C.; Liu, R. Soil microbial community compositions and metabolite profiles of *Achnatherum inebrians* affect phytoremediation potential in Cd contaminated soil. *J. Hazard. Mater.* **2023**, *459*, 132280. [CrossRef] [PubMed]
9. Schloter, M.; Dilly, O.; Munch, J.C. Indicators for evaluating soil quality. *Agric. Ecosyst. Environ.* **2003**, *98*, 255–262. [CrossRef]
10. Sessitsch, A.; Kuffner, M.; Kidd, P. The role of plant-associated bacteria in the mobilization and phytoextraction of trace elements in contaminated soils. *Soil Biol. Biochem.* **2013**, *60*, 182–194. [CrossRef]
11. Hema, T.G.; Getha, K.; Tan, G.; Sahira, H.L.; Syamil, A.M.; Fairuz, M.Y. Actinobacteria isolates from tin tailings and forest soil for bioremediation of heavy metals. *J. Trop. For. Sci.* **2014**, *26*, 153–162.
12. Sathya, A.; Vijayabharathi, R.; Gopalakrishnan, S. Plant growth-promoting actinobacteria: A new strategy for enhancing sustainable production and protection of grain legumes. *3 Biotech* **2017**, *7*, 102. [CrossRef]
13. Chanclud, E.; Morel, J.B. Plant hormones: A fungal point of view. *Mol. Plant Pathol.* **2016**, *17*, 1289–1297. [CrossRef]
14. Zhao, X.; Huang, J.; Lu, J.; Sun, Y. Study on the influence of soil microbial community on the long-term heavy metal pollution of different land use types and depth layers in mine. *Ecotoxicol. Environ. Saf.* **2019**, *170*, 218–226. [CrossRef] [PubMed]
15. Korenblum, E.; Dong, Y.; Szymanski, J. Rhizosphere microbiome mediates systemic root metabolite exudation by root-to-root signaling. *Proc. Natl. Acad. Sci. USA* **2020**, *117*, 3874–3883. [CrossRef]
16. Wang, Y.; Ren, W.; Li, Y.; Xu, Y.; Teng, Y.; Christie, P.; Luo, Y. Nontargeted metabolomic analysis to unravel the impact of di (2-ethylhexyl) phthalate stress on root exudates of alfalfa (*Medicago sativa*). *Sci. Total Environ.* **2019**, *646*, 212–219. [CrossRef] [PubMed]
17. Zu, Y.; Li, Y.; Christian, S.; Laurent, L.; Liu, F. Accumulation of Pb, Cd, Cu and Zn in plants and hyperaccumulator choice in Lanping lead-zinc mine area, China. *Environ. Int.* **2004**, *30*, 567–576. [CrossRef]
18. Tian, W.; Zhang, C.Q.; Qiao, P.; Milne, R. Diversity of culturable ericoid mycorrhizal fungi of *Rhododendron decorum* in Yunnan, China. *Mycologia* **2011**, *103*, 703–709. [CrossRef] [PubMed]
19. Gong, J.; Wang, C.; Wang, J.; Yang, Y.; Kong, X.; Liu, J.; Yi, Y. Integrative study of transcriptome and microbiome to reveal the response of Rhododendron decorum to cadmium stress. *Ecotoxicol. Environ. Saf.* **2024**, *280*, 116536. [CrossRef] [PubMed]
20. Tang, M.; Liu, J.; Hou, W.P.; Stubbendieck, R.M.; Xiong, H.; Jin, J.; Gong, J.Y.; Cheng, C.; Tang, X.X.; Liu, Y.L.; et al. Structural variability in the bulk soil, rhizosphere, and root endophyte fungal communities of *Themeda japonica* plants under different grades of karst rocky desertification. *Plant Soil* **2022**, *475*, 105–122. [CrossRef]
21. Hou, W.; Wang, J.; Nan, Z.; Christensen, M.J.; Xia, C.; Chen, T.; Niu, X. *Epichloë* gansuensis endophyte-infection alters soil enzymes activity and soil nutrients at different growth stages of *Achnatherum inebrians*. *Plant Soil* **2020**, *455*, 227–240. [CrossRef]
22. Hackenberger, D.K.; Palijan, G.; Lončarić, Ž.; Jovanović Glavaš, O.; Hackenberger, B.K. Influence of soil temperature and moisture on biochemical biomarkers in earthworm and microbial activity after exposure to propiconazole and chlorantraniliprole. *Ecotoxicol. Environ. Saf.* **2018**, *148*, 480–489. [CrossRef]

23. He, D.; Cui, J.; Gao, M. Effects of soil amendments applied on cadmium availability, soil enzyme activity, and plant uptake in contaminated purple soil. *Sci. Total Environ.* **2019**, *654*, 1364–1371. [CrossRef] [PubMed]
24. Giller, K.E.; Witter, E.; Steve, P. McGrath. Heavy metals and soil microbes. *Soil Biol. Biochem.* **2009**, *41*, 2031–2037. [CrossRef]
25. Khan, J.; Wang, L.; Zhu, L.; Wang, J. Individual and combined effects of enrofloxacin and cadmium on soil microbial biomass and the ammonia-oxidizing functional gene. *Sci. Total Environ.* **2018**, *624*, 900–907. [CrossRef]
26. Boampong, E.W. Uptake of Pb and Cd by Plant Species: A Review. *Int. J. Innov. Sci. Res. Technol.* **2020**, *5*, 306–310. [CrossRef]
27. Yang, L.P.; Zhu, J.; Wang, P. Effect of Cd on growth, physiological response, Cd subcellular distribution and chemical forms of *Koelreuteria paniculata*. *Ecotoxicol. Environ. Saf.* **2018**, *160*, 10–18. [CrossRef] [PubMed]
28. Cao, Y.; Du, P.; Zhang, J.; Ji, J.; Xu, J.; Liang, B. Dopamine alleviates cadmium stress in apple trees by recruiting beneficial microorganisms to enhance the physiological resilience revealed by high-throughput sequencing and soil metabolomics. *Hortic. Res.* **2023**, *10*, uhad112. [CrossRef]
29. Arfania, H.; Samadi, A.; Asadzadeh, F.; Sepehr, E.; Jaisi, D. Distribution of phosphorous pools in western river sediments of the Urmia Lake basin, Iran. *Environ. Sci. Pollut. Res. Int.* **2018**, *25*, 1614–11625. [CrossRef]
30. Rafiq, M.T.; Aziz, R.; Yang, X.; Xiao, W.; Rafiq, M.K.; Ali, B.; Li, T. Cadmium phytoavailability to rice (*Oryza sativa* L.) grown in representative Chinese soils. A model to improve soil environmental quality guidelines for food safety. *Ecotoxicol. Environ. Saf.* **2014**, *103*, 101–107. [CrossRef] [PubMed]
31. Hussain, B.; Ashraf, M.N.; Shafeeq-Ur-Rahman; Abbas, A.; Li, J.; Farooq, M. Cadmium stress in paddy fields: Effects of soil conditions and remediation strategies. *Sci. Total Environ.* **2021**, *754*, 142188. [CrossRef]
32. Li, Y.; Liang, L.; Li, W.; Ashraf, U.; Ma, L.; Tang, X. ZnO nanoparticlebased seed priming modulates early growth and enhances physio-biochemical and metabolic profiles of fragrant rice against cadmium toxicity. *J. Nanobiotechnol.* **2021**, *19*, 75. [CrossRef]
33. Li, W.; Xu, B.; Song, Q.; Liu, X.; Xu, J.; Brookes, P.C. The identification of 'hotspots' of heavy metal pollution in soil-rice systems at a regional scale in eastern China. *Sci. Total Environ.* **2014**, *472*, 407–420. [CrossRef]
34. Cui, J.; Wang, W.; Peng, Y. Effects of simulated Cd deposition on soil Cd availability, microbial response, and crop Cd uptake in the passivation-remediation process of Cd-contaminated purple soil. *Sci. Total Environ.* **2019**, *683*, 782–792. [CrossRef]
35. Dong, M.; Feng, R.; Wang, R.; Sun, Y.; Ding, Y.; Xu, Y.; Fan, Z.; Guo, J. Inoculation of Fe/Mn-oxidizing bacteria enhances Fe/Mn plaque formation and reduces Cd and As accumulation in Rice Plant tissues. *Plant Soil* **2016**, *404*, 75–83. [CrossRef]
36. Aponte, H.; Medina, J.; Butler, B.; Meier, S.; Cornejo, P.; Kuzyakov, Y. Soil quality indices for metal(loid) contamination: An enzymatic perspective. *Land Degrad. Dev.* **2020**, *31*, 2700–2719. [CrossRef]
37. Benbi, D.K.; Brar, K.; Toor, A.S.; Singh, P. Total and labile pools of soil organic carbon in cultivated and undisturbed soils in Northern India. *Geoderma* **2015**, *237–238*, 149–158. [CrossRef]
38. Jia, X.; Zhao, Y.; He, Y. Glomalin-related soil protein in the rhizosphere of *Robinia pseudoacacia* L. seedlings under higher air temperature combined with Cd-contaminated soil. *Eur. J. Soil Sci.* **2018**, *69*, 634–645. [CrossRef]
39. Kabuyah, R.N.T.M.; van Dongen, B.E.; Bewsher, A.D.; Robinson, C.H. Decomposition of lignin in wheat straw in a sand-dune grassland. *Soil Biol. Biochem.* **2012**, *45*, 128–131. [CrossRef]
40. Xiao, Z.; Duan, C.; Li, S. The microbial mechanisms by which long-term heavy metal contamination affects soil organic carbon levels. *Chemosphere* **2023**, *340*, 139770. [CrossRef]
41. Adetunji, A.; Lewu, F.; Mulidzi, R.; Ncube, B. The biological activities of β-glucosidase, phosphatase and urease as soil quality indicators: A review. *J. Soil Sci. Plant Nutr.* **2017**, *17*, 794–807. [CrossRef]
42. Bai, Y.; Li, F.; Yang, G.; Shi, S.; Dong, F.; Liu, M.; Hai, J. Meta-analysis of experimental warming on soil invertase and urease activities. *Acta Agric. Scand. Sect. B—Soil Plant Sci.* **2018**, *68*, 104–109. [CrossRef]
43. Wang, Y.; Nie, M.; Diwu, Z.; Chang, F.; Nie, H.; Zhang, B.; Yin, Q. Toxicity evaluation of the metabolites derived from the degradation of phenanthrene by one of a soil ubiquitous PAHs-degrading strain *Rhodococcus qingshengii* FF. *J. Hazard. Mater.* **2021**, *415*, 125657. [CrossRef]
44. Pan, J.; Yu, L. Effects of Cd or/and Pb on soil enzyme activities and microbial community structure. *Ecol. Eng.* **2011**, *37*, 1889–1894. [CrossRef]
45. Ru, S.; Xing, J.; Su, D. Rhizosphere cadmium speciation and mechanisms of cadmium tolerance in different oilseed rape species. *J. Plant Nutr.* **2006**, *29*, 921–932. [CrossRef]
46. Pan, X.; Zhang, S.; Zhong, Q. Effects of soil chemical properties and fractions of Pb, Cd, and Zn on bacterial and fungal communities. *Sci. Total Environ.* **2020**, *715*, 136904. [CrossRef] [PubMed]
47. Xu, Y.; Seshadri, B.; Sarkar, B.; Wang, H.; Rumpel, C.; Sparks, D. Biochar modulates heavy metal toxicity and improves microbial carbon use efficiency in soil. *Sci. Total Environ.* **2018**, *621*, 148–159. [CrossRef] [PubMed]
48. Guo, H.; Nasir, M.; Lv, J.; Dai, Y.; Gao, J. Understanding the variation of microbial community in heavy metals contaminated soil using high throughput sequencing. *Ecotoxicol. Environ. Saf.* **2017**, *144*, 300–306. [CrossRef]
49. Liu, H.; Wang, C.; Xie, Y.; Luo, Y.; Sheng, M.; Xu, F.; Xu, H. Ecological responses of soil microbial abundance and diversity to cadmium and soil properties in farmland around an enterprise-intensive region. *J. Hazard. Mater.* **2020**, *392*, 122478. [CrossRef]
50. Meyer, A.; Lipson, D.; Martin, A.; Schadt, C.; Schmidt, S. Molecular and metabolic characterization of cold-tolerant alpine soil *Pseudomonas sensu stricto*. *Appl. Environ. Microb.* **2004**, *70*, 483–489. [CrossRef] [PubMed]
51. Wu, B.; Hou, S.; Peng, D.; Wang, Y.; Wang, C.; Xu, F.; Xu, H. Response of soil micro-ecology to different levels of cadmium in alkaline soil. *Ecotoxicol. Environ. Saf.* **2018**, *166*, 116–122. [CrossRef]

2. Hou, D.; Lin, Z.; Wang, R. Cadmium Exposure-Sedum alfredii Planting Interactions Shape the Bacterial Community in the Hyperaccumulator Plant Rhizosphere. *Appl. Environ. Microbiol.* **2018**, *84*, e02797-17. [CrossRef] [PubMed]
3. Li, H.; Shen, Y.; He, Y.; Gao, T.; Li, G.; Zuo, M. Effects of heavy metals on bacterial community structures in two lead–zinc tailings situated in northwestern China. *Arch. Microbiol.* **2022**, *204*, 78. [CrossRef] [PubMed]
4. Li, X.; Meng, D.; Li, J.; Yin, H.; Liu, H.; Liu, X. Response of soil microbial communities and microbial interactions to long-term heavy metal contamination. *Environ. Pollut.* **2017**, *231*, 908–917. [CrossRef] [PubMed]
5. Narendrula-Kotha, R.; Nkongolo, K.K. Bacterial and fungal community structure and diversity in a mining region under long-term metal exposure revealed by metagenomics sequencing. *Ecol. Genet. Genom.* **2017**, *2*, 13–24. [CrossRef]
6. Arif, S.; Nacke, H.; Schliekmann, E.; Reimer, A.; Arp, G.; Hoppert, M. Composition and niche-specific characteristics of microbial consortia colonizing Marsberg copper mine in the *Rhenish massif*. *Biogeosciences* **2022**, *19*, 4883–4902. [CrossRef]
7. Van Bruggen, A.H.C.; Semenov, A.M. In search of biological indicators for soil health and disease suppression. *Appl. Soil Ecol.* **2012**, *15*, 13–24. [CrossRef]
8. Rastogi, G.; Barua, S.; Sani, R.K.; Peyton, B.M. Investigation of microbial populations in the extremely metal-contaminated Coeur d'Alene River sediments. *Microb. Ecol.* **2011**, *62*, 1–13. [CrossRef]
9. Eo, J.; Park, K. Long-term effects of imbalanced fertilization on the composition and diversity of soil bacterial community. *Agric. Ecosyst. Environ.* **2016**, *231*, 176–182. [CrossRef]
0. Fierer, N.; Lauber, C.L.; Ramirez, K.S.; Zaneveld, J.; Bradford, M.A.; Knight, R. Comparative metagenomic, phylogenetic and physiological analyses of soil microbial communities across nitrogen gradients. *ISME J.* **2012**, *6*, 1007–1017. [CrossRef]
1. Islam, Z.F.; Cordero, P.R.F.; Feng, J.; Chen, Y.J.; Bay, S.K.; Jirapanjawat, T. Two Chloroflexi classes independently evolved the ability to persist on atmospheric hydrogen and carbon monoxide. *ISME J.* **2019**, *13*, 1801–1813. [CrossRef]
2. Will, C.; Thürmer, A.; Wollherr, A.; Nacke, H.; Herold, N.; Schrumpf, M. Horizon-specific bacterial community composition of German grassland soils, as revealed by pyrosequencing-based analysis of 16S rRNA genes. *Appl. Environ. Microbiol.* **2010**, *76*, 6751–6759. [CrossRef] [PubMed]
3. Fernández, N.; Sierra-Alvarez, R.; Field, J.A.; Amils, R.; Sanz, J. Microbial community dynamics in a chemolithotrophic denitrification reactor inoculated with methanogenic granular sludge. *Chemosphere* **2008**, *70*, 462–474. [CrossRef] [PubMed]
4. Zhao, F.; Saleem, M.; Xie, Z.; Wei, X.; He, T.; He, G. Sensitive or tolerant functional microorganisms under cadmium stress: Suggesting potential specific interaction network characteristics in the rhizosphere system of karst potato. *Environ. Sci. Pollut. Res. Int.* **2023**, *30*, 55932–55947. [CrossRef] [PubMed]
5. Chen, G.; Fan, J.; Liu, R.; Zeng, G.; Chen, A.; Zou, Z. Removal of Cd (II), Cu (II) and Zn (II) from aqueous solutions by live *Phanerochaete chrysosporium*. *Environ. Technol.* **2012**, *33*, 2653–2659. [CrossRef]
6. Li, C.; Huang, H.; Gu, X.; Zhong, K.; Yin, J.; Mao, J.; Zhang, C. Accumulation of heavy metals in rice and the microbial response in a contaminated paddy field. *J. Soils Sediments* **2024**, *24*, 644–656. [CrossRef]
7. James, T.Y.; Pelin, A.; Bonen, L.; Ahrendt, S.; Sain, D.; Corradi, N.; Stajich, J. Shared signatures of parasitism and phylogenomics unite Cryptomycota and microsporidia. *Curr. Biol.* **2013**, *23*, 548–1553. [CrossRef]
8. Gundacker, C.; Gencik, M.; Hengstschläger, M. The relevance of the individual genetic background for the toxicokinetics of two significant neurodevelopmental toxicants: Mercury and lead. *Mutat. Res.* **2010**, *705*, 130–140. [CrossRef]
9. Mohapatra, B.R.; Gould, W.D.; Dinardo, O.; Koren, D.W. Tracking the prokaryotic diversity in acid mine drainage-contaminated environments: A review of molecular methods. *Miner. Eng.* **2011**, *24*, 709–718. [CrossRef]
0. Anjitha, K.S.; Sameena, P.P.; Puthur, J.T. Functional aspects of plant secondary metabolites in metal stress tolerance and their importance in pharmacology. *Plant Stress* **2021**, *2*, 100038. [CrossRef]
1. Li, X.; Wang, S.; Guo, L.; Huang, L. Effect of cadmium in the soil on growth, secondary metabolites and metal uptake in *Salvia miltiorrhiza*. *Toxicol. Environ. Chem.* **2013**, *95*, 1525–1538. [CrossRef]
2. Xiong, H.; Duan, C.; Xinxiang, A.; Chen, M. Response of Scutellarin Content to Heavy Metals in *Erigeron breviscapus*. *Int. J. Environ. Sci. Dev.* **2013**, *4*, 277–281. [CrossRef]
3. Haichar, F.Z.; Marol, C.; Berge, O.; Rangel-Castro, J.I.; Prosser, J.I.; Balesdent, J.; Heulin, T.; Achouak, W. Plant host habitat and root exudates shape soil bacterial community structure. *ISME J.* **2008**, *2*, 1221–1230. [CrossRef]
4. Chaparro, J.M.; Badri, D.V.; Vivanco, J.M. Rhizosphere microbiome assemblage is affected by plant development. *ISME J.* **2014**, *8*, 790–803. [CrossRef]
5. Lareen, A.; Burton, F.; Schäfer, P. Plant root-microbe communication in shaping root microbiomes. *Plant Mol. Biol.* **2016**, *90*, 575–587. [CrossRef]
6. Santolini, M.; Barabási, A.L. Predicting perturbation patterns from the topology of biological networks. *Proc. Natl. Acad. Sci. USA* **2018**, *115*, E6375–E6383. [CrossRef] [PubMed]
7. Du, L.; Zhong, S.; Luo, K.; Yang, S.; Xia, J.; Chen, Q. Effect of metal pollution on the distribution and co-occurrence pattern of bacterial, archaeal and fungal communities throughout the soil profiles. *Chemosphere* **2023**, *315*, 137692. [CrossRef]
8. Mougi, A.; Kondoh, M. Diversity of interaction types and ecological community stability. *Science* **2012**, *337*, 349–351. [CrossRef]
9. Wagg, C.; Schlaeppi, K.; Banerjee, S.; Kuramae, E.E.; van der Heijden, M.G.A. Fungal-bacterial diversity and microbiome complexity predict ecosystem functioning. *Nat. Commun.* **2019**, *10*, 4841. [CrossRef] [PubMed]

80. Li, D.; Zhou, C.; Wu, Y.; An, Q.; Zhang, J.; Fang, Y.; Li, J.Q.; Pan, C. Nanoselenium integrates soil-pepper plant homeostasis by recruiting rhizosphere beneficial microbiomes and allocating signaling molecule levels under Cd stress. *J. Hazard. Mater.* **2022**, *432*, 128763. [CrossRef]
81. Hanrahan-Tan, D.G.; Henderson, L.; Kertesz, M.A.; Lilje, O. The Effects of Nitrogen and Phosphorus on Colony Growth and Zoospore Characteristics of Soil Chytridiomycota. *J. Fungi* **2022**, *8*, 341. [CrossRef]

horticulturae

Article

Identification of Key Candidate Genes Involved in Aluminum Accumulation in the Sepals of *Hydrangea macrophylla*

Shuwen Luo †, Ying Li †, Yibing Wan, Youwei Fan, Chun Liu and Suxia Yuan *

State Key Laboratory of Vegetable Biobreeding, Institute of Vegetables and Flowers, Chinese Academy of Agricultural Sciences, Beijing 100081, China; luoshuw2024@163.com (S.L.)
* Correspondence: yuansuxia@caas.cn
† These authors contributed equally to this work.

Abstract: *Hydrangea macrophylla* (*H. macrophylla*), a species in the genus *Hydrangea* in the family Hydrangeaceae, is widely valued for its ornamental qualities in both domestic and international markets. Notably, *H. macrophylla* is known for its ability to accumulate aluminum (Al). Moreover, aluminum ions (Al^{3+}) participate in sepal bluing. However, the underlying mechanisms of Al accumulation in the sepals remain unclear. In this study, we utilized transcriptome data from two cultivars to identify genes associated with Al accumulation. In total, 154 differentially expressed isoforms between the CK and Tr groups in the sepals of both cultivars were screened. Through gene enrichment analysis and similarity identification in the CDS (coding sequence) region, 43 differentially expressed genes were identified, including 30 upregulated and 13 downregulated genes, in the sepals of the Al treatment group. Further analysis revealed that seven of these upregulated genes are related to Al accumulation in sepals. Among the seven, the gene *HmALS3.1* was identified as a potential key player in Al transport within the sepals of *H. macrophylla*. This study lays the groundwork for further exploration into the mechanisms by which *HmALS3.1* regulates Al accumulation in *H. macrophylla*.

Keywords: *Hydrangea macrophylla*; sepal; Al accumulation; *Hm*ALS3

1. Introduction

Al is not an essential element for plant growth and development, and it is toxic to most plants, causing stunted growth, weak stalks, leaf chlorosis, and in severe cases, leaf desiccation and plant death [1]. However, moderate amounts of Al have been shown to be beneficial to the growth and development of certain plants [2,3]. For instance, moderate Al levels can promote root system development [4], enhance the absorption of trace elements, and improve photosynthesis in plants [1,5–7]. Some plants can accumulate high levels of Al^{3+} in their aboveground tissues without exhibiting toxic symptoms; these species are commonly referred to as Al-accumulating plants. Al-accumulating plants are specifically defined as those containing more than 1000 mg/kg dry weight of Al in their stems and leaves [8], including species like *H. macrophylla*, *Fagopyrum esculentum*, and *Camellia sinensis* [9,10].

H. macrophylla accumulates significant levels of Al^{3+} in its aboveground tissues [11]. Leaves and sepals are the primary organs for Al accumulation in these tissues. Al content in leaves can reach up to 5 mg per g of dry weight over several months [11]. Al content in red sepals ranges from 0 to 10 μg/g fresh weight, while in blue sepals, it exceeds 40 μg/g fresh weight [12,13]. In *H. macrophylla*, appropriate Al concentrations promote the growth and development of roots, stems, and leaves, as well as the absorption of essential nutrients, including nitrogen (N), potassium (K), phosphorus (P), calcium (Ca), and magnesium (Mg) [14]. Moreover, Al alters flower color from pink to blue, thereby enhancing their ornamental value [1,15]. Al^{3+} is critical for the blue pigmentation of sepals [13,16,17]. Given that *H. macrophylla* is an Al-accumulating plant, it likely possesses a specialized mechanism for Al accumulation. As a result, the mechanisms of Al absorption and transport in *H. macrophylla* have become prominent research topics.

Organic acids can form complexes with Al^{3+} to mitigate its toxicity in *H. macrophylla*. In the rhizosphere, Al^{3+} forms complexes with oxalic acid secreted by the root system, which partially prevents Al^{3+} from entering root cells. Conversely, in the xylem and leaves, Al^{3+} is absorbed and transported by forming complexes with citric acid [18–20]. Additionally, several genes related to Al accumulation have been identified. Some ABC transporter genes, including the multidrug and toxic compound extrusion (MATE) gene and the NRAMP aluminum transporter, are highly expressed in the roots and leaves of *H. macrophylla* under Al stress [21,22]. These transporters mediate the transport of organic acids, participate in the absorption, transport, and storage of Al^{3+}, and play a crucial role in signaling, thereby enhancing plant resistance to Al stress [21,22]. Moreover, two aquaporin genes, *H. macrophylla* vacuolar Al transporter (*HmVALT*) and *H. macrophylla* plasma membrane Al transporter 1 (*HmPALT1*), have been associated with Al transport. *HmPALT1* is responsible for transporting Al^{3+} from outside the cell into the cytoplasm, while *HmVALT* transports Al^{3+} to the vacuole [23]. However, the precise mechanisms of Al^{3+} absorption, transport, and storage in *H. macrophylla* remain unclear.

In this study, transcriptome data from the sepals of the CK group (plants without Al treatment) and the Tr group (plants with Al treatment) in two cultivars were analyzed. The identification of key candidate genes involved in Al accumulation in sepals was examined. These results lay the foundation for elucidating the Al accumulation mechanism in *H. macrophylla* and provide a significant theoretical reference for studying the mechanism of Al accumulation in Al-accumulating plants.

2. Materials and Methods

2.1. Plant Materials

Two cultivars of *H. macrophylla*, 'Bailmer' (Endless Summer™) and 'Duro', were used in this study and were cultivated in a greenhouse at the Institute of Vegetables and Flowers, Chinese Academy of Agricultural Sciences. Following flower bud formation, the plants of both cultivars were subjected to a low temperature (2 °C~6 °C) for approximately six weeks. Subsequently, the plants were returned to the greenhouse for flowering.

Half of the plants from each cultivar (Tr group) were grown in a mixed substrate (peat: perlite = 7:3, *v*/*v*). A 400 mL solution of aluminum sulfate ($Al_2(SO_4)_3 \cdot 18H_2O$ dissolved in water) was applied to the substrate at the concentration of 6 g/L every seven to ten days, starting from the flower bud formation stage until flowering. While the remaining half, without Al treatment, cultivated in the same mixed substrate served as the control (CK group). Under Al sulfate treatment, the sepal color changed from pink to blue in 'Bailmer', while in 'Duro', it changed only slightly and did not turn blue [16]. Sepals at three different developmental stages (S1, S2, S3) from the Tr and CK groups of 'Bailmer' and 'Duro' (Figure 1) [16] were sampled on April 28, 2020 and May 19, 2020, respectively, and used in each experiment.

Figure 1. Sepals at three different developmental stages. S1, Stage 1; S2, Stage 2; S3, Stage 3. Reprinted with permission from Ref. [16]. 2022, Suxia Yuan.

2.2. *Al Content Measurement*

Initially, all samples were dried to a constant weight. Each 0.2–0.4 g sample was digested and soaked in 5 mL of nitric acid (≥99.8%) overnight. The treated samples were placed in a constant temperature drying oven for 2 h at 80 °C, 2 h at 120 °C, and 4 h at 160 °C, followed by cooling to room temperature. The cooled solution was heated to evaporate the acid, after which the digestive solution was transferred into a 25 mL volumetric flask. The inner tank and cap were rinsed three times with a small amount of 1% nitric acid solution. The rinsed solution was combined in the volumetric flask, diluted to the mark with 1% nitric acid, and thoroughly mixed to determine the Al content. The Al content was determined using inductively coupled plasma mass spectrometry (ICP-MS) following the methods for the determination of multiple elements in food (National Food Safety Standards, GB 5009.268-2016, China) [24]. The samples were characterized by element-specific mass numbers (mass-to-charge ratio, m/z). The detection wavelength was set at 396.152 nm.

2.3. *Transcriptome Sequencing, Annotation*

In prior research, a full-length transcriptome database was constructed using the PacBio Sequel II platform, incorporating the flowers, pedicels, buds, leaves, stems, and roots of 'Bailmer' [25]. The database contained 72,848 high-quality isoforms, of which 67,941 were annotated using the NCBI Non-Redundant (Nr) database (http://www.ncbi.nlm.nih.gov, accessed on 2 December 2020), Swiss-Prot protein database (http://www.expasy.ch/sprot, accessed on 4 December 2020), Kyoto Encyclopedia of Genes and Genomes (KEGG) database (http://www.genome.jp/kegg, accessed on 7 December 2020), and COG/KOG database (http://www.ncbi.nlm.nih.gov/COG, accessed on 9 December 2020), with an E-value threshold of 1e-5. These PacBio Iso-Seq data served as the reference transcriptome for *H. macrophylla*. cDNA libraries from 36 samples, including three biological replicates, were constructed from sepals at the S1, S2, and S3 stages of CK and Tr groups in 'Bailmer' and 'Duro' (Figure 1). These 36 cDNA libraries were sequenced using the Illumina HiSeq™ 4000 [16]. Between 89.45% and 92.36% of the high-quality clean reads from 'Bailmer', and 83.06% to 85.97% from 'Duro', were mapped to the isoforms from the reference transcriptome. Gene abundances were calculated and normalized to reads per kilobase per million mapped reads (RPKM). The mapped isoforms were annotated using the NR, Swiss-Prot, KEGG, and COG/KOG databases, as previously described.

2.4. *Analysis of Differentially Expressed Genes (DEIs)*

Differentially expressed genes (DEIs) between the CK and Tr groups were identified. The threshold values for screening DEIs were a false discovery rate (FDR) ≤0.05 and |log2 (foldchange)| ≥ 1. DEIs were annotated using the NR, KEGG, COG, and GO databases, as described for isoform annotation.

2.5. *Quantitative Real-Time Fluorescent PCR (qRT-PCR) Analysis*

The RNA isolation kit is a quick RNA isolation kit (Hua Yue Yang, Beijing, China). First-strand cDNA was synthesized using HiScript III RT SuperMix for qPCR (+gDNA wiper) according to the manufacturer's instructions. Primer pairs for qRT-PCR were designed using Primer Premier v5.0 software (Premier Biosoft Int., Palo Alto, CA, USA) (Table S1). The specificity and efficiency of the primers were confirmed using 1.5% agarose gel electrophoresis. All reactions were conducted in 96-well plates using the CFX96 Real-Time System (Bio-Rad, Hercules, CA, USA). qRT-PCR analyses were conducted using Taq Pro Universal SYBR qPCR Master Mix (Vazyme, Nanjing, China). Each PCR reaction (20 μL) contained 5 μL of cDNA (100–200 ng/μL), 0.4 μL of each forward and reverse primer (10 pmol/μL), 10 μL of 2 × Taq Pro Universal SYBR qPCR Master Mix, and 4.2 μL of ddH_2O. Amplification reactions were performed under the following conditions: 95 °C for 30 s, followed by 40 cycles of 95 °C for 10 s and 60 °C for 30 s. A melting curve was generated for each PCR to examine the amplification specificity of primers and the presence of reaction contaminants. The melting curve was gained by heating the amplification products under the following conditions: 65–95 °C with a temperature increment of 0.5 °C every 5 s. The *HmActin* gene was used as the internal control for normalization. The primer sequences were *HmActin*-F (5′-GCCTGCCATGTATGTTGCCATC-3′) and *HmActin*-R (5′-CGGAATCCAGCACAATACCAGTTG-3′) [26]. The relative expression levels of the target genes were calculated using the $2^{-\Delta\Delta CT}$ method [27] relative to the internal control.

2.6. Protein Interaction Analysis

Protein interactions were predicted using the online software STRING (functional protein association networks; https://string-db.org/, accessed on 24 March 2024).

2.7. Data Analysis

Data are expressed as means ± standard deviations (SDs) of three replicates. Data analysis was conducted using Microsoft Office Excel 2010 and IBM SPSS Statistics 24.0. Bar graphs were generated using Origin 9.0.

3. Results

3.1. Al Contents in the Sepals

The Al content in sepals from the Tr groups was higher than that in sepals from the CK groups at all three stages (S1, S2, and S3) in both 'Bailmer' and 'Duro'. Under Al treatment, sepals accumulated a substantial amount of Al, particularly at S1. The Al content in sepals was 2260.67 µg/g dry weight (DW) for 'Bailmer' and 1204.67 µg/g DW for 'Duro', which were approximately 2.90 and 5.47 times higher than that of the CK group (779.67 and 220.33 µg/g DW), respectively (Figure 2).

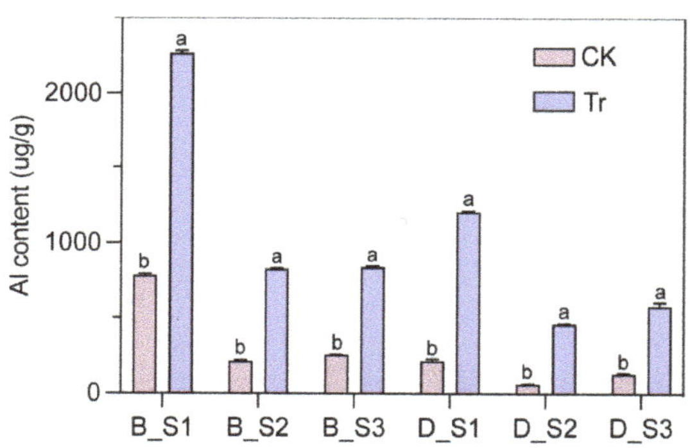

Figure 2. Al content in sepals at three different developmental stages (S1, S2, S3) in 'Bailmer' and 'Duro'. B and D represent 'Bailmer' and 'Duro', respectively. Different lowercase letters indicate significant differences between two treatments at $p < 0.05$ according to Duncan's multiple range test.

3.2. Analysis of Illumina Transcriptome Data

In our previous study, 72,848 high-quality, full-length isoforms were obtained through full-length transcriptome sequencing. Of these, 71,617 isoforms were expressed in the Illumina-sequenced samples, and a total of 67,932 isoforms were annotated in at least one database. Of these, 44,208 (65.08%), 66,896 (98.47%), 32,980 (48.54%), and 55,845 (82.21%) isoforms were annotated according to the KOG, NR, KEGG, and Swiss-Prot databases, respectively. Across the four databases, 27,077 isoforms were annotated in all of them (Figure 3).

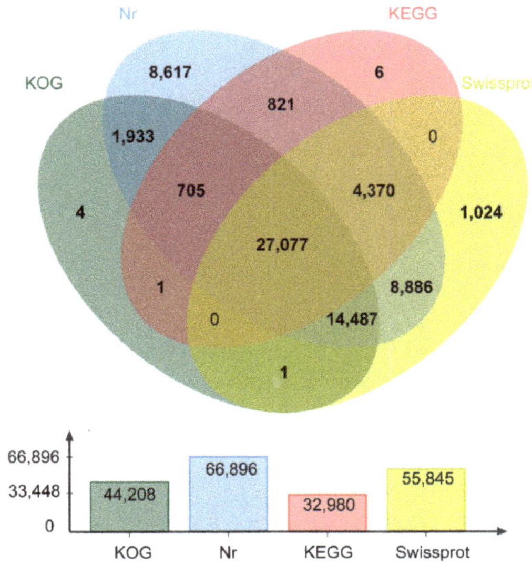

Figure 3. Venn diagram of NR, KOG, KEGG, and Swiss-Prot function annotations.

3.3. *Identification of Key Candidate Genes Mediating Aluminum Accumulation in Sepals*

Differentially expressed isoforms (DEIs) between the CK group and the Tr group were identified at each of the three stages (S1–S3) in every cultivar (Figure 4). Considering all three developmental stages (S1–S3) of sepals, a total of 6575 DEIs were obtained in 'Bailmer', of which 2972 isoforms were upregulated and 3603 were downregulated. In total, 788 DEIs were identified in 'Duro', of which 426 were upregulated and 362 were downregulated. Among these DEIs, 84 upregulated isoforms and 70 downregulated isoforms were identified in both cultivars (Figures 5 and S1).

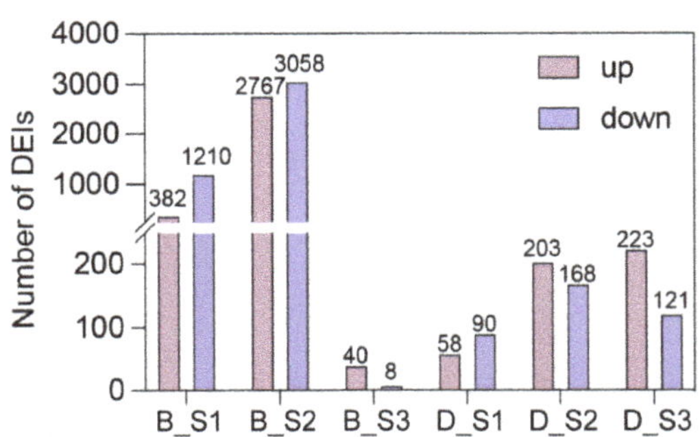

Figure 4. The number of differentially expressed isoforms in the comparison of treatment versus control and three different stages (S1, S2, and S3). B and D represent 'Bailmer' and 'Duro', respectively.

According to GO annotation, the 154 DEIs (84 upregulated isoforms and 70 downregulated isoforms) were classified into three categories: biological process, cellular component, and molecular function (Figure 6). Among these, 55 upregulated isoforms and 39 downregulated isoforms were significantly enriched ($p \leq 0.05$). Isoforms with over 99% similarity in the CDS region were identified as the same gene, resulting in 30 upregulated genes and 13 downregulated genes (Table S2). Of these, seven upregulated and two downregulated genes were related to stress resistance, thirteen upregulated and six downregulated genes were involved in metabolism, six upregulated and two downregulated genes were involved in both metabolism and ion transport, and the remaining (four upregulated and three downregulated genes) were transporter genes.

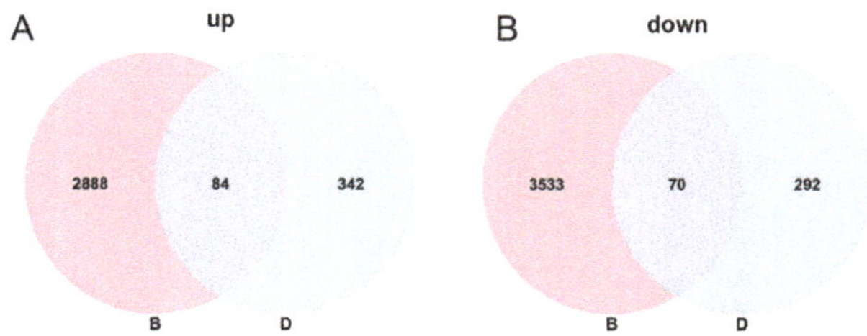

Figure 5. Venn diagram of DEIs between two cultivars. (**A**) Upregulated expression genes under Al treatment in 'Bailmer' and 'Duro'. (**B**) Downregulated expression genes under Al treatment in 'Bailmer' and 'Duro'. B and D represent 'Bailmer' and 'Duro', respectively.

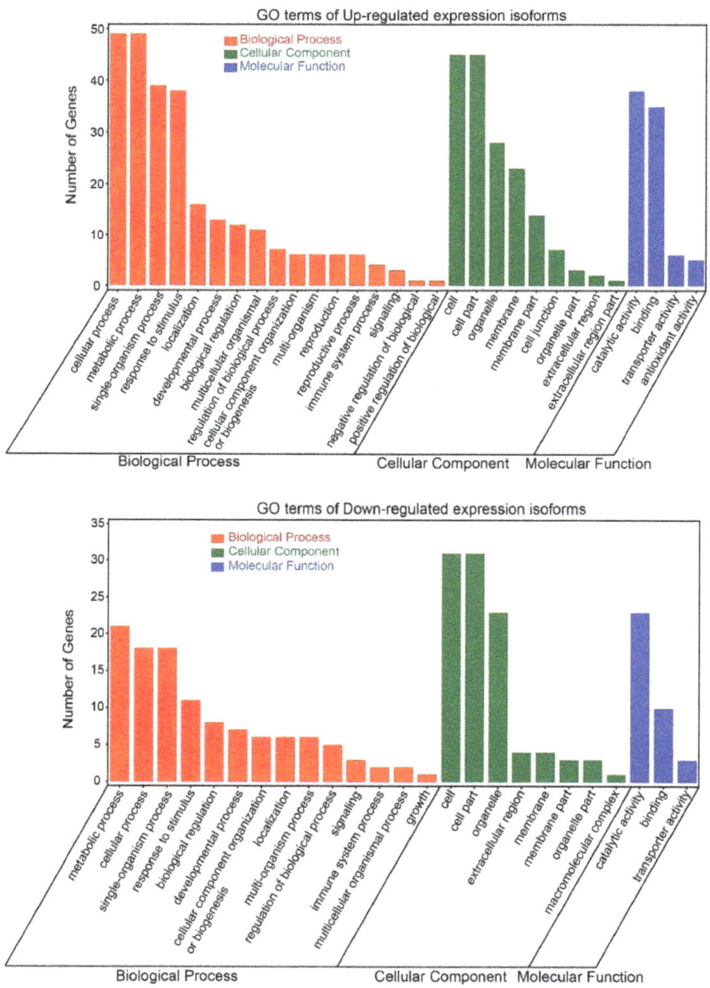

(**A**)

(**B**)

Figure 6. GO enrichment analysis of DEIs. (**A**) GO terms of upregulated expression isoforms. (**B**) GO terms of downregulated expression isoforms.

3.4. Expression Verification of Key Candidate Genes Using qRT-PCR

The relative expression of 43 differentially expressed genes (30 upregulated and 13 downregulated) was verified using qRT-PCR. Seven genes, including *HmDDS* (*Dammarenediol II synthase-like*), *HmDHK* (*1,2-dihydroxy-3-keto-5-methylthiopentene dioxygenase 1*), *HmALS3.1* (*protein Aluminum sensitive 3*), *HmABAH4* (*Abscisic acid 8′-hydroxylase*), *HmMSMO1-1* (*Methylsterol monooxygenase 1-1*), *HmPAP8* (*Purple acid phosphatase 8-like*), and *HmSULTR3;1* (*Sulfate transporter 3.1-like*), were upregulated across all three stages of sepals in both cultivars under Al treatment (Figure 7). Additionally, the specificity and efficiency of PCR amplification were confirmed. PCR amplifications for each primer pair showed a single and desired amplicon of the expected size on a 1.5% agarose gel (Figure 8A), and a single peak in melt curves (Figure 8B). This pattern was consistent with the Al content in the sepals and the gene expression profile based on transcriptome data for the two cultivars (Figures 2 and 5). These seven genes were identified as being related to Al accumulation in sepals.

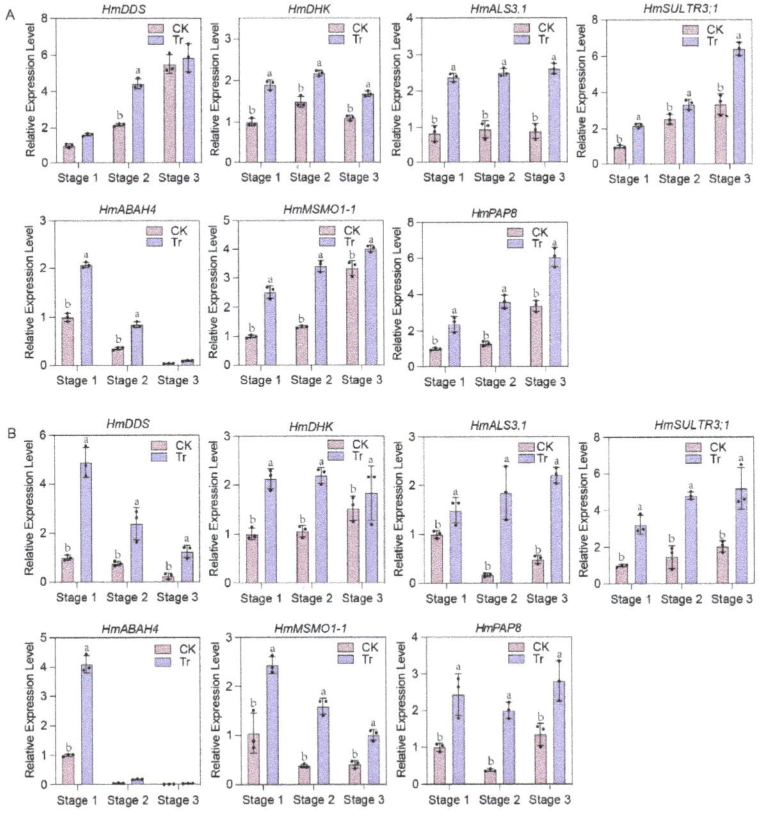

Figure 7. Expression of seven genes between two treatments in 'Bailmer' (**A**) and 'Duro' (**B**). Different lowercase letters indicate significant differences between two treatments at $p \leq 0.05$ according to Duncan's multiple range test.

According to gene functions (Table S2), HmDDS and HmMSMO1-1 are associated with organic substance biosynthesis; HmDHK, HmABAH4, and HmPAP8 are primarily involved in macromolecule metabolism and transition metal ion binding; and HmALS3.1 and HmSULTR3;1 are transporters responsible for metal ion and sulfate transport, respectively.

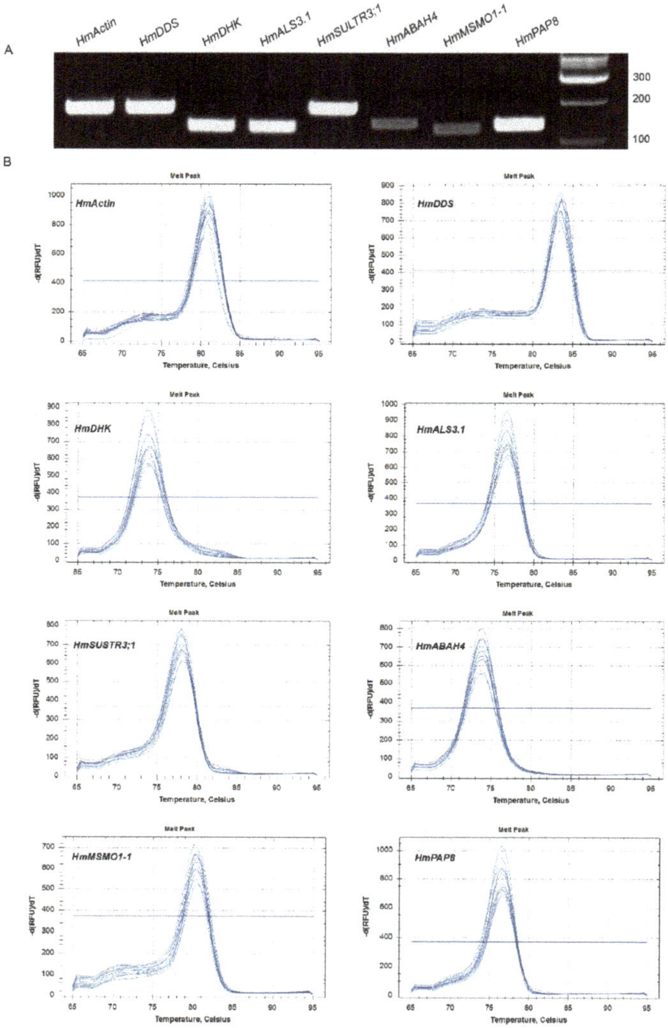

Figure 8. Specificity of qRT-PCR amplification for *HmAction* and seven candidate genes. (**A**) Amplification of a specific PCR product for each gene through agarose gel. (**B**) Melting curves for each gene.

3.5. Protein Interaction Analysis

Among the seven genes, only the HmALS3.1 gene was found to have an interaction network with several proteins (Figure 9). HmALS3.1, as the central protein, was closely linked to the ATP-binding cassette (ABC) transporter protein family. The only experimentally verified interaction with the HmALS3.1 protein is ABCI17, which belongs to the ABC transporter superfamily. Among the predicted interactions, ABCB27, ABCI17, and LPR2 are identified as neighborhood proteins to HmALS3.1. Through automated and unsupervised text mining, nine proteins (ABCB27, ABCI1, ABCI17, ALMT1, Q9LJC4_ARATH, DTX42, MRS2-10, STOP1, and STOP2) were frequently mentioned and were the first shell of interactors with HmALS3.1.

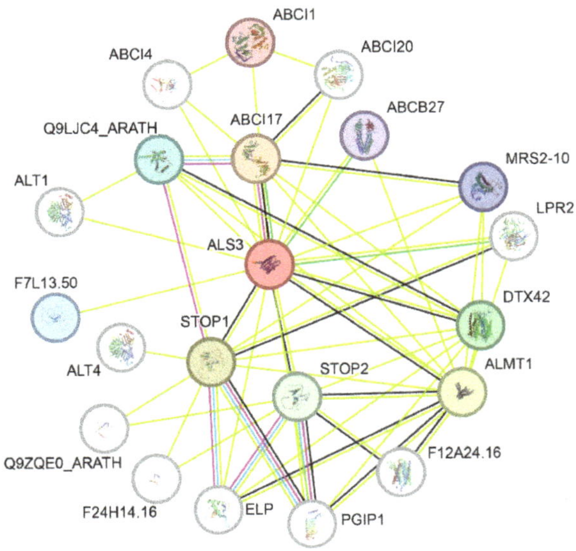

Figure 9. Protein interaction network diagram of HmALS3.1. The blue line indicates interactions known from curated databases; the pink line indicates experimentally determined interactions; the green line represents text mining-based interactions; and the black line denotes co-expression interactions.

4. Discussion

H. macrophylla is famous for its sepal color changes under different cultivation conditions. The sepal color is pink-red when the plant is grown in neutral to alkaline soils and blue when grown in acidic soils. The main reason for sepals bluing is that Al^{3+} is absorbed by roots in acidic soil and transported to sepals to participate in the formation of blue complex [28,29]. In fact, not every cultivar with a pink-red sepal color has blue sepals in acidic soils with Al^{3+} because the bluing capability of the sepal color was positively correlated with the proportion of delphinidin-3-glucoside among all the anthocyanins in sepals [16,30]. In this experiment, two cultivars with different sepal color bluing capabilities were utilized. The sepal color of 'Bailmer' varied significantly from pink to blue under Al treatment, whereas 'Duro' exhibited slightly and did not turn blue from pink [16]. However, both cultivars had higher Al content in sepals under Al treatment. Consequently, we analyzed the transcriptome data from the two cultivars to elucidate how *H. macrophylla* responded to Al.

To identify genes associated with Al accumulation, transcriptome data from two cultivars of *H. macrophylla* were used to screen 43 differentially expressed genes. Additionally, seven genes associated with Al accumulation were identified using the qRT-PCR technique: *HmDDS*, *HmDHK*, *HmALS3.1*, *HmABAH4*, *HmMSMO1-1*, *HmPAP8*, and *HmSULTR3;1* (Figure 7). In the research, we failed to obtain hydroponically cultivated flowering plants, so all the plants were grown in the mixed substrate containing peat and perlite (7:3, by volume). The lower Al^{3+} level in peat might affect gene expression. The seven genes screened exhibited significant differences in the expression level at all three developmental stages between the CK and Tr groups in both cultivars.

Except for *HmALS3.1*, the other six genes have not been reported in the literature as being associated with Al stress. In the experiments, Al^{3+} in the form of $Al_2(SO_4)_3$ was applied to the substrate; thus, the addition of sulfate ions may have significantly increased the expression of *HmSULTR3;1* [31].

Aluminum sensitive 3 (*ALS3*) encodes a half-size ABC protein, which belongs to the large family of ATP-binding cassette (ABC) proteins [32]. Studies have shown that the *ALS3* gene functions in Al resistance across various plant species. In *Arabidopsis thaliana*, the expression of *AtALS3* was regulated by Al signal in roots and shoots [32,33]. Homologs of ALS3 genes required for Al resistance have also been identified in *Rhododendron yunnanense* [34], soybean [35], and buckwheat [36]. The potential role of this protein is to transfer Al^{3+} from sensitive root tissues to less sensitive shoots, thereby reducing Al toxicity [32,37]. Plant roots can also tolerate Al^{3+} internally by sequestering it into vacuoles [38,39].

Simulations using the Arabidopsis database (Figure 9) revealed that some proteins interacting with HmALS3.1 are also involved in plant Al resistance. Among them, two organic acid transporters were mentioned, ALMT1 and DTX42. Aluminum-activated malate transporter 1 (ALMT1) was identified as an Al-activated malate efflux transporter, secreting malate to chelate Al^{3+} in the rhizosphere, thus preventing Al from sensitive root cells [40–42]. The expression of *ALMT1* plays an important role in improving Al resistance in wheat [43], *Arabidopsis* [44], and apple [45]. Although DTX42 belonging to the MATE efflux family has not yet been verified as related to Al response, such as DTX30 from the same family, it indirectly modulates citrate exudation to promote Al tolerance by modulating auxin levels in root [46]. The HmALS3.1 gene was also closely associated with the other ATP-binding cassette (ABC) transporter proteins. The ABC transporter family is divided into eight subfamilies based on the evolutionary relationships of conserved regions, with ABCB transporter proteins being the second-largest subfamily [47]. ABCB transporter proteins mediate the transport of heavy metal ions, including cadmium, lead, and Al, and enhance plant tolerance to heavy metals [32,48]. Among them, ABCB27 (ALS1) may play a role in Al intracellular movement [32,49]. Additionally, some ABCI transporter proteins also respond to Al stress. In soybeans, GsABCI1 transgenic plants exhibited resistance to Al treatment through ion translocation or alterations in root components [50]. The ABCI17 (sensitive to aluminum rhizotoxicity 1 (STAR1)) can interact with STAR2 to form a complex as a bacterial-type ABC transporter, which transports UDP-Glc and is required for Al tolerance in rice [39]. HmABCI17 from *H. macrophylla* has been shown to enhance Al tolerance in yeast [51]. Q9LJC4_ARATH also belongs to the ABC transporter family, but it has not been verified as being related to the Al response yet. Research has shown that the transcription factor (TF) sensitive to proton rhizotoxicity1 (STOP1) regulates *ALS3*, *AtALMT1*, and *AtMATE* expression under Al stress [52,53]. Additionally, the STOP1-independent NO signaling pathway and STOP1-dependent regulation in the phosphoinositide (PI) signaling pathway are involved in the regulation of PGIP1 expression under Al stress [54]. To some extent, STOP2 can substitute for STOP1 in activating the transcription of genes regulated by STOP1, including *ALS3* and *AtMATE* [55,56]. In Arabidopsis, ALS3 and STAR1 form the ATP-binding cassette (ABC) transporter complex located in the vacuolar membrane. ALS3/STAR1 promotes the degradation of the Al resistance transcription factor STOP1 to regulate ALMT1 expression [57]. Two metal transporters are also mentioned in the protein interaction network, magnesium transporter 1 (MRS2-10) and low phosphate root 2 (LPR2). MRS2-10 has shown the activity to transport Al in Arabidopsis [58]. These proteins mentioned by the protein network may become the key points in the future study on Al accumulation in *H. macrophylla*.

Additionally, other Al resistance genes have been identified in various plants, including those from the MATE family first described in sorghum [59]. Subsequently, VrMATE19 and VrMATE30 were found to play a potential role in Al tolerance in mung beans [60]. In soybeans, the *GmMATE13* and *GmMATE75* genes enhanced Al tolerance in the plant through citrate secretion [61]. Moreover, the nbatural resistance-associated macrophage protein 1 (Nrat1) gene may be a useful tool for enhancing Al tolerance in rice and *Arabidopsis* [62]. A knockout mutation in the Nrat1 gene resulted in decreased Al uptake and increased binding of Al^{3+} to the cell wall [63]. In *H. macrophylla*, two aquaporin genes, *HmVALT* and *HmPALT1*, were found to be associated with Al transport.

In our study, *HmALS3.1* was identified as being closely related to Al accumulation in sepals of *H. macrophylla*, while the functions of the remaining six genes require further investigation. At present, the specific functions and regulatory mechanisms of *HmALS3.1* participating in Al accumulation are being studied in our laboratory.

5. Conclusions

This study first reports the identification of *HmALS3.1* as a key gene related to Al accumulation in the sepals of *H. macrophylla*. By analyzing the transcriptome data from the CK group and the Al treatment group of the two cultivars of *H. macrophylla*, 43 differentially expressed genes (30 upregulated and 13 downregulated under the Al treatment) in sepals were identified. Further research indicated that seven genes upregulated across all three stages under Al treatment in both cultivars, including *HmDDS*, *HmDHK*, *HmALS3.1*, *HmABAH4*, *HmMSMO1-1*, *HmPAP8*, and *HmSUSTR3;1*. Among the seven, the gene *HmALS3.1* was identified as a potential key player in Al transport within the sepals of *H. macrophylla*. This study lays the groundwork for further exploration into the mechanisms by which *HmALS3.1* regulates Al accumulation in *H. macrophylla*.

Supplementary Materials: The following supporting information can be downloaded at https://www.mdpi.com/article/10.3390/horticulturae10111180/s1: Figure S1: Heat maps of expression of 84 upregulated and 70 downregulated isoforms based on transcriptome data; Table S1: Primers used for qRT-PCR in this study; and Table S2: Annotation of 96 DEIs related to aluminum accumulation in sepals of hydrangea.

Author Contributions: Conceptualization, S.Y.; methodology, S.Y.; validation, Y.F.; formal analysis, Y.L.; investigation, Y.L.; resources, Y.L. and Y.W.; data curation, S.L.; writing—original draft preparation, S.L.; writing—review and editing, S.Y.; visualization, S.L.; supervision, C.L.; project administration, S.Y.; funding acquisition, C.L. All authors have read and agreed to the published version of the manuscript.

Funding: This research was funded by the Central Public-interest Scientific Institution Basal Research Fund (grant number: IVF-BRF2022017 and IVF-BRF2023009), the Science and Technology Innovation Program of the Chinese Academy of Agricultural Sciences (grant number: CAAS-ASTIP-2022-IVFCAAS and CAAS-ASTIP-2023-IVFCAAS), the National Flower Improvement Center, and the Key Laboratory of Biology and Genetic Improvement of Flower Crops (North China), Ministry of Agriculture and Rural Affairs, China.

Data Availability Statement: The RNA-Seq data generated in this study are available at the SRA Archive (https://submit.ncbi.nlm.nih.gov/subs/sra/, accessed on 10 May 2020), with accession number PRJNA1125644.

Conflicts of Interest: The authors declare no conflicts of interest.

References

1. Ma, J.F. Role of organic acids in detoxification of aluminum in higher plants. *Plant Cell Physiol.* **2000**, *41*, 383–390. [CrossRef] [PubMed]
2. Mortvedt, J.J.; Asher, C.J. Beneficial Elements, Functional Nutrients, and Possible New Essential Elements. *Micronutr. Agric.* **1991**, *4*, 703–723.
3. Famoso, A.N.; Clark, R.T.; Shaff, J.E.; Craft, E.; McCouch, S.R.; Kochian, L.V. Development of a Novel Aluminum Tolerance Phenotyping Platform Used for Comparisons of Cereal Aluminum Tolerance and Investigations into Rice Aluminum Tolerance Mechanisms. *Plant Physiol.* **2010**, *153*, 1678–1691. [CrossRef]
4. Xu, Q.S.; Wang, Y.; Ding, Z.T.; Song, L.B.; Li, Y.S.; Ma, D.X.; Wang, Y.; Shen, J.Z.; Jia, S.S.; Sun, H.W.; et al. Aluminum induced metabolic responses in two tea cultivars. *Plant Physiol. Biochem.* **2016**, *101*, 162–172. [CrossRef] [PubMed]
5. Bojórquez-Quintal, E.; Escalante-Magaña, C.; Echevarría-Machado, I.; Martínez-Estévez, M. Aluminum, a Friend or Foe of Higher Plants in Acid Soils. *Front. Plant Sci.* **2017**, *8*, 1767. [CrossRef]
6. Muhammad, N.; Zvobgo, G.; Zhang, G.P. A review: The beneficial effects and possible mechanisms of aluminum on plant growth in acidic soil. *J. Integr. Agric.* **2019**, *18*, 1518–1528. [CrossRef]
7. Chauhan, D.K.; Yadav, V.; Vaculik, M.; Gassmann, W.; Pike, S.; Arif, N.; Singh, V.P.; Deshmukh, R.; Sahi, S.; Tripathi, D.K. Aluminum toxicity and aluminum stress-induced physiological tolerance responses in higher plants. *Crit. Rev. Biotechnol.* **2021**, *41*, 715–730. [CrossRef]
8. Schmitt, M.; Watanabe, T.; Jansen, S. The effects of aluminium on plant growth in a temperate and deciduous aluminium accumulating species. *AoB Plants* **2016**, *8*, plw065. [CrossRef] [PubMed]
9. Jansen, S.; Broadley, M.R.; Robbrecht, E.; Smets, E. Aluminum hyperaccumulation in angiosperms: A review of its phylogenetic significance. *Bot. Rev.* **2002**, *68*, 235–269. [CrossRef]

10. Sun, L.; Zhang, M.; Liu, X.; Mao, Q.; Shi, C.; Kochian, L.V.; Liao, H. Aluminium is essential for root growth and development c tea plants (*Camellia sinensis*). *J. Integr. Plant Biol.* **2020**, *62*, 984–997. [CrossRef]
11. Ma, J.F.; Hiradate, S.; Nomoto, K.; Iwashita, T.; Matsumoto, H. Internal detoxification mechanism of Al in hydrangea–Identification of Al form in the leaves. *Plant Physiol.* **1997**, *113*, 1033–1039. [CrossRef] [PubMed]
12. Chen, H.; Wang, X.; Xu, L.F. Identification and Bioinformatics Analysis of ABC Transporter Gene Family in Hydrangea unde Aluminum Stress. *Mol. Plant Breed.* **2022**, *13*, 1–15. [CrossRef]
13. Ito, D.; Shinkai, Y.; Kato, Y.; Kondo, T.; Yoshida, K. Chemical Studies on Different Color Development in Blue- and Red-Colorec Sepal Cells of. *Biosci. Biotechnol. Biochem.* **2009**, *73*, 1054–1059. [CrossRef] [PubMed]
14. Osaki, M.; Watanabe, T.; Tadano, T. Beneficial effect of aluminum on growth of plants adapted to low pH soils. *Soil Sci. Plant Nut* **1997**, *43*, 551–563. [CrossRef]
15. Peng, J.; Dong, X.; Xue, C.; Liu, Z.; Cao, F. Exploring the Molecular Mechanism of Blue Flower Color Formation in *Hydrange macrophylla* cv. "Forever Summer". *Front. Plant Sci.* **2021**, *12*, 585665. [CrossRef]
16. Yuan, S.X.; Qi, H.; Yang, S.N.; Chu, Z.Y.; Zhang, G.T.; Liu, C. Role of delphinidin-3-glucoside in the sepal blue color chang among *Hydrangea macrophylla* cultivars. *Sci. Hortic.* **2023**, *313*, 111902. [CrossRef]
17. Liang, C.Y.; Rengasamy, K.P.; Huang, L.M.; Hsu, C.C.; Jeng, M.F.; Chen, W.H.; Chen, H.H. Assessment of violet-blue colo formation in *Phalaenopsis* orchids. *BMC Plant Biol.* **2020**, *20*, 212. [CrossRef]
18. Naumann, A.; Horst, W.J. Effect of aluminium supply on aluminium uptake, translocation and blueing of *Hydrangea macrophyll* (Thunb.) Ser. cultivars in a peatclay substrate. *J. Hortic. Sci. Biotechnol.* **2003**, *78*, 463–469. [CrossRef]
19. Hotta, H.; Wang, Q.; Fukuda, M.; Aizawa, S.; Umemura, T.; Sekizawa, K.; Tsunoda, K.I. Identification of aluminum species in ar aluminum-accumulating plant, hydrangea (*Hydrangea macrophylla*), by electrospray ionization mass spectrometry. *Anal. Sci.* **200**8 *24*, 795–798. [CrossRef]
20. Dong, B.; Meng, D.; Song, Z.; Cao, H.; Du, T.; Qi, M.; Wang, S.; Xue, J.; Yang, Q.; Fu, Y. CcNFYB3-CcMATE35 and LncRNA CcLTCS-CcCS modules jointly regulate the efflux and synthesis of citrate to enhance aluminium tolerance in pigeon pea. *Plan Biotechnol. J.* **2024**, *22*, 181–199. [CrossRef]
21. Wilkens, S. Structure and mechanism of ABC transporters. *F1000Prime Rep.* **2015**, *7*, 14. [CrossRef] [PubMed]
22. Aryal, B.; Laurent, C.; Geisler, M. Learning from each other: ABC transporter regulation by protein phosphorylation in plan and mammalian systems. *Biochem. Soc. Trans.* **2015**, *43*, 966–974, Erratum in *Biochem. Soc. Trans.* **2016**, *44*, 663–673. [CrossRef [PubMed]
23. Negishi, T.; Oshima, K.; Hattori, M.; Kanai, M.; Mano, S.; Nishimura, M.; Yoshida, K. Tonoplast- and plasma membrane-localizec aquaporin-family transporters in blue hydrangea sepals of aluminum hyperaccumulating plant. *PLoS ONE* **2012**, *7*, e4318 [CrossRef] [PubMed]
24. *GB 5009.268–2016*; National Food Safety Standard. Determination of Multi-elements in Food. General Administration of Quality Supervision, Inspection and Quarantine of the People's Republic of China: Beijing, China, 2016. (In Chinese)
25. Qi, H.; Zhang, G.T.; Chu, Z.Y.; Liu, C.; Yuan, S.X. Identification of Seven Key Structural Genes in the Anthocyanin Biosynthesis Pathway in Sepals of *Hydrangea macrophylla*. *Curr. Issues Mol. Biol.* **2022**, *44*, 4167–4180. [CrossRef]
26. Zhang, G.T.; Yuan, S.X.; Qi, H.; Chu, Z.Y.; Liu, C. Identification of Reliable Reference Genes for the Expression of *Hydrangea macrophylla* 'Bailmer' and 'Duro' Sepal Color. *Horticulturae* **2022**, *8*, 835. [CrossRef]
27. Livak, K.J.; Schmittgen, T.D. Analysis of relative gene expression data using real-time quantitative PCR and the $2^{-\Delta\Delta CT}$ method *Methods* **2001**, *25*, 402–408. [CrossRef]
28. Rahmati, R.; Hamid, R.; Ghorbanzadeh, Z.; Jacob, F.; Azadi, P.; Zeinalabedini, M.; Karimi Farsad, L.; Kazemi, M.; Ebrahimi, M.A. Shahinnia, F.; et al. Comparative Transcriptome Analysis Unveils the Molecular Mechanism Underlying Sepal Colour Changes under Acidic pH Substratum in *Hydrangea macrophylla*. *Int. J. Mol. Sci.* **2022**, *23*, 15428. [CrossRef]
29. Ito, T.; Oyama, K.I.; Yoshida, K. Direct Observation of Hydrangea Blue-Complex Composed of 3-*O*-Glucosyldelphinidin, Al^{3+} and 5-*O*-Acylquinic Acid by ESI-Mass Spectrometry. *Molecules* **2018**, *23*, 1424. [CrossRef]
30. Yoshida, K.; Ito, D.; Miki, N.; Kondo, T. Single-cell analysis clarifies mosaic color development in purple hydrangea sepal. *New Phytol.* **2021**, *229*, 3549–3557. [CrossRef]
31. Zhang, H.J.; Hao, X.Y.; Zhang, J.J.; Wang, L.; Wang, Y.C.; Li, N.N.; Guo, L.N.; Ren, H.Z.; Zeng, J.M. Genome-wide identification of SULTR genes in tea plant and analysis of their expression in response to sulfur and selenium. *Protoplasma* **2022**, *259*, 127–140 [CrossRef]
32. Larsen, P.B.; Cancel, J.; Rounds, M.; Ochoa, V. *Arabidopsis ALS1* encodes a root tip and stele localized half type ABC transporter required for root growth in an aluminum toxic environment. *Planta* **2007**, *225*, 1447–1458. [CrossRef] [PubMed]
33. Sadhukhan, A.; Agrahari, R.K.; Wu, L.J.; Watanabe, T.; Nakano, Y.; Panda, S.K.; Koyama, H.; Kobayashi, Y. Expression genome wide association study identifies that phosphatidylinositol-derived signalling regulates ALUMINIUM SENSITIVE3 expression under aluminium stress in the shoots of *Arabidopsis thaliana*. *Plant Sci.* **2021**, *302*, 110711. [CrossRef] [PubMed]
34. Xu, Y.X.; Lei, Y.S.; Huang, S.X.; Zhang, J.; Wan, Z.Y.; Zhu, X.T.; Jin, S.H. Combined de novo transcriptomic and physiological analyses reveal RyALS3-mediated aluminum tolerance in *Rhododendron yunnanense* Franch. *Front. Plant Sci.* **2022**, *13*, 951003 [CrossRef]
35. Agrahari, R.K.; Kobayashi, Y.; Borgohain, P.; Panda, S.K.; Koyama, H. Aluminum-Specific Upregulation of *GmALS3* in the Shoots of Soybeans: A Potential Biomarker for Managing Soybean Production in Acidic Soil Regions. *Agronomy* **2020**, *10*, 1228. [CrossRef]

36. Reyna-Llorens, I.; Corrales, I.; Poschenrieder, C.; Barcelo, J.; Cruz-Ortega, R. Both aluminum and ABA induce the expression of an ABC-like transporter gene (*FeALS3*) in the Al-tolerant species. *Environ. Exp. Bot.* **2015**, *111*, 74–82. [CrossRef]
37. Larsen, P.B.; Tai, C.Y.; Kochian, L.V.; Howell, S.H. Arabidopsis mutants with increased sensitivity to aluminum. *Plant Physiol.* **1996**, *110*, 743–751. [CrossRef]
38. Huang, C.F.; Yamaji, N.; Ma, J.F. Knockout of a Bacterial-Type ATP-Binding Cassette Transporter Gene, *AtSTAR1*, Results in Increased Aluminum Sensitivity in Arabidopsis. *Plant Physiol.* **2010**, *153*, 1669–1677. [CrossRef]
39. Huang, C.F.; Yamaji, N.; Mitani, N.; Yano, M.; Nagamura, Y.; Ma, J.F. A Bacterial-Type ABC Transporter Is Involved in Aluminum Tolerance in Rice. *Plant Cell* **2009**, *21*, 655–667. [CrossRef]
40. Kochian, L.V.; Piñeros, M.A.; Liu, J.P.; Magalhaes, J.V. Plant Adaptation to Acid Soils: The Molecular Basis for Crop Aluminum Resistance. *Annu. Rev. Plant Biol.* **2015**, *66*, 571–598. [CrossRef]
41. Ma, J.F.; Chen, Z.C.; Shen, R.F. Molecular mechanisms of Al tolerance in gramineous plants. *Plant Soil* **2014**, *381*, 1–12. [CrossRef]
42. Peng, F.C.; Yuan, M.; Zhou, L.; Zheng, B.Q.; Wang, Y. Identification and Analysis of Aluminum-Activated Malate Transporter Gene Family Reveals Functional Diversification in Orchidaceae and the Expression Patterns of *Dendrobium catenatum* Aluminum-Activated Malate Transporters. *Int. J. Mol. Sci.* **2024**, *25*, 9662. [CrossRef] [PubMed]
43. Ligaba, A.; Dreyer, I.; Margaryan, A.; Schneider, D.J.; Kochian, L.; Piñeros, M. Functional, structural and phylogenetic analysis of domains underlying the Al sensitivity of the aluminum-activated malate/anion transporter, TaALMT1. *Plant J.* **2013**, *76*, 766–780. [CrossRef]
44. Wang, J.Q.; Yu, X.F.; Ding, Z.J.; Zhang, X.K.; Luo, Y.P.; Xu, X.M.; Xie, Y.; Li, X.X.; Yuan, T.; Zheng, S.J.; et al. Structural basis of ALMT1-mediated aluminum resistance in *Arabidopsis*. *Cell Res.* **2022**, *32*, 89–98. [CrossRef]
45. Lu, J.; Liu, X.; Ma, Q.J.; Kang, H.; Liu, Y.J.; Hao, Y.J.; You, C.X. Molecular cloning and functional characterization of the Aluminum-activated malate transporter gene. *Sci. Hortic.* **2019**, *244*, 208–217. [CrossRef]
46. Upadhyay, N.; Kar, D.; Datta, S. A multidrug and toxic compound extrusion (MATE) transporter modulates auxin levels in root to regulate root development and promotes aluminium tolerance. *Plant Cell Environ.* **2020**, *43*, 745–759. [CrossRef]
47. Lee, M.; Choi, Y.; Burla, B.; Kim, Y.Y.; Jeon, B.; Maeshima, M.; Yoo, J.Y.; Martinoia, E.; Lee, Y. The ABC transporter AtABCB14 is a malate importer and modulates stomatal response to CO_2. *Nat. Cell Biol.* **2008**, *10*, 1217–1223. [CrossRef]
48. Devi, R.; Goyal, P.; Verma, B.; Hussain, S.; Chowdhary, F.; Arora, P.; Gupta, S. A transcriptome-wide identification of ATP-binding cassette (ABC) transporters revealed participation of ABCB subfamily in abiotic stress management of *Glycyrrhiza glabra* L. *BMC Genomics* **2024**, *25*, 315. [CrossRef]
49. Fang, C.; Wu, J.; Liang, W. Systematic Investigation of Aluminum Stress-Related Genes and Their Critical Roles in Plants. *Int. J. Mol. Sci.* **2024**, *25*, 9045. [CrossRef]
50. Wen, K.; Pan, H.T.; Li, X.A.; Huang, R.; Ma, Q.B.; Nian, H. Identification of an ATP-Binding Cassette Transporter Implicated in Aluminum Tolerance in Wild Soybean (*Glycine soja*). *Int. J. Mol. Sci.* **2021**, *22*, 13264. [CrossRef]
51. Chen, S.S.; Qi, X.Y.; Feng, J.; Chen, H.J.; Qin, Z.Y.; Wang, H.D.; Deng, Y.M. Biochemistry and transcriptome analyses reveal key genes and pathways involved in high-aluminum stress response and tolerance in hydrangea sepals. *Plant Physiol. Biochem.* **2022**, *185*, 268–278. [CrossRef]
52. Sawaki, K.; Sawaki, Y.; Zhao, C.R.; Kobayashi, Y.; Koyama, H. Specific transcriptomic response in the shoots of *Arabidopsis thaliana* after exposure to Al rhizotoxicity: - Potential gene expression biomarkers for evaluating Al toxicity in soils. *Plant Soil* **2016**, *409*, 131–142. [CrossRef]
53. Sawaki, Y.; Iuchi, S.; Kobayashi, Y.; Kobayashi, Y.; Ikka, T.; Sakurai, N.; Fujita, M.; Shinozaki, K.; Shibata, D.; Kobayashi, M.; et al. STOP1 Regulates Multiple Genes That Protect Arabidopsis from Proton and Aluminum Toxicities. *Plant Physiol.* **2009**, *150*, 281–294. [CrossRef] [PubMed]
54. Agrahari, R.K.; Enomoto, T.; Ito, H.; Nakano, Y.; Yanase, E.; Watanabe, T.; Sadhukhan, A.; Iuchi, S.; Kobayashi, M.; Panda, S.K.; et al. Expression GWAS of *PGIP1* Identifies STOP1-Dependent and STOP1-Independent Regulation of *PGIP1* in Aluminum Stress Signaling in *Arabidopsis*. *Front. Plant Sci.* **2021**, *12*, 774687. [CrossRef] [PubMed]
55. Kobayashi, Y.; Ohyama, Y.; Kobayashi, Y.; Ito, H.; Iuchi, S.; Fujita, M.; Zhao, C.R.; Tanveer, T.; Ganesan, M.; Kobayashi, M.; et al. STOP2 Activates Transcription of Several Genes for Al- and Low pH-Tolerance that Are Regulated by STOP1 in. *Mol. Plant* **2014**, *7*, 311–322. [CrossRef]
56. Sadhukhan, A.; Kobayashi, Y.; Iuchi, S.; Koyama, H. Synergistic and antagonistic pleiotropy of STOP1 in stress tolerance. *Trends Plant Sci.* **2021**, *26*, 1014–1022. [CrossRef]
57. Fan, N.; Li, X.B.; Xie, W.X.; Wei, X.; Fang, Q.; Xu, J.Y.; Huang, C.F. Modulation of external and internal aluminum resistance by ALS3-dependent STAR1-mediated promotion of STOP1 degradation. *New Phytol.* **2024**, *244*, 511–527. [CrossRef]
58. Ishijima, S.; Manabe, Y.; Shinkawa, Y.; Hotta, A.; Tokumasu, A.; Ida, M.; Sagami, I. The homologous Arabidopsis MRS2/MGT/CorA-type Mg channels, AtMRS2-10 and AtMRS2-1 exhibit different aluminum transport activity. *Biochim. Biophys. Acta BBA-Biomembr.* **2018**, *1860*, 2184–2191. [CrossRef]
59. Magalhaes, J.V.; Liu, J.; Guimaraes, C.T.; Lana, U.G.P.; Alves, V.M.C.; Wang, Y.H.; Schaffert, R.E.; Hoekenga, O.A.; Piñeros, M.A.; Shaff, J.E.; et al. A gene in the multidrug and toxic compound extrusion (MATE) family confers aluminum tolerance in sorghum. *Nat. Genet.* **2007**, *39*, 1156–1161. [CrossRef]

60. Singh, D.; Tripathi, A.; Mitra, R.; Bhati, J.; Rani, V.; Taunk, J.; Singh, D.; Yadav, R.K.; Siddiqui, M.H.; Pal, M. Genome-wide identification of MATE and ALMT genes and their expression profiling in mungbean (*Vigna radiata* L.) under aluminium stress. *Ecotoxicol. Environ. Saf.* **2024**, *280*, 116558. [CrossRef]
61. Gao, P.X.; Han, R.R.; Xu, H.; Wei, Y.M.; Yu, Y.X. Identification of *MATE* Family and Characterization of *GmMATE1* and *GmMATE75* in Soybean's Response to Stress. *Int. J. Mol. Sci.* **2024**, *25*, 3711. [CrossRef]
62. Li, J.Y.; Liu, J.P.; Dong, D.K.; Jia, X.M.; McCouch, S.R.; Kochian, L.V. Natural variation underlies alterations in Nramp aluminum transporter (NRAT1) expression and function that play a key role in rice aluminum tolerance. *Proc. Natl. Acad. Sci. USA* **2014**, *111*, 6503–6508. [CrossRef]
63. Ofoe, R.; Thomas, R.H.; Asiedu, S.K.; Wang-Pruski, G.; Fofana, B.; Abbey, L. Aluminum in plant: Benefits, toxicity and tolerance mechanisms. *Front. Plant Sci.* **2023**, *13*, 1085998. [CrossRef]

horticulturae

MDPI

rticle

Transcriptomic Analysis Reveals Calcium and Ethylene Signaling Pathway Genes in Response to Cold Stress in *Cinnamomum camphora*

o Bi [1,2,3], Lingmei Shao [1], Tong Xu [1], Hao Du [1,3] and Danqing Li [4,*]

1 Department of Horticulture, College of Agriculture and Biotechnology, Zhejiang University, Hangzhou 310058, China; 0621973@zju.edu.cn (B.B.)
2 Crop Station, Agriculture and Rural Bureau of Qingtian County, Lishui 323900, China
3 ZJU-Hangzhou Global Scientific and Technological Innovation Centre, Hangzhou 311200, China
4 Department of Landscape Architecture, School of Civil Engineering and Architecture, Zhejiang Sci-Tech University, Hangzhou 310018, China
* Correspondence: danqingli@zju.edu.cn

Abstract: *Cinnamomum camphora* is one of the most dominant broad-leaved evergreen trees in tropical and subtropical regions. Understanding its response to cold stress is crucial for enhancing its resilience to climate changes and expanding the cultivation range. Cold stress response is a vital strategy for plants to withstand cold stress, typically involving transcriptional changes across various pathways. In this study, RNA-Sequencing (RNA-Seq) was conducted on the leaves of *C. camphora* subjected to different cold stress treatments (0 h, 2 h, and 12 h). Transcriptome analyses revealed that short-term cold stress treatment rapidly induced the upregulation of genes associated with calcium and ethylene signaling pathways, including *GLR2.7*, *CaM*, *CPK7*, and *ERF1/3/4/5/7*. Subsequently, 12 h cold response treatment further activated genes related to the cold response, jasmonic acid signaling pathways, and the negative regulation of cellular biosynthetic processes, such as *CBF2* and *CBF4*. Notably, ERFs emerged as the most differentially expressed transcription factors in this study. A total of 133 ERF family members from *C. camphora* were identified through phylogenetic analysis, and these ERFs were classified into 12 clusters. Many of these ERFs are likely to play pivotal roles in the cold response of *C. camphora*, especially *ERF1/3/4/5/7*. These findings offer novel insights into the mechanisms underlying the cold response and present valuable candidates for further research, advancing our understanding of plant responses to cold stress.

Keywords: *Cinnamomum camphora*; cold stress; calcium signaling; ethylene; ERF family; transcriptomics

1. Introduction

Low temperatures are a major environmental stress that restricts plant growth, development, and geographical distribution, further impacting flowering and yield in the subsequent growing season [1–3]. In tropical and subtropical regions, 0–15 °C chilling temperatures are the primary cold stress. Upon exposure to chilling temperatures, plants can develop enhanced freezing tolerance through a process known as cold acclimation [4]. To withstand chilling stress, plants have evolved various regulatory mechanisms, including multiple signal transduction pathways [5,6], hormone homeostasis [7], osmoprotectants [8], and reactive oxygen species (ROS) scavenging [9]. In recent years, research on the cold stress response process of evergreen woody plants has garnered increasing attention due to their significant economic, ornamental, and ecological value, as well as the impact of severe climate abnormal changes.

The regulatory mechanisms underlying the cold stress response in model plants have been extensively investigated. Numerous studies have demonstrated that the perception and transduction of cold signals are two pivotal steps in the plant cold response [10]. Low temperatures may alter the structure of the plasma membrane, leading to cytoskeletal rearrangement, a process involving the perception of ion channels, transport proteins, and membrane-localized protein kinases [11]. In plant cells, calcium signaling acts as a secondary messenger and is an early signaling molecule in the cold response [12]. Under cold stress conditions, calcium enters the cells through calcium channels and is detected using calcium sensors, which then triggers the expression of downstream genes [13]. Various calcium sensors have been identified in plants, such as CALMODULIN (CaM), CALMODULIN-LIKE (CML), CALCINEURIN B-LIKE PROTEIN (CBL), CALCIUM DEPENDENT PROTEIN KINASE (CPK), and CALCIUM DEPENDENT PROTEIN KINASE (CDPK) [14]. Furthermore, secondary signal transduction is mediated by the calcium binding protein calmodulin. For example, CALMODULIN-BINDING TRANSCRIPTION ACTIVATOR (CAMTA) is a calcium-loaded CaM-dependent transcription factor [15]. Cold-induced calcium signaling also promotes the production of ROS and the Mitogen-activated protein kinase (MAPK) cascade [16,17]. Ultimately, these pathways further regulate the downstream cold response cascade involving ICE-CBF-COR.

Phytohormones also play significant signaling roles in regulating the cold response. Increasing evidence has shown that ethylene is integral to the plant cold response, with cold stress affecting ethylene biosynthesis and signal transduction [18]. For instance, exposure to 4 °C can promote ethylene release in grape (*Vitis vinifera*) seedlings, and ethylene positively modulates the cold response in apples (*Malus domestica*) [19,20]. ETHYLENE RESPONSIVE FACTOR (ERF) family members are crucial in mediating plant cold acclimation [21]. ERF102 and ERF103 are indispensable for cold acclimation in *Arabidopsis thaliana* [22]. The ERF family members contain an APETALA2/ERF (AP2/ERF) domain [23]. This family is further divided into two major subfamilies: the ERF subfamily and the DEHYDRATION RESPONSE ELEMENT/DEHYDRATION-RESPONSIVE ELEMENT BINDING PROTEIN (CBF/DREB) subfamily. The role of CBF/DREB in the cold response is well-established. Simultaneously, ERFs are likely key hubs in the ethylene-mediated cold response. ERFs rely on ethylene signal-mediated cold response pathways to regulate plant freezing tolerance [20]. They can initiate the expression of downstream *CBF* and *COLD-REGULATED* (*COR*) genes, thereby influencing plant freezing tolerance [24]. In summary, ethylene plays a crucial regulatory role in the cold response, yet the underlying molecular mechanisms remain unclear.

Cinnamomum camphora is a broad-leaved evergreen tree belonging to the Lauraceae family, renowned as a prominent component in tropical and subtropical urban landscapes and valued for its significant economic and ornamental contributions [25]. The Lauraceae family includes approximately 45 genera and around 2500 to 3000 species. In contrast to the widespread distribution of *C. camphora*, most Lauraceae species are confined to tropical regions [25,26]. Li et al. (2023) [26] identified the genes responsible for cold acclimation in *C. camphora* through comparative genomic analysis between *C. camphora* and its close relative *C. kanehrae*, revealing the factors contributing to its dominance in subtropical urban landscapes. Studying the cold response mechanisms of *C. camphora* offers valuable theoretical insights and potential candidate genes to enhance the environmental adaptability of plants within the Lauraceae family. In this study, RNA-Sequencing (RNA-Seq) was conducted on *C. camphora* leaves subjected to different cold stress treatments. Transcriptome analysis revealed differentially expressed genes (DEGs) related to the cold response under both 2 h and 12 h cold stress responses. Our findings elucidate key pathways and genes involved in the cold response of *C. camphora*, providing a substantial number of candidate genes for genetic improvement to enhance cold resilience and expand the cultivation range of Lauraceae plants.

2. Materials and Methods

2.1. Plant Materials and Cold Stress Treatments

Branches with leaves were harvested from three 30-year-old *C. camphora* trees located in Hangzhou (30° N, 120° E), Zhejiang Province, China. Branches with leaves at 25 °C were subjected to different cold stress treatments (0 h, 2 h, and 12 h) in a chamber and were divided into three groups: CK (plants with no treatment), CA2 (2-h treatment at 4 °C), and CA12 (12-h treatment at 4 °C). After different cold stress treatments in branches of *C. camphora*, we randomly selected three mature *C. camphora* leaves from each of the branches as a biological replicate. Three biological replicates were collected, rapidly frozen in liquid nitrogen, and subsequently stored at −80 °C until further analysis.

2.2. RNA Extraction

The total RNA from *C. camphora* leaves was extracted using the RNAprep Pure Plant Plus Kit (DP441, Tiangen, Beijing, China). The RNA concentration and purity were quantified using a Nanodrop ND-1000 spectrophotometer (Isogen Life Science, Utrecht, The Netherlands). The quality and integrity of the RNA were evaluated using 1% agarose gel electrophoresis.

2.3. Transcriptome Analysis

RNA-Seq was conducted on *C. camphora* leaves subjected to various cold stress treatments (0 h, 2 h, and 12 h). Three biological replicates for each treatment were used to construct cDNA libraries following the Illumina standard operating procedure. Then, RNA-seq was performed on the Illumina X-ten platform with 150-bp paired-end reads to obtain raw data. We downloaded RNA-seq data from the NCBI BioProject database under the accession number PRJNA1000241 (https://www.ncbi.nlm.nih.gov/bioproject/PRJNA1000241, accessed on 4 July 2024), which was uploaded by our team during the publication of the *C. camphora* genome. Genome assembly and gene annotations were downloaded from figshare (https://doi.org/10.6084/m9.figshare.20647452.v1, accessed on 4 July 2024). RNA-seq reads were filtered with FASTP (v.0.12.4) [27], mapped to genome assemblies using HiSAT2 (v.2.2.0) [28], and analyzed for transcripts per kilobase of exon model per million mapped reads (TPM) with StringTie (v.2.1.4) [29]. Transcription factors were identified using PlantTFDB v5.0 (https://planttfdb.gao-lab.org/, accessed on 10 July 2024).

DEGs were identified using RStudio with the DESeq2 R package. Genes with $|\log2(\text{fold change})| > 1$ and Padj < 0.01 were classified as DEGs. The accuracy of the RNA-Seq was validated by quantitative real-time PCR (qRT-PCR). Gene ontology (GO) enrichment analysis and weighted gene co-expression network analysis (WGCNA) were conducted using RStudio, employing the topGO and WGCNA R packages, respectively.

2.4. Identification of ERF Proteins in *C. camphora*

To identify ERF proteins in *C. camphora*, HMM Search and BLAST screening were conducted using TBtools v2.110 [30]. Specifically, the domain of the ERF family, the AP2/ERF domain (PF00847), was downloaded as an hmm file from the Pfam database (http://pfam.xfam.org/, accessed on 10 July 2024). This was used as a seed for identifying ERF proteins against all protein sequences of *C. camphora* using the Simple HMM Search function in TBtools v2.110 (E-value threshold 0.05). BLAST screening (E-value threshold 1×10^{-5}) was conducted to identify ERF proteins in *C. camphora* based on ERF protein sequences from *Arabidopsis* (derived from the TAIR database). The sequences obtained from both methods were analyzed using the CDD search tool from the NCBI website (https://www.ncbi.nlm.nih.gov/Structure/cdd/wrpsb.cgi, accessed on 12 July 2024), and the results were visualized with TBtools v2.110. Ultimately, only those sequences featuring a single AP2 domain were selected to identify all ERF proteins in *C. camphora* [31].

2.5. Phylogenetic Analysis

To clarify the evolutionary relationship of ERF proteins in *C. camphora*, a phylogenetic tree was constructed that includes all ERF proteins from both *Arabidopsis* and *C. camphora*. According to the study by Nakano et al. (2006) [31], the ERF protein sequences of *Arabidopsis* were downloaded from the TAIR database (http://www.arabidopsis.org/, accessed on 12 July 2024). Subsequently, multiple sequence alignment of ERF proteins from *Arabidopsis* and *C. camphora* was performed using MAFFT v7.467 with default parameters [32]. FastTree v2.1.12 was used to construct maximum-likelihood phylogenetic trees employing the JTT+CAT model [33]. MEGA 7.0 was utilized for visualizing and enhancing the phylogenetic trees [34].

2.6. Reverse Transcription and qRT-PCR Analysis

RNA (1 μg) was reverse transcribed into cDNA using the PrimeScript™ RT Reagent Kit with gDNA Eraser (TaKaRa, Kyoto, Japan). The cDNA was diluted 20-fold with deionized water and utilized as the template for qRT-PCR, following the protocol established by Wang et al. (2020) [35].

The primers used for qRT-PCR, listed in Table S1, were designed using the Primer-BLAST tool from NCBI (https://www.ncbi.nlm.nih.gov/tools/primer-blast/index.cgi, accessed on 20 July 2024), with a product length ranging from 200 to 300 bp, an annealing temperature of approximately 60 °C, and a GC content of 40% to 60%. *UBC22* was used as a reference gene for normalizing gene expression levels [36]. qRT-PCR was conducted on the CFX Connect™ Real-Time PCR Detection System (Bio-Rad, Hercules, CA, USA) using the TB Green® Premix Ex Taq enzyme (TaKaRa, Kyoto, Japan). The reaction system consisted of 4 μL of 20-fold diluted cDNA, 5 μL of TB Green, and 0.5 μL 10 μmol/L each of forward and reverse primers. The reaction program was as follows: 95 °C for 2 min; then 95 °C for 5 s, 55 °C for 30 s, repeated for 39 cycles; followed by a melting curve analysis of 95 °C for 5 s, 65 °C for 5 s, and 95 °C for 5 s. The temperature increments used between 65 °C and 95 °C were set at 0.5 °C per 5 s to check the specificity of the reaction. Finally, relative gene expression levels for each sample were evaluated using the $2^{-\Delta\Delta Ct}$ method [37].

2.7. Statistical Analyses

Statistical analyses were carried out utilizing SPSS 23 (IBM Corporation, Armonk, NY, USA). One-way ANOVA and Duncan's multiple range test were employed to compare differences among various samples, with the significance threshold set at 0.05 (denoted by different letters). All data comprised a minimum of three biological replicates.

3. Results

3.1. Functional Enrichment Analysis of DEGs between Different Cold Stress Treatments

Mature leaves of *C. camphora* that underwent different cold stress treatments were randomly selected for phenotype observation. The results showed that there were no significant phenotype changes in mature leaves of *C. camphora* after different cold stress treatments, although water-soaked lesions increased with prolonged exposure to cold stress (Figure S1).

Furthermore, we obtained a total of 83.13 Gb raw data, and the average mapping rate was over 90% as calculated by HiSAT2 (v.2.2.0) [28]. RNA-Seq analysis revealed a total of 34,918 genes expressed in at least one sample [with TPM > 2]. Furthermore, a total of 1377 DEGs were identified across the different cold stress treatments, as illustrated in Figure S2. To investigate the key pathways and genes involved in the cold response of *C. camphora* under different cold stress treatments, GO enrichment analysis was conducted on DEGs comparing the different treatments, including CK vs. CA2, CA2 vs. CA12, and CA12 vs. CK (Figure 1). The results indicated that DEGs following 2 h cold stress treatment (CK vs. CA2) were predominantly enriched in the biological processes of "calcium ion transport", "calcium ion transmembrane transport", "ethylene-activated signaling pathway", and "cellular response to hypoxia". Compared to the 2 h cold stress treatment, after 12 h cold stress treatment DEGs (CA2 vs. CA12) were primarily involved in the biological processes of "regulation of signal transduction", "response to cold", and "regulation of the jasmonic acid-mediated signaling pathway". The DEGs in the comparison between CK and CA12 were mainly enriched in the biological processes of "response to cold", "negative regulation of cellular biosynthetic", and "cellular response to oxygen levels".

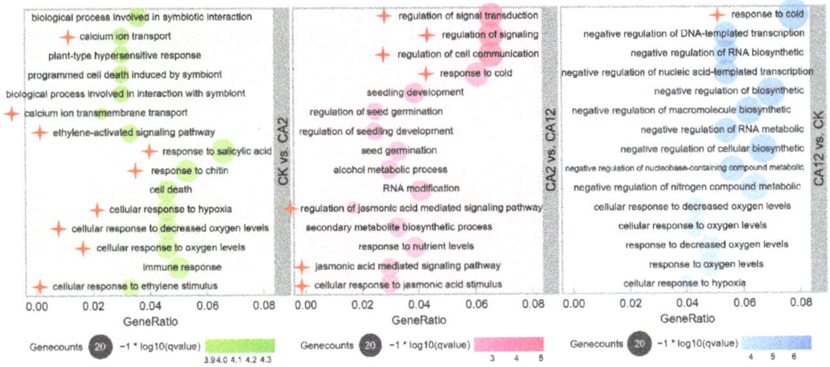

Figure 1. Enriched gene ontology (GO) terms of differentially expressed genes under different cold stress treatments in *Cinnamomum camphora*. The figure shows the top 15 results of biological processes of GO enrichment analysis between different treatments. The red signs highlight the biological processes that may be related to the *C. camphora* cold response. CK, plants without any treatment; CA2, 2-h treatment under 4 °C cold stress; CA12, 12-h treatment under 4 °C cold stress.

3.2. WGCNA—Revealed Key Pathways and Gene Responses to Cold

To identify key stage-specific modules, module-stage association analyses were conducted using all 1448 DEGs through WGCNA. In total, six distinct co-expression modules were identified (Figure 2A). Notably, MEblack and MEmidnightblue were highly correlated with the CA2 stage (correlation coefficients of 0.86 and 0.78, respectively, $p < 0.01$). Moreover, MEblue and MEgreenyellow showed a significant positive correlation with the CA12 stage (correlation coefficients of 0.95 and 0.73, respectively, $p < 0.05$). Based on the correlation between modules and stages, MEblack, MEmidnightblue, MEblue, and MEgreenyellow were identified as key modules related to the cold response.

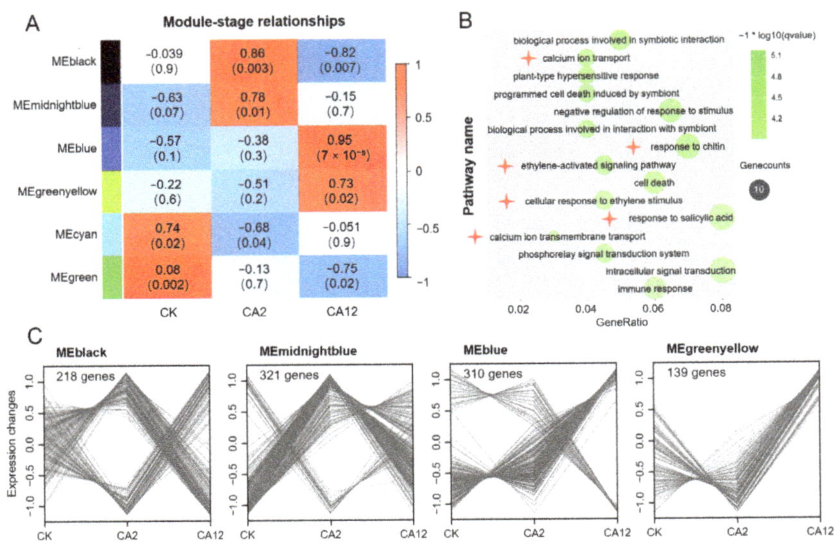

Figure 2. Weighted gene co-expression network analysis of *Cinnamomum camphora* under different cold stress treatments. (**A**) Module-stage relationships. CK, plants without any treatment; CA2, 2-h treatment under 4 °C cold stress; CA12, 12-h treatment under 4 °C cold stress. (**B**) Enriched gene ontology terms of differentially expressed genes in four key modules (MEblack, MEmidnightblue,

MEblue, and MEgreenyellow). The red signs highlight the biological processes that may be related to the *C. camphora* cold response. (**C**) Gene expression patterns of four key modules. The expression values were Z-scaled $\log_2$(TPM+1).

GO enrichment analysis was conducted on all DEGs within these four key cold response-related modules (Figure 2B). The results indicated that the DEGs within cold response-related modules were predominantly enriched in the biological processes of "calcium ion transport", "calcium ion transmembrane transport", "ethylene-activated signaling pathway", and "cellular response to ethylene stimulus". This is consistent with the findings for DEGs following 2 h cold stress treatment (CK vs. CA2). Additionally, there were 218, 321, 310, and 139 genes in MEblack, MEmidnightblue, MEblue, and MEgreenyellow, respectively (Figure 2C). The gene expression patterns of MEblack and MEmidnightblue exhibited two distinct trends: one where DEGs were significantly upregulated during the CA2 stage and then downregulated back to CK stage levels during the CA12 stage; and another where DEGs were significantly downregulated during the CA2 stage and subsequently restored during the CA12 stage. The gene expression patterns of MEblue were characterized by two main trends: most DEGs showed a significant upregulation with the cold response, while a smaller portion displayed a significant downregulation. For MEgreenyellow, the gene expression pattern predominantly showed a significant upregulation with the cold response.

3.3. Impact of Cold Stress Treatment on Calcium Signaling-Related Gene Expression

The results of GO enrichment analyses and WGCNA consistently highlighted that transcriptional changes during cold stress in *C. camphora* were predominantly associated with the calcium and ethylene signaling pathways. Consequently, we initially visualized the stage-specific expression patterns of DEGs involved in the calcium signaling pathway (Figure 3A). The results indicated that genes encoding proteins associated with calcium ion transport were significantly upregulated at both the CA2 and CA12 stages, such as *CYCLIC NUCLEOTIDE GATED CHANNEL* (*CNGC*) and *GLUTAMATE RECEPTOR* (*GLR*) (Figure 3B). Genes encoding calcium-sensing proteins, including *CaM* and *CML*, were predominantly upregulated at the CA2 stage. Additionally, genes encoding calcium-binding sensory proteins, such as *CPK*, were significantly upregulated at both the CA2 and CA12 stages. Furthermore, the calcium sensor target gene *CIPK* was predominantly upregulated at the CA12 stage.

In addition to genes directly involved in calcium signaling, those associated with MAPK pathways were also markedly upregulated during cold stress treatment, such as *MAPK/ERK KINASE KINASE* (*MAPKKK*) and *MITOGEN-ACTIVATED PROTEIN KINASE* (*MPK*) (Figure 3B). Similarly, cold-responsive genes such as *CBF* were significantly upregulated at both the CA2 and CA12 stages, with a notable increase at the CA12 stage.

To further validate the expression levels of genes related to the calcium signaling pathway, qRT-PCR analyses were conducted (Figure 3C). The qRT-PCR results were in concordance with the RNA-Seq. Overall, calcium signaling pathway genes were rapidly upregulated at the CA2 stage, whereas cold response and MAPK pathway genes exhibited delayed upregulation compared to the calcium signaling pathway genes. These results suggest that the calcium signaling pathway may play a crucial role in the early cold response of *C. camphora* leaves.

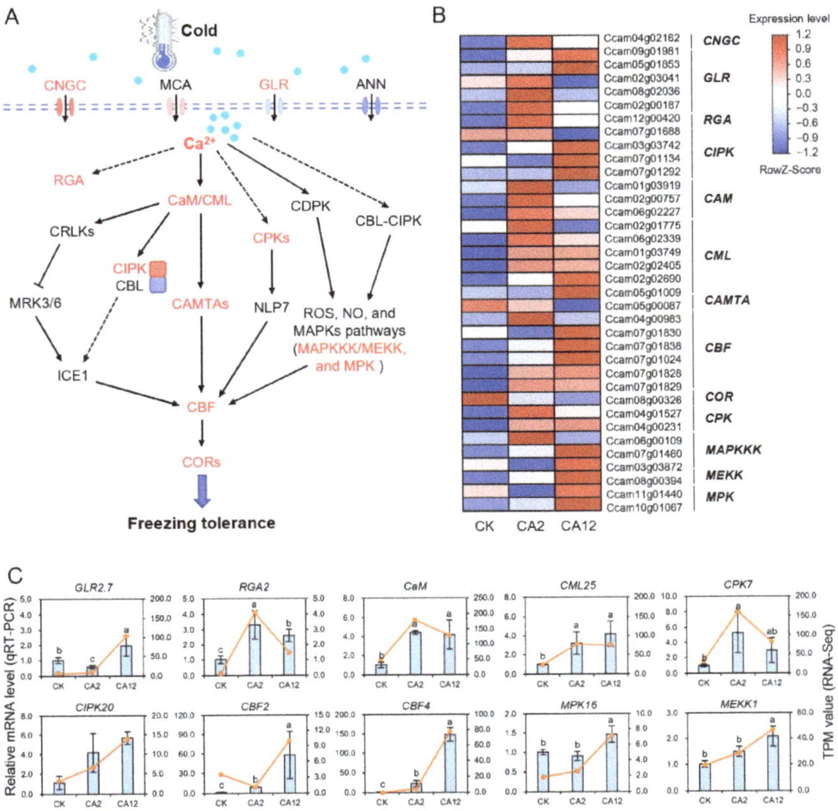

Figure 3. Expression patterns of differentially expressed genes (DEGs) involved in the calcium signaling pathway. (**A**) Calcium signaling-driven plant response to cold stress. The genes showing evidently different expression patterns are highlighted. (**B**) Heatmap of DEGs in the calcium signaling pathway of *Cinnamomum camphora* under different treatments. CK, plants without any treatment; CA2, 2-h treatment under 4 °C cold stress; CA12, 12-h treatment under 4 °C cold stress. The expression values were Z-scaled $\log_2$(TPM+1). (**C**) Quantitative real-time PCR expression levels of genes involved in the calcium signaling pathway. Error bars represent ± SE of three biological replicates. Different letters indicate significant differences at $p < 0.05$, following statistical analysis by one-way ANOVA. Gene names are listed in Table S4.

3.4. Alterations in Gene Expression Related to Ethylene Biosynthesis and Signaling Pathways

Under cold stress treatment, significant changes were observed in genes associated with the ethylene biosynthesis and signaling pathways (Figure 4A). Specifically, the key regulatory gene in ethylene biosynthesis, *ACC SYNTHASE* (*ACS*), was markedly upregulated at both the CA2 and CA12 stages, with a particularly pronounced increase at the CA2 stage (Figure 4B). ERF family genes, which play a positive regulatory role in the ethylene signaling pathway, exhibit a variety of expression patterns (Figure 4B). The 36 differentially expressed ERF family genes can be broadly categorized into four expression patterns. The first pattern demonstrates a decrease in expression following cold stress treatment, with downregulation observed at the CA2 or CA12 stages. The second pattern reveals consistently high expression levels at both the CA2 and CA12 stages. The third pattern is characterized by a significant upregulation after 2 h cold stress treatment (CA2 stage). The fourth pattern exhibits substantial upregulation following a prolonged cold stress treatment (CA12 stage).

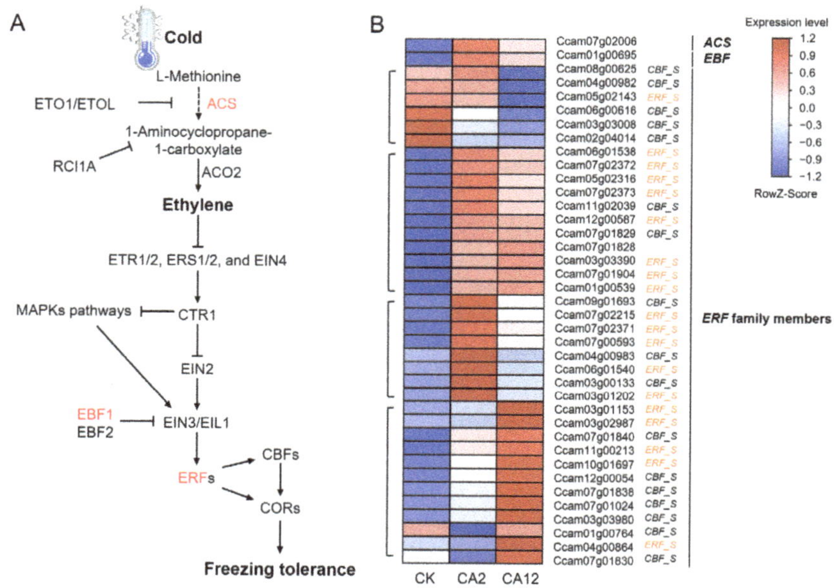

Figure 4. Expression patterns of differentially expressed genes (DEGs) involved in the ethylene biosynthesis and signaling pathway. (**A**) Diagram of the ethylene-mediated cold stress response in plants; (**B**) Heatmap of DEGs in the ethylene biosynthesis and signaling pathway of *Cinnamomum camphora* under different treatments. The expression values were Z-scaled $\log_2$(TPM+1). CK, plants without any treatment; CA2, 2-h treatment under 4 °C cold stress; CA12, 12-h treatment under 4 °C cold stress. CBF_S, C-REPEAT BINDING FACTOR subfamily; ERF_S, ERF DOMAIN PROTEIN subfamily. Gene names are listed in Table S4.

3.5. Phylogenetic Characterization of ERF Gene Family Members in *C. camphora*

To investigate the impact of cold stress treatment on the expression levels of transcription factors, the number of differentially expressed transcription factors in the four cold response-related modules was calculated (Figure 5A). Among these, the ERF, MYB DOMAIN PROTEIN (MYB), and WRKY DNA-BINDING PROTEIN (WRKY) transcription factor families exhibited the highest differential expression ratios in *C. camphora*.

Given that the ERF family is a large gene family of transcription factors and part of the AP2/ERF superfamily [38], the proteins of the ERF family are characterized by containing a single AP2/ERF domain. A total of 133 ERF genes from *C. camphora* were identified using Simple HMM Search and BLAST, with ERF proteins from Arabidopsis. Phylogenetic analysis was conducted to identify ERF genes in *C. camphora* (Figure 5B). The phylogenetic tree includes 133 of the AtERF protein sequences and 133 of the CcERF protein sequences, categorized into 12 main clusters. These 12 clusters were named according to the results by Nakano et al. (2006) [31] as I to X, VI-like (VI-L), and Xb-like (Xb-L). Clusters A-1 to A-6 belong to the CBF/DREB subfamily, while clusters B-1 to B-6 belong to the ERF subfamily. Additional detailed information on the CcERF proteins is available in Table S3.

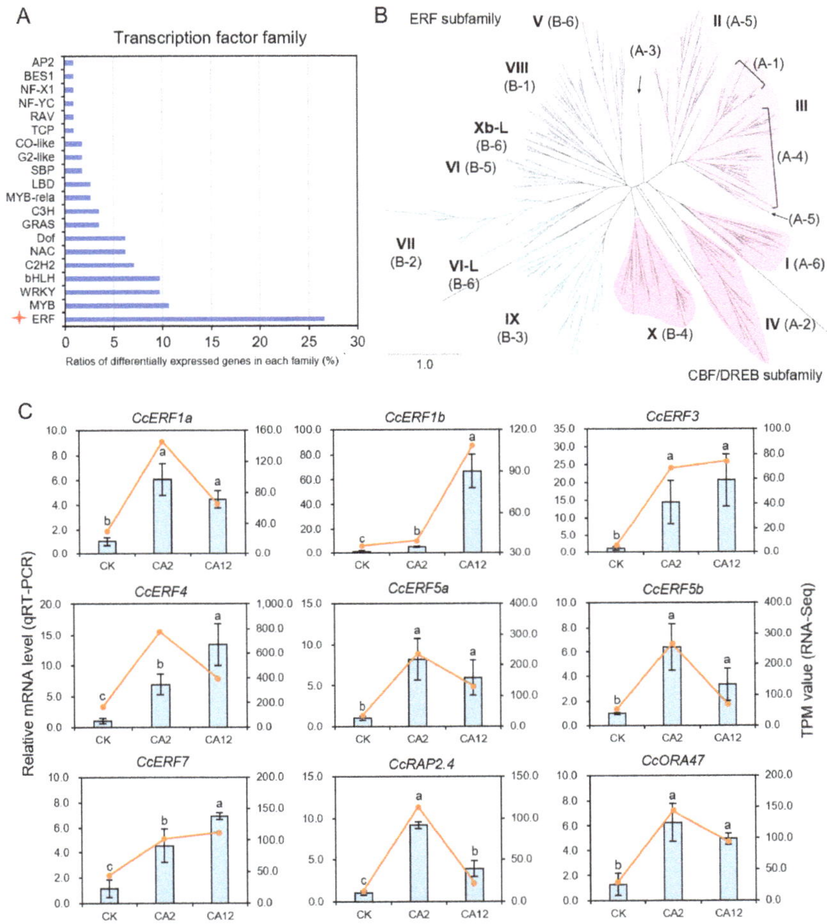

Figure 5. Differentially expressed transcription factor families and phylogenetic analysis of ERF proteins. (**A**) Transcription factor family analysis of differentially expressed genes (DEGs) in four key modules (MEblack, MEmidnightblue, MEblue, and MEgreenyellow). We have marked the transcription factor family with the highest ratio, ERF, using red signs. (**B**) The phylogenetic analysis of ERF family proteins in *Arabidopsis thaliana* and *Cinnamomum camphora*. (**C**) Quantitative real-time PCR expression levels of ERF subfamily genes. Different letters indicate significant differences at $p < 0.05$, following statistic analysis by one-way ANOVA. Gene names are listed in Table S4.

qRT-PCR was performed on nine ERF subfamily members, and the results were consistent with the RNA-Seq data (Figure 5C). All nine ERF subfamily genes were markedly upregulated following 2 h cold stress treatment. Moreover, considering the phylogenetic analysis and the annotation of DEGs identified as ERF (Figure 4B), it was found that the upregulation of ERF subfamily genes occurred earlier than that of CBF/DREB subfamily genes.

4. Discussion

C. camphora is a widely distributed evergreen broadleaf tree, valued for its significant ornamental and economic value. In addition to *C. camphora*, many other species in the Lauraceae family possess considerable practical value. However, their limited cold tolerance restricts these species primarily to tropical regions, thereby constraining the development and utilization of Lauraceae family resources. The extensive distribution of

C. camphora may be attributed to its unique cold response mechanisms. Under cold stress, plants transmit cold signals via various regulatory pathways, initiating the expression of downstream genes that subsequently regulate multiple transcriptional and metabolic processes, ultimately enhancing freezing tolerance. This study aims to explore the key regulatory pathways and genes involved in the cold stress response of *C. camphora* by conducting RNA-Seq on leaves subjected to different cold stress treatments.

Through GO enrichment analysis and WGCNA, we identified four key modules related to the cold response. Notably, genes in the MEblack and MEmidnightblue modules were predominantly upregulated after 2 h of 4 °C treatment, while genes in the MEblue and MEgreenyellow modules showed upregulation after 12 h of 4 °C treatment (Figure 2). Further analysis revealed that the calcium and ethylene signaling pathways are the earliest activated in the *C. camphora* cold response (Figure 2). Conversely, transcriptional changes in the salicylic acid, jasmonic acid, and ICE-CBF-COR cascades become prominent after 12 h cold stress treatment (Figure 6). Consistent with our findings, signaling pathways involved in the perception and transmission of cold stress signals, such as the calcium signaling pathway, have been identified in the early gene expression profiles of many plants [38,39].

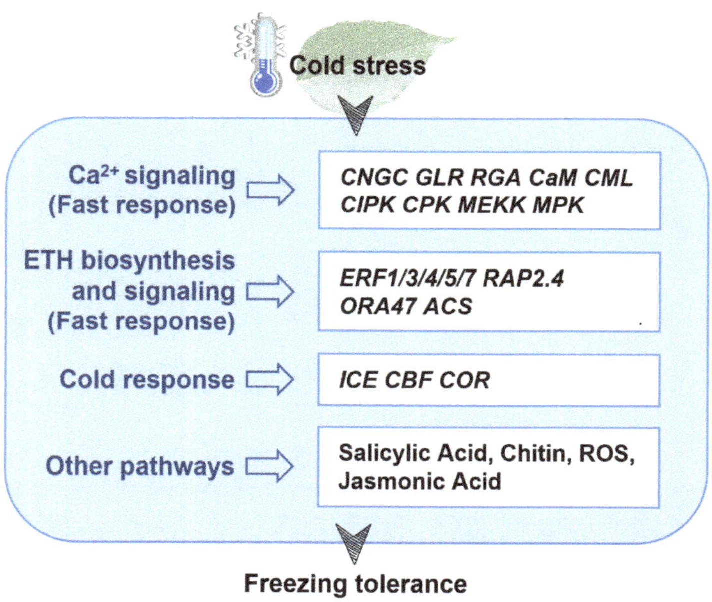

Figure 6. Key pathways and genes associated with the cold response process of *Cinnamomum camphora* following cold stress treatment. Gene names are listed in Table S4.

Under cold stress, calcium rapidly influxes into plant cells, triggering molecular mechanisms that enable plants to adapt to the environment [40]. Exogenous calcium can enhance the cold resistance of maize (*Zea mays*) and tea (*Camellia sinensis*) [41,42]. Calcium channels and sensors have been widely studied in the cold stress response. In rice (*Oryza sativa*), transcriptional activation and phosphorylation of OsCNGC9 confer enhanced chilling tolerance [43]. Cold acclimation elevates the expression of tomato (*Solanum lycopersicum*) *GLR3.3* and *GLR3.5*, subsequently enhancing chilling tolerance [44]. In *C. camphora*, the calcium channel genes *CNGC* and *GLR* were rapidly upregulated after 2 h of 4 °C treatment (Figure 3), suggesting that these genes may play a positive regulatory role in the cold response of *C. camphora*.

Additionally, various calcium sensors in *C. camphora* also respond promptly to cold stress, with significant upregulation of *CaM*, *CML*, *CPK*, and *RGA* (Figure 3). Among these,

CaM is one of the most essential calcium sensors. CaM plays a crucial role in positively regulating cold resistance in tea plants [45]. In rice, *OsCPK24* is upregulated in response to cold and positively regulates chilling tolerance [46]. Moreover, the CaM4/PATL1 complex, CML42, and CIPK13 positively regulate the expression of downstream *CBF* genes [47–49]. In *C. camphora*, the upregulation of genes in the MAPK pathway and the CBF-COR module occurs later than that of calcium channel and sensor genes, possibly because the calcium signaling pathway is more upstream in the cold response process. Based on these findings, we speculate that calcium channels and sensor proteins in *C. camphora* may enhance freezing tolerance by modulating downstream CBF genes.

Ethylene also plays a crucial role in plant stress resistance. Cold stress significantly induces the expression of the apple ethylene synthesis gene *MdACS1*, which positively modulates the apple's responses to cold stress [20]. Similarly, a 4 °C treatment significantly induces the expression of the *Gossypium hirsutum ACS* gene [50]. In apple and grape seedlings, exposure to 4 °C rapidly triggers ethylene release, whereas blocking ethylene signaling reduces the sensitivity of grape seedlings to cold stress [19,20]. Cold storage also induces the expression of the *LeACS2* gene and promotes ethylene synthesis in tomato fruit [51]. Conversely, ethylene negatively regulates cold tolerance in *Medicago truncatula* and *Arabidopsis* [52,53]. In this study, 2 h of 4 °C treatment rapidly induced the expression of the *C. camphora ACS1* gene (Figure 4), suggesting that ethylene may positively regulate the response of *C. camphora* to cold stress.

ERF family members in the ethylene signaling pathway have been identified as a key regulator in plant responses to cold stress [21]. A 4 °C treatment rapidly induces the expression of *MdERF1B*, which encodes an ethylene signal activator, in apple seedlings. Citrus *ERF1*, *RsERF40*, *BpERF13*, and *MdERF1B* have been demonstrated to enhance cold tolerance [20,54–56]. Moreover, *ERF102*, *ERF103*, and *ERF105* are essential for cold acclimation responses in *Arabidopsis* [22,57]. In grapes, the expression of *VaERF057* and *VaERF092* is inhibited by the ethylene biosynthesis inhibitor, thereby diminishing their roles in cold tolerance [19,58]. In this study, ERF family genes showed significant differential expression following cold treatment, especially *ERF1/3/4/5/7* (Figure 4). Phylogenetic analysis indicated that most ERF subfamily genes were rapidly upregulated after 2 h of 4 °C exposure, whereas CBF/DREB subfamily genes were mainly upregulated after 12 h of 4 °C exposure. In tomato, TERF2/LeERF2 is a transcription factor that acts upstream of *CBF*, mediating cold tolerance by feedback-regulating ethylene biosynthesis genes [24]. Therefore, we hypothesize that ERFs such as ERF1/3/4/5/7 in *C. camphora* may similarly function upstream of *CBF* genes, responding rapidly to cold stimuli and regulating the expression of downstream *CBF* genes, ultimately enhancing cold tolerance. The diverse roles of ERF subfamily members in cold stress responses merit further investigation.

5. Conclusions

In this study, transcriptome analysis revealed that genes involved in calcium signaling, ethylene biosynthesis and signaling, and cold response pathways were upregulated following cold stress treatment. The 2 h cold stress treatment rapidly activated gene expression related to calcium signaling and ethylene biosynthesis and signaling pathways. In contrast, the 12 h cold stress treatment further activated processes related to the cold response, jasmonic acid signaling pathways, and the negative regulation of cellular biosynthetic processes. Additionally, this study preliminarily identified 133 ERF family members in *C. camphora*, which were divided into 12 clusters. qRT-PCR results indicated that many ERFs likely play a crucial role in the cold response of *C. camphora* leaves, especially *ERF1/3/4/5/7*. This study suggested that the response of *C. camphora* to cold stress is closely related to the calcium and ethylene signaling pathways, providing a wealth of candidate genes for further research on the cold response in *C. camphora*.

Supplementary Materials: The following supporting information can be downloaded at: https://www.mdpi.com/article/10.3390/horticulturae10090995/s1, Table S1: Primers used in qRT-PCR; Table S2: Gene Ontology (GO) enrichment analysis of RNA-Seq data in the cold response of *Cinnamomum camphora*; Table S3: Details of the ERF family members in *Cinnamomum camphora*; Table S4: Gene name; Figure S1: Phenotype observation of the leaves of *Cinnamomum camphora* under different cold stress treatments; Figure S2: Numbers of up- and down-regulated DEGs in comparisons between different cold stress treatments of *Cinnamomum camphora*; Figure S3: The raw figure of phylogenetic analysis of ERF family proteins in Arabidopsis thaliana and *Cinnamomum camphora*.

Author Contributions: Conceptualization: B.B. and D.L.; writing—original draft preparation, B.B.; investigation, D.L., B.B., L.S. and T.X.; form analysis and visualization, B.B., L.S. and T.X.; writing—review and editing, H.D. and D.L.; funding acquisition, D.L. All authors have read and agreed to the published version of the manuscript.

Funding: This research was funded by the China Postdoctoral Science Foundation (No. 2021M692790) and the Zhejiang Sci-Tech University Start-up Fund (No. 24052162-Y).

Data Availability Statement: The original contributions presented in this study are included in the article/Supplementary Materials, further inquiries can be directed to the corresponding authors.

Conflicts of Interest: The authors declare no conflicts of interest.

References

1. Ding, Y.; Yang, S. Surviving and thriving: How plants perceive and respond to temperature stress. *Dev. Cell* **2022**, *57*, 947–958. [CrossRef] [PubMed]
2. Hasanuzzaman, M.; Nahar, K.; Hossain, M.S.; Mahmud, J.A.; Rahman, A.; Inafuku, M.; Oku, H.; Fujita, M. Coordinated Actions of Glyoxalase and Antioxidant Defense Systems in Conferring Abiotic Stress Tolerance in Plants. *Int. J. Mol. Sci.* **2017**, *18*, 200. [CrossRef] [PubMed]
3. Sedaghathoor, S.; Baladeh, M.K.; Piri, S. A study on chilling hardiness of three persimmon genotypes by electrolyte leakage parameter. *Technol. Hortic.* **2023**, *3*, 10. [CrossRef]
4. Liu, C.T.; Wang, W.; Mao, B.G.; Chu, C.C. Cold stress tolerance in rice: Physiological changes, molecular mechanism, and future prospects. *Yi Chuan = Hered.* **2018**, *40*, 171–185. [CrossRef]
5. Guo, X.; Liu, D.; Chong, K. Cold signaling in plants: Insights into mechanisms and regulation. *J. Integr. Plant Biol.* **2018**, *60*, 745–756. [CrossRef]
6. Liu, Q.; Ding, Y.; Shi, Y.; Ma, L.; Wang, Y.; Song, C.; Wilkins, K.A.; Davies, J.M.; Knight, H.; Knight, M.R.; et al. The calcium transporter ANNEXIN1 mediates cold-induced calcium signaling and freezing tolerance in plants. *EMBO J.* **2021**, *40*, e104559. [CrossRef]
7. Shao, L.; Xu, T.; Wang, X.; Zhang, R.; Wang, X.; Ren, Z.; Zhang, J.; Xia, Y.; Li, D. Integrative Comparative Assessment of Cold Acclimation in Evergreen and Deciduous Iris Species. *Antioxidants* **2022**, *11*, 977. [CrossRef]
8. Wingler, A.; Tijero, V.; Müller, M.; Yuan, B.; Munné-Bosch, S. Interactions between sucrose and jasmonate signalling in the response to cold stress. *BMC Plant Biol.* **2020**, *20*, 176. [CrossRef]
9. You, J.; Chan, Z. ROS Regulation During Abiotic Stress Responses in Crop Plants. *Front. Plant Sci.* **2015**, *6*, 1092. [CrossRef]
10. Nykiel, M.; Gietler, M.; Fidler, J.; Prabucka, B.; Labudda, M. Abiotic Stress Signaling and Responses in Plants. *Plants* **2023**, *12*, 3405. [CrossRef]
11. Sangwan, V.; Orvar, B.L.; Beyerly, J.; Hirt, H.; Dhindsa, R.S. Opposite changes in membrane fluidity mimic cold and heat stress activation of distinct plant MAP kinase pathways. *Plant J. Cell Mol. Biol.* **2002**, *31*, 629–638. [CrossRef] [PubMed]
12. Yang, T.; Poovaiah, B.W. Calcium/calmodulin-mediated signal network in plants. *Trends Plant Sci.* **2003**, *8*, 505–512. [CrossRef] [PubMed]
13. Sanders, D.; Pelloux, J.; Brownlee, C.; Harper, J.F. Calcium at the crossroads of signaling. *Plant Cell* **2002**, *14*, S401–S417. [CrossRef] [PubMed]
14. Kim, M.C.; Chung, W.S.; Yun, D.J.; Cho, M.J. Calcium and calmodulin-mediated regulation of gene expression in plants. *Mol. Plant* **2009**, *2*, 13–21. [CrossRef]
15. Liu, J.; Lenzoni, G.; Knight, M.R. Design Principle for Decoding Calcium Signals to Generate Specific Gene Expression Via Transcription. *Plant Physiol.* **2020**, *182*, 1743–1761. [CrossRef] [PubMed]
16. Yuan, P.; Yang, T.; Poovaiah, B.W. Calcium Signaling-Mediated Plant Response to Cold Stress. *Int. J. Mol. Sci.* **2018**, *19*, 3896. [CrossRef]
17. Lv, X.; Li, H.; Chen, X.; Xiang, X.; Guo, Z.; Yu, J.; Zhou, Y. The role of calcium-dependent protein kinase in hydrogen peroxide, nitric oxide and ABA-dependent cold acclimation. *J. Exp. Bot.* **2018**, *69*, 4127–4139. [CrossRef]
18. Huang, J.; Zhao, X.; Bürger, M.; Chory, J.; Wang, X. The role of ethylene in plant temperature stress response. *Trends Plant Sci.* **2023**, *28*, 808–824. [CrossRef]

19. Sun, X.; Zhao, T.; Gan, S.; Ren, X.; Fang, L.; Karungo, S.K.; Wang, Y.; Chen, L.; Li, S.; Xin, H. Ethylene positively regulates cold tolerance in grapevine by modulating the expression of ETHYLENE RESPONSE FACTOR 057. *Sci. Rep.* **2016**, *6*, 24066. [CrossRef]
20. Wang, Y.; Jiang, H.; Mao, Z.; Liu, W.; Jiang, S.; Xu, H.; Su, M.; Zhang, J.; Wang, N.; Zhang, Z.; et al. Ethylene increases the cold tolerance of apple via the MdERF1B-MdCIbHLH1 regulatory module. *Plant J. Cell Mol. Biol.* **2021**, *106*, 379–393. [CrossRef]
21. Ritonga, F.N.; Ngatia, J.N.; Wang, Y.; Khoso, M.A.; Farooq, U.; Chen, S. AP2/ERF, an important cold stress-related transcription factor family in plants: A review. *Physiol. Mol. Biol. Plants Int. J. Funct. Plant Biol.* **2021**, *27*, 1953–1968. [CrossRef] [PubMed]
22. Illgen, S.; Zintl, S.; Zuther, E.; Hincha, D.K.; Schmülling, T. Characterisation of the ERF102 to ERF105 genes of Arabidopsis thaliana and their role in the response to cold stress. *Plant Mol. Biol.* **2020**, *103*, 303–320. [CrossRef] [PubMed]
23. Sakuma, Y.; Liu, Q.; Dubouzet, J.G.; Abe, H.; Shinozaki, K.; Yamaguchi-Shinozaki, K. DNA-binding specificity of the ERF/AP2 domain of Arabidopsis DREBs, transcription factors involved in dehydration- and cold-inducible gene expression. *Biochem. Biophys. Res. Commun.* **2002**, *290*, 998–1009. [CrossRef] [PubMed]
24. Zhang, Z.; Huang, R. Enhanced tolerance to freezing in tobacco and tomato overexpressing transcription factor TERF2/LeERF2 is modulated by ethylene biosynthesis. *Plant Mol. Biol.* **2010**, *73*, 241–249. [CrossRef] [PubMed]
25. Li, H.; Liu, B.; Davis, C.C.; Yang, Y. Plastome phylogenomics, systematics, and divergence time estimation of the Beilschmiedia group (Lauraceae). *Mol. Phylogenet. Evol.* **2020**, *151*, 106901. [CrossRef]
26. Li, D.Q.; Lin, H.Y.; Wang, X.Y.; Bi, B.; Gao, Y.; Shao, L.M.; Zhang, R.L.; Liang, Y.W.; Xia, Y.P.; Zhao, Y.P.; et al. Genome and whole-genome resequencing of elucidate its dominance in subtropical urban landscapes. *BMC Biol.* **2023**, *21*, 192. [CrossRef]
27. Chen, S.F.; Zhou, Y.Q.; Chen, Y.R.; Gu, J. fastp: An ultra-fast all-in-one FASTQ preprocessor. *Bioinformatics* **2018**, *34*, 884–890. [CrossRef]
28. Kim, D.; Paggi, J.M.; Park, C.; Bennett, C.; Salzberg, S.L. Graph-based genome alignment and genotyping with HISAT2 and HISAT-genotype. *Nat. Biotechnol.* **2019**, *37*, 907–915. [CrossRef]
29. Kovaka, S.; Zimin, A.V.; Pertea, G.M.; Razaghi, R.; Salzberg, S.L.; Pertea, M. Transcriptome assembly from long-read RNA-seq alignments with StringTie2. *Genome Biol.* **2019**, *20*, 278. [CrossRef]
30. Chen, C.; Chen, H.; Zhang, Y.; Thomas, H.R.; Frank, M.H.; He, Y.; Xia, R. TBtools: An Integrative Toolkit Developed for Interactive Analyses of Big Biological Data. *Mol. Plant* **2020**, *13*, 1194–1202. [CrossRef]
31. Nakano, T.; Suzuki, K.; Fujimura, T.; Shinshi, H. Genome-wide analysis of the ERF gene family in Arabidopsis and rice. *Plant Physiol.* **2006**, *140*, 411–432. [CrossRef] [PubMed]
32. Katoh, K.; Standley, D.M. MAFFT multiple sequence alignment software version 7: Improvements in performance and usability. *Mol. Biol. Evol.* **2013**, *30*, 772–780. [CrossRef] [PubMed]
33. Price, M.N.; Dehal, P.S.; Arkin, A.P. FastTree: Computing large minimum evolution trees with profiles instead of a distance matrix. *Mol. Biol. Evol.* **2009**, *26*, 1641–1650. [CrossRef] [PubMed]
34. Kumar, S.; Stecher, G.; Li, M.; Knyaz, C.; Tamura, K. MEGA X: Molecular Evolutionary Genetics Analysis across Computing Platforms. *Mol. Biol. Evol.* **2018**, *35*, 1547–1549. [CrossRef] [PubMed]
35. Wang, X.; Li, D.; Zhang, D.; Shi, X.; Wu, Y.; Qi, Z.; Ding, H.; Zhu, K.; Xia, Y.; Zhang, J. Improving crucial details and selecting the optimal model for evaluating the chilling requirement of Paeonia lactiflora Pall. at low latitudes during four winters. *Sci. Hortic.* **2020**, *265*, 109175. [CrossRef]
36. Johnson, N.; Rodriguez Diaz, D.; Ganapathy, S.; Bass, J.S.; Kutchan, T.M.; Khan, A.L.; Flavier, A.B. Evaluation of reference genes for qRT-PCR studies in the colchicine producing *Gloriosa superba* L. *Plant Biotechnol. Rep.* **2023**, *17*, 561–571. [CrossRef]
37. Livak, K.J.; Schmittgen, T.D. Analysis of relative gene expression data using real-time quantitative PCR and the 2(-Delta Delta C(T)) Method. *Methods* **2001**, *25*, 402–408. [CrossRef]
38. Li, W.; Fu, Y.; Lv, W.; Zhao, S.; Feng, H.; Shao, L.; Li, C.; Yang, J. Characterization of the early gene expression profile in Populus ussuriensis under cold stress using PacBio SMRT sequencing integrated with RNA-seq reads. *Tree Physiol.* **2022**, *42*, 646–663. [CrossRef]
39. Wang, J.; Yang, Y.; Liu, X.; Huang, J.; Wang, Q.; Gu, J.; Lu, Y. Transcriptome profiling of the cold response and signaling pathways in Lilium lancifolium. *BMC Genom.* **2014**, *15*, 203. [CrossRef]
40. Ren, H.; Zhang, Y.; Zhong, M.; Hussian, J.; Tang, Y.; Liu, S.; Qi, G. Calcium signaling-mediated transcriptional reprogramming during abiotic stress response in plants. *Theor. Appl. Genet.* **2023**, *136*, 210. [CrossRef]
41. Zhang, Q.; Liu, Y.; Yu, Q.; Ma, Y.; Gu, W.; Yang, D. Physiological changes associated with enhanced cold resistance during maize (*Zea mays*) germination and seedling growth in response to exogenous calcium. *Crop Pasture Sci.* **2020**, *71*, 529–538. [CrossRef]
42. Chen, S.; Wang, L.; Kang, R.; Liu, C.; Xing, L.; Wu, S.; Wang, Z.; Wu, C.; Zhou, Q.; Zhao, R.J.H. Exogenous Calcium Alleviates the Photosynthetic Inhibition and Oxidative Damage of the Tea Plant under Cold Stress. *Horticulturae* **2024**, *10*, 666. [CrossRef]
43. Wang, J.; Ren, Y.; Liu, X.; Luo, S.; Zhang, X.; Liu, X.; Lin, Q.; Zhu, S.; Wan, H.; Yang, Y.; et al. Transcriptional activation and phosphorylation of OsCNGC9 confer enhanced chilling tolerance in rice. *Mol. Plant* **2021**, *14*, 315–329. [CrossRef] [PubMed]
44. Li, H.; Jiang, X.; Lv, X.; Ahammed, G.J.; Guo, Z.; Qi, Z.; Yu, J.; Zhou, Y. Tomato GLR3.3 and GLR3.5 mediate cold acclimation-induced chilling tolerance by regulating apoplastic H(2) O(2) production and redox homeostasis. *Plant Cell Environ.* **2019**, *42*, 3326–3339. [CrossRef] [PubMed]
45. Ye, K.; Shen, W.; Zhao, Y. External application of brassinolide enhances cold resistance of tea plants (*Camellia sinensis* L.) by integrating calcium signals. *Planta* **2023**, *258*, 114. [CrossRef]

46. Liu, Y.; Xu, C.; Zhu, Y.; Zhang, L.; Chen, T.; Zhou, F.; Chen, H.; Lin, Y. The calcium-dependent kinase OsCPK24 functions in cold stress responses in rice. *J. Integr. Plant Biol.* **2018**, *60*, 173–188. [CrossRef] [PubMed]
47. Sun, Q.; Huang, R.; Zhu, H.; Sun, Y.; Guo, Z. A novel Medicago truncatula calmodulin-like protein (MtCML42) regulates cold tolerance and flowering time. *Plant J. Cell Mol. Biol.* **2021**, *108*, 1069–1082. [CrossRef]
48. Chu, M.; Li, J.; Zhang, J.; Shen, S.; Li, C.; Gao, Y.; Zhang, S. AtCaM4 interacts with a Sec14-like protein, PATL1, to regulate freezing tolerance in Arabidopsis in a CBF-independent manner. *J. Exp. Bot.* **2018**, *69*, 5241–5253. [CrossRef]
49. Ma, X.; Gai, W.X.; Li, Y.; Yu, Y.N.; Ali, M.; Gong, Z.H. The CBL-interacting protein kinase CaCIPK13 positively regulates defence mechanisms against cold stress in pepper. *J. Exp. Bot.* **2022**, *73*, 1655–1667. [CrossRef]
50. Li, J.; Zou, X.; Chen, G.; Meng, Y.; Ma, Q.; Chen, Q.; Wang, Z.; Li, F. Potential Roles of 1-Aminocyclopropane-1-carboxylic Acid Synthase Genes in the Response of Gossypium Species to Abiotic Stress by Genome-Wide Identification and Expression Analysis. *Plants* **2022**, *11*, 1524. [CrossRef]
51. Zhao, R.; Xie, H.; Lv, S.; Zheng, Y.; Yu, M.; Shen, L.; Sheng, J. LeMAPK4 participated in cold-induced ethylene production in tomato fruit. *J. Sci. Food Agric.* **2013**, *93*, 1003–1009. [CrossRef] [PubMed]
52. Zhao, M.; Liu, W.; Xia, X.; Wang, T.; Zhang, W.H. Cold acclimation-induced freezing tolerance of Medicago truncatula seedlings is negatively regulated by ethylene. *Physiol. Plant.* **2014**, *152*, 115–129. [CrossRef] [PubMed]
53. Shi, Y.; Tian, S.; Hou, L.; Huang, X.; Zhang, X.; Guo, H.; Yang, S. Ethylene signaling negatively regulates freezing tolerance by repressing expression of CBF and type-A ARR genes in Arabidopsis. *Plant Cell* **2012**, *24*, 2578–2595. [CrossRef]
54. Ma, Y.; Zhang, L.; Zhang, J.; Chen, J.; Wu, T.; Zhu, S.; Yan, S.; Zhao, X.; Zhong, G. Expressing a Citrus ortholog of Arabidopsis ERF1 enhanced cold-tolerance in tobacco. *Sci. Hortic.* **2014**, *174*, 65–76. [CrossRef]
55. Li, C.; Mao, B.; Wang, K.; Xu, L.; Fan, L.; Wang, Y.; Li, Y.; Ma, Y.; Wang, L.; Liu, L. RsERF40 contributes to cold stress tolerance and cell expansion of taproot in radish (*Raphanus sativus* L.). *Hortic. Res.* **2023**, *10*, uhad013. [CrossRef]
56. Lv, K.; Li, J.; Zhao, K.; Chen, S.; Nie, J.; Zhang, W.; Liu, G.; Wei, H. Overexpression of an AP2/ERF family gene, BpERF13, in birch enhances cold tolerance through upregulating CBF genes and mitigating reactive oxygen species. *Plant Sci. Int. J. Exp. Plant Biol.* **2020**, *292*, 110375. [CrossRef] [PubMed]
57. Bolt, S.; Zuther, E.; Zintl, S.; Hincha, D.K.; Schmülling, T. ERF105 is a transcription factor gene of Arabidopsis thaliana required for freezing tolerance and cold acclimation. *Plant Cell Environ.* **2017**, *40*, 108–120. [CrossRef]
58. Sun, X.; Zhang, L.; Wong, D.C.J.; Wang, Y.; Zhu, Z.; Xu, G.; Wang, Q.; Li, S.; Liang, Z.; Xin, H. The ethylene response factor VaERF092 from Amur grape regulates the transcription factor VaWRKY33, improving cold tolerance. *Plant J. Cell Mol. Biol.* **2019**, *99*, 988–1002. [CrossRef]

horticulturae

MDPI

Review

Emerging Ornamental Plant Diseases and Their Management Trends in Northern Italy

Maria Lodovica Gullino [1], Domenico Bertetti [2], Massimo Pugliese [2,*] and Angelo Garibaldi [2]

1 Department of Civil, Chemical and Environmental Engineering, University of Genoa, Via Balbi, 5, 16126 Genova, Italy; marialodovica.gullino@unito.it
2 Interdepartmental Centre for Innovation in the Agri-Environmental Field Agroinnova, University of Turin, Largo Paolo Braccini, 2, 10095 Grugliasco, Italy; domenico.bertetti@unito.it (D.B.); angelo.garibaldi@unito.it (A.G.)
* Correspondence: massimo.pugliese@unito.it

Abstract: The ornamental plant sector is characterized by the production of a large variety of genera, species and cultivars that are much more numerous than those of other agricultural production sectors. Many countries throughout the world are involved in an intensive exchange of potted plants, cut flowers and propagation material. This intense trade exchange favors the introduction of the causal agents of new diseases on farms, in parks, along tree-lined avenues and in city gardens. Global warming can favor plant pathogens that thrive under high temperatures. Moreover, the interaction between the ongoing increase in temperature and in the CO_2 concentration has caused a significant increase in the disease severity of many pathosystems. The numerous reports of new plant pathogens on ornamental plants in Italy in recent years fall into this context. In plant pathology research, living labs incorporate the complexities and variability of natural conditions, and they can thus be used to conduct experiments and test hypotheses. A private garden, located in the hamlet of Bariola (Piedmont, Biella province, northern Italy), has become an ideal living lab that is used to monitor the evolution of the phytosanitary situation of ornamental plants. The results obtained in this living lab are reported hereafter. Moreover, new trends in disease prevention and management, such as the adoption of appropriate prevention practices, water and fertilization management and use of environmentally friendly methods to reduce pesticide use as part of an integrated pest management approach, are also examined.

Keywords: climate changes; emerging diseases; environmental factors; living lab; parks and gardens

1. Introduction

The ornamental plant sector is characterized by the production of a large variety of genera, species and cultivars that are much more numerous than those of other agricultural production sectors. This is in part due to the growing demand for the novelty of increasingly competent and demanding buyers, as evidenced by the numerous floral events that are frequently held in a large variety of locations. Although these requests are often influenced by trends and fads, hybridizers, growers and nurserymen are obliged to constantly search for "novelties" for the market. The introduction of new species and cultivars mainly depends on the importation of plants and propagative material from abroad, which may be produced in geographically distant locations with very different climatic conditions. Some parasites, imported together with the new hosts, may find environmental conditions that are more favorable for their development in the new cultivation sites, or imported plants may encounter parasites that are already present in the new country to which they had not previously been exposed.

The ornamental plant industry is dynamic, and trends within this industry can change rapidly. The global demand for ornamental plants is generally increasing, and it is driven by the urbanization phenomenon and a growing interest in home and garden esthetics. The online sales of ornamental plants are increasing, and consumers are being provided with more and easier access to a wider variety of plants. In 2022, European Union countries, including Italy, imported 4.5 billion potted plants and 1.58 billion fresh cut flowers, all of which came from non-European countries, such as Kenya, Colombia and Ethiopia [1]. This facilitates the entry of new pathogens introduced with imported plants, and the rising volume of imports increases the risk of diseases.

Italian floriculture products have reached a value of more than EUR 1.2 billion, and Italy is in second place in Europe for their production after The Netherlands [2]. However, it is difficult to find public funding for research in the ornamental sector, although it is needed to improve its competitiveness. This research may involve, for example, the selection of widespread native species that grow spontaneously in Italian areas and have esthetic characteristics that could enrich the production of plants, flowers and cut fronds. Some examples of this kind of research are those studies that have been conducted on genera known for their rusticity, such as *Euphorbia* and *Verbascum*, which are suitable for low-maintenance gardens. The selection of genotypes of already cultivated species also helps to improve their production and/or to expand it, as in the case of hellebore, eucalyptus and viburnum. Research also provides a valid contribution through susceptibility trials, which have been carried out on several cultivars and hybrids of flowering plants against some of the most dangerous plant pathogens that can threaten such crops: *Fusarium oxysporum* f. sp. *chrysanthemi* on African daisy (*Dimorphotheca* sp.) and *Gerbera jamesonii*, *Colletotrichum acutatum* and *Erysiphe azaleae* on azalea [3] are some examples.

New plant pathogens, the causal agents of emerging diseases, can also appear on ornamental herbaceous, shrubby and arboreal species grown in public and private greenery, of which Italy is particularly rich. Indeed, there are numerous historic gardens, tree-lined avenues and city gardens, that is, from large parks to the greenery of a neighborhood, for which the recent pandemic caused by COVID-19 increased the need for citizens to feel healthy. The management of this large green heritage is further complicated in Italy and in many other European countries by the stringent limitations on the use of pesticides, which are prohibited or minimized in sensitive areas, such as public parks, sports grounds, hospitals and schools.

This article focuses on the emerging fungal and bacterial pathogens detected in northern Italy over the past twenty years, whose appearance is often linked to ongoing climate changes. Furthermore, it highlights the importance of so-called "living labs" as early warning systems for monitoring the evolving phytosanitary landscape and for testing appropriate sustainable control strategies. The activities carried out in one such lab are summarized.

2. Factors That Affect the Emergence of New Diseases

The ornamental plant market is extremely lively, and many countries throughout the world are involved in an intensive exchange of potted plants, cut flowers and propagation material. The production of ornamental plants requires large investments, in part due to the costs of labor and the heating of greenhouses and tunnels, especially in countries with cold climates. Unrooted cuttings are often produced in third-world countries, where the production costs are significantly lower and the weather conditions are more favorable.

The propagative material is transferred from these countries to the traditional hubs for ornamental production (such as The Netherlands and Denmark, in Europe), and it eventually reaches consumers all over the world. However, this movement of plant material has facilitated the spread of a variety of plant pathogens, including *Phytophthora* spp. such as *P. ramorum* and the fearsome *formae speciales* of *Fusarium oxysporum* [4,5]. Moreover, this intense trade exchange favors the introduction of the causal agents of new diseases and/or brings newly imported hosts into contact with plant pathogens already present in a territory. A highly significant aspect derives from theories developed regarding invasive alien plants [6]. Similarly, there is a risk that certain ornamental species, when introduced into new production contexts, may become naturalized and invasive, facing limited control from local natural enemies—including fungal pathogens. Examples include butterfly bush (*Buddleja davidii*) [7] and Japanese spirea (*Spiraea japonica*) [8].

Some pathogens find temperatures in the new cultivation sites that are more favorable for their propagation, and this in turn can lead to more severe attacks. Among the various environmental factors, temperature plays a significant role in the onset of disease. Global warming, on the one hand, can favor plant pathogens that thrive in high temperatures, as well as the survival of pathogens in the winter periods; on the other hand, it can weaken plants, especially the mesophilic species, thereby making them more susceptible to infection. Temperature variations can alter the development cycles of both plants and pathogens and can favor the development of some diseases at certain times of the year in which they normally would not appear. The climate changes recorded in recent years have caused more frequent and intense extreme weather events (prolonged heat waves, intense droughts, late frosts, torrential rains) than in the past, and these events can cause stress (thermal, water and light imbalances) that weakens the plants and makes them more vulnerable to diseases. Rises in temperatures can also determine an increase in the number of generations per year of dangerous insects, such as bark beetles. The occurrence of intense storms can promote, even indirectly, the proliferation of insects that are normally present [9]. An example of this is the "Vaia storm": the numerous trunks that fell onto the ground represented a very abundant source of nourishment for the bark beetle (*Ips typographus*), whose proliferation compromises thousands of spruce trees.

The increase in the concentration of atmospheric carbon dioxide (CO_2), as a result of the increase in human activities, has led to an increase in atmospheric temperatures. The impact of climate change on plant health can be studied through three different but complementary approaches: monitoring the evolution of the phytopathological situation in specific geographic areas; using models to hypothesize future scenarios; using special research climate cells (phytotrons) that are able to modify the main environmental parameters, including the temperature and the CO_2 concentration [10]. For example, experimental tests carried out in a controlled atmosphere have shown that the interaction between the increase in temperature and in the CO_2 concentration has caused a significant increase in the severity of many pathosystems, such as *Pelargonium–Puccinia pelargonii* [11]. A variation in just one of the studied climatic parameters can cause more variable effects. The same studies underlined that the host–parasite–environmental parameter interaction is very complex and variable, and it can also have consequences on the adopted control methods [12,13].

3. New or Re-Emerging Diseases, with Special Focus on the Mediterranean Area

The increase in temperature caused by climate change will continue to determine certain consequences in northern European countries, especially in the wintertime, and will lead to the expansion of areas that are climatically suitable for a large variety of crops. However, in southern European countries, the consequences will instead mainly concern the summer period, and they will lead to increased water shortages and to extreme events, such as heatwaves, storms and droughts [14]. Climatic data recorded in Italy in the last few decades have outlined a dangerous trend. The significant increase in summer temperatures, the reduction in rainfall, especially in flat and hilly areas, the early mild spring temperatures, which induce plants to develop leaves and flowers earlier, and the increased damage caused by late frosts are all causing particularly negative consequences on agricultural crops [15]. Ornamental plants grown on farms, and in nurseries, gardens and parks, can suffer from similar consequences.

4. Italy as a Hot Spot for the Appearance of New Diseases

The numerous reports on the presence of new plant pathogens on ornamental plants in recent years fall into this context [3]. Among these, special attention has been paid to pathogens that are favored by high temperatures, such as *Fusarium oxysporum* (Table 1), whose optimal temperature, which is between 25 and 30 °C, depends on its specific form.

Table 1. First reports of *Fusarium oxysporum* on ornamental plants in northern Italy in the last 20 years. The *formae speciales* are listed.

Host	*Forma Specialis*	Region	Year
Gerbera jamesonii	*Chrysanthemi*	Liguria	2004
Osteospermum spp.	*Chrysanthemi*	Liguria	2004
Lewisia cotyledon	*Unidentified*	Piedmont	2005
Cereus peruvianus monstruosus	*Opuntiarum*	Liguria	2011
Crassula ovata	*Crassulae* *	Liguria	2011
Papaver nudicaule	*Papaveris* *	Liguria	2012
Echeveria agavoides	*Echeveriae* *	Liguria	2013
Cereus marginatus var. *cristata*	*Opuntiarum*	Liguria	2014
Echeveria tolimanensis	*Echeveriae* *	Liguria	2015
Euphorbia mammillaris var. *variegata*	*Opuntiarum*	Liguria	2015
Astrophytum myriostigma	*Opuntiarum*	Liguria	2016
Cereus peruvianus florida	*Opuntiarum*	Liguria	2016
Lavandula × *allardii*	*Lavandulae* *	Liguria	2016
Mammillaria zeilmanniana	*Opuntiarum*	Liguria	2016
Rudbeckia fulgida	*Chrysanthemi*	Piedmont	2017
Sulcorebutia heliosa	*Opuntiarum*	Liguria	2019
Mammillaria painteri	*Opuntiarum*	Liguria	2020
Sulcorebutia rauschii	*Opuntiarum*	Liguria	2020

* New forma specialis.

The most favorable conditions for the causal agents of powdery mildew (Table 2) are warm and humid weather and temperatures ranging from 18 to 25 °C.

Table 2. The powdery mildews reported on ornamental plants for the first time in northern Italy in the last 20 years. The pathogen proliferation observed in the Bariola garden aligns with the increase in the average temperatures registered by the weather station of Oropa and Tollegno (Figures 1 and 2).

Host	Pathogen	Region	Detected in the Bariola Garden	Year
Akebia quinata	*Oidium* sp.	Piedmont	Yes	2004
Aquilegia flabellata	*Erysiphe aquilegiae*	Piedmont	Yes	2004
Lonicera caprifolium	*Oidium* subgenus *Pseudoidium*	Piedmont	Yes	2004
Photinia × *fraserii*	*Podosphaera leucotricha*	Piedmont		2005
Potentilla fruticosa	*Podosphaera aphanis* var. *aphanis*	Piedmont	Yes	2005
Veronica spicata	*Golovinomyces orontii*	Piedmont	Yes	2006
Coreopsis lanceolata	*Podosphaera fusca*	Piedmont	Yes	2007

Table 2. *Cont.*

Host	Pathogen	Region	Detected in the Bariola Garden	Year
Lamium galeobdolon	*Golovinomyces orontii*	Piedmont	Yes	2007
Petunia × *hybrida*	*Golovinomyces orontii*	Piedmont	Yes	2007
Argyranthemum frutescens	*Golovinomyces cichoracearum*	Liguria		2008
Calendula officinalis	*Podosphaera xanthii*	Piedmont	Yes	2008
Hedera helix	*Erysiphe heraclei*	Liguria		2008
Rudbeckia fulgida	*Golovinomyces cichoracearum*	Piedmont		2008
Cleome hassleriana	*Erysiphe cruciferarum*	Piedmont		2009
Cornus florida	*Erysiphe pulchra*	Piedmont	Yes	2009
Gerbera jamesonii	*Golovinomyces cichoracearum*	Piedmont		2010
Phlox drummondii	*Podosphaera* sp.	Piedmont	Yes	2011
Verbascum blattaria	*Golovinomyces cichoracearum*	Piedmont		2011
Aster novi-belgii	*Golovinomyces cichoracearum*	Piedmont	Yes	2012
Campanula rapunculoides	*Golovinomyces orontii*	Piedmont	Yes	2012
Euphorbia aggregata	*Podosphaera* sp.	Liguria		2012
Euphorbia inermis	*Podosphaera* sp.	Liguria		2012
Euphorbia perdorfiana	*Podosphaera* sp.	Liguria		2012
Euphorbia susannae	*Podosphaera* sp.	Liguria		2012
Oenothera biennis	*Erysiphe* sp.	Piedmont	Yes	2012
Phlox paniculata	*Golovinomyces magnicellulatus*	Piedmont	Yes	2016
Thymus × *citriodorus* "Aureus"	*Golovinomyces biocellatus*	Liguria		2016
Verbascum nigrum "Album"	*Golovinomyces cichoracearum*	Piedmont	Yes	2016
Campanula glomerata	*Golovinomyces orontii*	Piedmont		2018
Echinacea purpurea	*Golovinomyces cichoracearum*	Piedmont	Yes	2018
Abelmoschus manihot	*Golovinomyces orontii*	Piedmont		2019
Lavandula stoechas "Blueberry Ruffles"	*Golovinomyces neosalviae*	Liguria		2020
Salvia nemorosa	*Golovinomyces biocellatus*	Piedmont	Yes	2021
Dianthus caryophyllus	*Erysiphe buhrii*	Liguria		2022
Phlox maculata	*Golovinomyces magnicellulatus*	Piedmont	Yes	2022

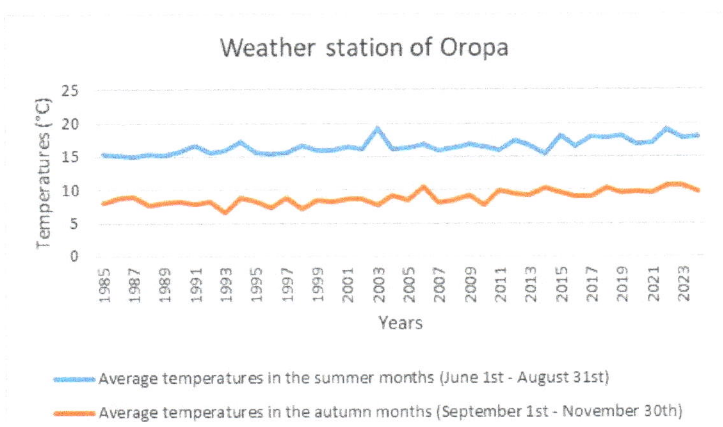

Figure 1. Trend of the average temperatures registered in the summer months (1 June–31 August) and in the autumn months (1 September–30 November), in the last 40 years by the weather station of Oropa (Biella province, northern Italy). The temperature rise is related to the numerous powdery mildews detected in the Bariola garden listed in Table 2.

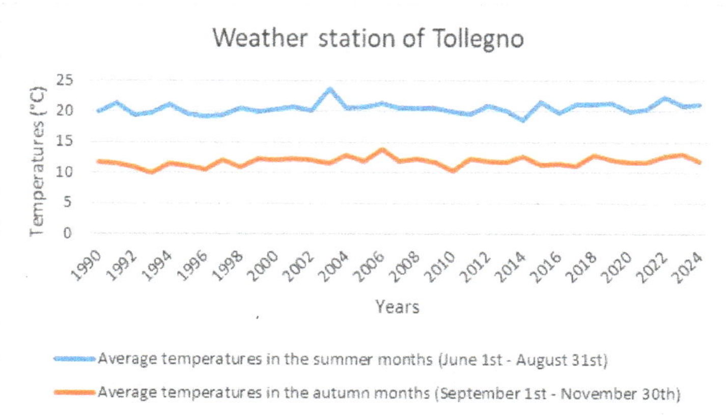

Figure 2. Trend of the average temperatures registered in the summer months (1 June–31 August) and in the autumn months (1 September–30 November), in the last 40 years by the weather station of Tollegno (Biella province, northern Italy). The temperature rise is related to the numerous powdery mildews detected in the Bariola garden listed in Table 2.

Among the emerging foliar plant pathogens, *Alternaria* spp. has been detected in recent years on numerous ornamental species [16], many of which are listed in Table 3.

Table 3. Leaf plant fungal pathogens reported on several ornamental plants for the first time in Italy in the Bariola living lab in the last 20 years.

Host	Pathogen	Year
Hydrangea macrophylla	*Phoma exigua*	2006
Clematis × *jackmanii*	*Phoma* sp.	2007
Hydrangea macrophylla	*Alternaria alternata*	2007
Hydrangea anomala	*Alternaria compacta*	2008
Cornus florida	*Botrytis cinerea*	2009
Platycodon grandiflorum	*Botrytis cinerea*	2009
Fuchsia × *hybrida*	*Phoma multirostrata*	2010
Rudbeckia fulgida	*Phoma* sp.	2010
Aquilegia flabellata	*Phoma aquilegiicola*	2011
Lupinus polyphyllus	*Pleiochaeta setosa*	2012
Saponaria officinalis	*Alternaria nobilis*	2013
Verbascum nigrum	*Phoma novae-verbascicola*	2013
Verbascum blattaria	*Phoma novae-verbascicola*	2014
Campanula glomerata	*Alternaria* sp.	2015
Campanula medium	*Stagonosporopsis trachelii*	2015
Rudbeckia fulgida	*Alternaria* sp.	2015
Salvia oxyphora	*Botrytis cinerea*	2015
Anemone japonica	*Botrytis cinerea*	2016
Campanula medium	*Alternaria alternata*	2016
Liquidambar styraciflua	*Colletotrichum kahawae*	2016
Salvia greggii	*Boeremia exigua var. linicola*	2016
Salvia leucantha	*Colletotrichum fioriniae*	2016
Campanula trachelium	*Stagonosporopsis trachelii*	2017
Ceratostigma willmottianum	*Alternaria alternata*	2019
Abelmoschus manihot	*Alternaria alternata*	2020
Coreopsis lanceolata	*Colletotrichum fuscum*	2020
Mahonia aquifolium	*Colletotrichum fioriniae*	2020
Phlox maculata	*Alternaria alternata*	2020
Aquilegia flabellata	*Plectosphaerella cucumerina*	2021
Campanula rapunculoides	*Stagonosporopsis trachelii*	2022
Hydrangea paniculata	*Alternaria alternata*	2023
Rhododendron arboreum	*Botrytis cinerea*	2024

The appearance of insect pests and plant pathogens on new tree hosts can lead to a reduction in growth, decay and even the death of the trees, which in turn affects the esthetics of the landscape and the human health of the citizens and/or can have negative economic impacts on forest plants grown in the parks and gardens of urban areas [17]. A significant example is that of *Geosmithia morbida* (with the walnut twig vector *Pityophthorus juglandis*), the causal agent of the thousand cankers disease, which can spread on *Juglans nigra*, on *Juglans* spp. and also on *Pterocarya fraxinifolia*, with the latter host being grown in several urban and historical parks [18]. Temperature increases and water stress phenomena also favor the spread of plant pathogens that are already present in Italian areas, such as *Anthostoma decipiens*, which causes the decline and death of hornbeam (*Carpinus betulus*). This pathogen has been constantly detected on hornbeam together with *Endothiella* sp., and it has been proved that both of these fungi are potential pathogens for other forest species diffused in gardens and parks [19].

Some endophytic fungi that are usually considered "non-pathogenic" can cause diseases on plants subjected to certain stress conditions, such as the occurrence of prolonged periods of drought. An example is that of *Tubakia dryina*, which, in conditions of excessive heat and drought, becomes parasitic and causes leaf spots, early leaf drop and crown decline on some ornamental broadleaf trees, such as oaks (*Quercus* spp.), horse chestnut (*Aesculus* spp.) and beech (*Fagus sylvatica*) [20,21].

5. The Bariola Garden

In plant pathology research, a living lab is a real-world environment (e.g., a farm, orchard or natural ecosystem) that is used to conduct experiments and test hypotheses. Unlike controlled laboratory settings, living labs incorporate the complexities and variability of natural conditions, thereby offering more realistic data and insights into disease dynamics and management strategies. This approach bridges the gap between laboratory findings and practical applications, and it allows researchers to evaluate the effectiveness of interventions under realistic field conditions.

Numerous herbaceous, shrubby and arboreal ornamental plants, both native and from all over the world, are currently being grown in a private garden located in the hamlet of Bariola, in the municipality of Campiglia Cervo (Piedmont, Biella province, northern Italy, 45.65236894890476, 8.006370027342234), at 850 m above sea level, on a surface of about 2.5 ha. This garden has been enriched each year with new species and cultivars, and, over the past 25 years, it has become an ideal living lab that is used to monitor the evolution of the phytosanitary situation of ornamental plants. The observations that have been carried out have made it possible to:

- investigate the impact of climate change on the prevalence and severity of plant diseases;
- compare the disease resistance of different varieties/cultivars under natural infection pressure conditions, considering such factors as the weather and soil conditions;
- evaluate the efficacy of new disease control methods (a novel fungicide, biocontrol agent or cultural practice) in reducing disease incidence and severity in a real-world agricultural setting.

This has led to the achievement of more reliable data than those that can be obtained in controlled lab experiments.

Numerous causal agents of powdery mildew, which has never been detected in a geographical area affected by their presence until a few decades ago (Table 2), are among the plant pathogens that have been reported on this site. Their appearance, in a Piedmont location known for its rainy climates and cool summers, on hosts that have been grown in the Bariola garden for many years (for example, on *Akebia quinata*, *Cornus florida*, *Lonicera caprifolium* and *Potentilla fruticosa*), supports the hypothesis that climate changes can favor the spread of these plant pathogens [12]. It is interesting to note that the appearance of the numerous powdery mildews reported on this site (Table 2) occurred during a period in which there was a gradual increase in the average temperatures in the summer months (1 June–31 August) and in the autumn months (1 September–30 November), as recorded by the nearby weather stations of Oropa (45°37′36.39″ N 7°58′52.47″ E) and Tollegno (45°35′27″ N 8°02′51″ E) (see Figures 1 and 2).

Some of the fungal pathogens reported on this site, such as *Erysiphe pulchra* on *Cornus florida*, were later also detected on floricultural farms, and they successively appeared on the "Daybreak", "Rainbow" and "Rubra" cultivars grown in nurseries in the same area. Among the numerous plant pathogens reported for the first time on ornamental hosts at this site (Tables 2–5), bacterial pathogens were also identified (Table 6), including *Pseudomonas syringae* pv. *viburnii*, the causal agent of bacterial blight in *Viburnum sargentii*. A subsequent survey also assessed the susceptibility of various *Viburnum* species and cultivars to the disease. Therefore, the Bariola garden can be considered one of the so-called "sentinel gardens and nurseries" that play a fundamental role in intercepting parasites (fungi, bacteria, insects, nematodes, etc.) that could appear on new hosts grown in such locations [22]. An example of these locations is the sentinel gardens created in China, which have allowed the presence of some dangerous parasites on important ornamental species to be reported in a timely manner [23–25]. In this context, the International Plant Sentinel Network (IPSN) brings together gardens, arboreta, the National Plant Protection Organisations (NPPOs), and plant pathologists who collaborate to form a network capable of timely detecting new and emerging plant pathogens. The IPSN was the first to detect, on a global scale, the presence of *Hymenoscyphus fraxineus* on hosts other than ash trees: *Phillyrea latifolia*, *P. angustifolia* and *Chionanthus virginicus* [26].

Table 4. Soil-borne plant pathogens reported on ornamental plants for the first time in Italy in the Bariola living lab in the last 20 years.

Host	Pathogen	Year
Heuchera sanguinea	*Rhizoctonia solani*	2007
Lupinus polyphyllus	*Verticillium dahliae*	2007
Aquilegia flabellata	*Rhizoctonia solani*	2009
Digitalis purpurea	*Rhizoctonia solani*	2009
Rudbeckia hirta	*Verticillium dahliae*	2012
Phlox paniculata	*Verticillium dahliae*	2014
Campanula trachelium	*Rhizoctonia solani*	2015
Lychnis coronaria	*Rhizoctonia solani*	2015
Campanula carpatica	*Rhizoctonia solani*	2018
Echinacea purpurea	*Rhizoctonia solani*	2019
Coreopsis lanceolata	*Phytopythium oedochilum*	2022
Anemone japonica	*Globisporangium sylvaticum*	2023

Table 5. Obligate plant pathogens reported on ornamental plants for the first time in Italy in the Bariola living lab in the last 20 years.

Host	Pathogen	Year
Fuchsia × hybrida	*Pucciniastrum circaeae*	2012
Digitalis purpurea	*Peronospora digitalidis*	2013
Impatiens walleriana	*Plasmopara obducens*	2013
Oenothera biennis	*Peronospora arthurii*	2018

Table 6. Bacterial plant pathogens reported on ornamental plants for the first time in Italy in the Bariola living lab in the last 20 years.

Host	Pathogen	Year
Phlox paniculata	*Pseudomonas cichorii*	2005
Viburnum sargentii	*Pseudomonas syringae* pv. *viburnii*	2005
Coreopsis lanceolata	*Pseudomonas cichorii*	2009
Verbena hybrida	*Erwinia* sp.	2011

6. New Trends in Disease Prevention and Management

Sustainable production practices of ornamental plants, including reducing pesticide use, conserving water and using eco-friendly packaging, have been emphasized more and more in recent years. Such trends have led to a profound change in disease management practices, even for intensive production systems. The use of healthy plant material and the adoption of appropriate prevention practices (good ventilation, reduction in relative humidity) in production systems, coupled with the use of tolerant or resistant varieties, when available, reduces the need for using pesticides, to which the resort has become very limited during the last few decades. All this takes on significant importance for protected crops, where closed environments create more favorable conditions for fungal and bacterial infections.

The transition from the use of chemical products to an integrated pest and disease pest management approach has been determined for many reasons: the risk of developing resistance to fungicides, phytotoxicity damage, high costs, detrimental effects for both the workers and for the environment [27,28]. Furthermore, there is a growing request from consumers for ornamental plants with no chemical residues and that are grown while respecting pollinating insects.

All these reasons explain the increasing recourse to the biological control of ornamentals crops in greenhouses that has been observed in Europe and in North America [29]. As far as biocontrol agents are concerned, Lecomte et al. listed 26 biological control products used for the control of Fusarium wilt and stated that a great deal of research is needed to learn how to best integrate biocontrol measures with other control methods [30]. Some microorganisms that are effective against plant pathogens have been registered in Italy for ornamental plants: *Ampelomyces quisqualis* and *Bacillus amyloliquefaciens* (powdery mildew agents), *Pythium oligandrum* (*Botrytis, Pythium, Rhizoctonia, Fusarium, Phytophthora, Sclerotinia* and *Verticillium* genera), *Trichoderma asperellum* (*Sclerotinia*), *Trichoderma asperellum* TV1 (root rot causal agents), *Coniothyrium minitans* (*Sclerotinia*). Encouraging results have been obtained in experimental trials from microorganisms that have antagonistic activity, such as a *Trichoderma* strain isolated from compost, which was able to control both *Phytophthora nicotianae* on *Skimmia japonica* and *P. cinnamomi* on azalea [31,32].

The increasing demand for environmentally responsible pest management methods requires the search for "alternative" control methods. The fungicide activity carried out by some salts, such as sodium and potassium bicarbonates, is of interest for this purpose. Some phosphites (currently registered in Italy as fungicides) and phosphates perform like resistance inducers, especially the latter against powdery mildew agents. Interesting results have been obtained for some calcium-based products, which have been found to make the plant cell wall stronger [33,34].

Prevention plays an essential role in parks and gardens, where respect for the ecological needs of species cannot be ignored, and which, for new parks, should already start at the project phase. Respect for ecological needs is the essential premise for the growth of vigorous plants that are less susceptible to disease. The essential starting point to prevent plant diseases is the use of healthy plants and propagative material of certain origin. In addition, landscapers should be aware of the main phytopathological problems and, if possible, orient their choices toward resistant or, at least, tolerant cultivars. In this regard, the numerous experimental tests that have specifically been conducted to evaluate the sensitivity of plants against the most important parasites have provided precise indications on how to detect resistant or tolerant cultivars. For example, a monitoring campaign carried out in the Burcina park (Biella province, northern Italy) established that most of the approximately 50 tested rhododendron cultivars were resistant or tolerant to *Pycnostysanus azaleae*, the causal agent of rhododendron bud rot. Similarly, interesting results have emerged for downy mildew on *Coleus* (*Solenostemon scutellarioides*) cultivars [35].

However, the cultural practices that can be implemented to modify environmental factors in parks and gardens are somewhat limited. Respect for the correct planting distance and the exploitation of the direction of the prevailing winds both help to reduce the moisture on the leaves of shrubs and hedges that are prone to fungal and bacterial pathogens. Water management, using drip irrigation systems and fertilization management, helps prevent the development of diseases. Moreover, if any new phytopathological problems appear, it is essential to eliminate the sources of infection as soon as possible. In this context, there is a growing need to train landscape architects, green technicians and specialized gardeners, who will have to be constantly updated in order to be able to recognize new plant pathogens. These professional figures will have to be trained through training courses, seminars, meetings and specialized reviews. Collaboration with laboratories specialized in diagnostic services can provide a fundamental contribution to improving plant disease protection, through traditional and molecular diagnostic methods.

Finally, certain actions (meetings, debates, and posters) undertaken to educate people and empower them with scientific knowledge are also very useful. Such actions allow the public to contribute constructively to the work of green technicians. In this context, a significant and innovative example of collaboration between the managers of public green areas and citizens is that of the so-called "Citizen science", whereby citizens can contribute, on a voluntary basis, to the collection of scientific data, using smartphones and dedicated apps, to monitor the health of public parks and report any emerging problems. Scientific observations involving nature lovers, made during "phytopathological walks" in home gardens and green areas near homes during the pandemic period, produced surprising results [36].

Thus, it is reasonable to expect that the ornamental sector will continue to be enriched, more than other sectors, with new plant pathogens. This should encourage collaboration between specialized technicians, research institutions, growers, plant lovers and users of greenery to help update a constantly evolving phytopathological framework, which could, in particular, be used to identify the most significant dangerous emergencies.

This is a complex challenge. However, the results from scientific research can provide innovative, rapid and specific solutions to defend one of the most important Italian agricultural production sectors as well as the Italian tree heritage in parks, public gardens and the gardens of historic mansions, of which there are many throughout Italy. In this context, molecular diagnostic methods—which have become increasingly sophisticated and precise in recent years [37]—can make a decisive contribution. These methods are continually evolving to become faster, more accurate and more cost-effective. Additionally, experimental trials carried out in controlled environment facilities (phytotrons) can simulate critical situations and provide valuable insights into the effects of ongoing climate change. Finally, the development of mathematical models enables the prediction of phytopathological trends in specific geographic areas, including urban environments [38].

Author Contributions: Writing—original draft preparation, M.L.G. and D.B.; writing—review and editing, M.L.G., D.B., M.P., and A.G.; supervision, M.L.G. and A.G.; funding acquisition, M.P. All authors have read and agreed to the published version of the manuscript.

Funding: This research was funded by the Fondazione Compagnia di San Paolo as a part of the "Biodiversità e Salute delle Piante (BIO-SAL)" project, by the University of Torino within the project Ricerca Locale (ex 60%) "Malattie delle piante e salute circolare: caratterizzazione, epidemiologia, diagnostica e strategie di lotta a patogeni emergenti" and by the European Union Next-Generation EU (PIANO NAZIONALE DI RIPRESA E RESILIENZA (PNRR)-MISSIONE 4 COMPONENTE 2, INVESTIMENTO 1.4-D.D. 1032 17/06/2022, CN00000022), Spoke 6, Task 6.1.1. This manuscript only reflects the authors' views and opinions, and neither the European Union nor the European Commission can be considered responsible for them.

Conflicts of Interest: The authors declare no conflicts of interest.

References

1. Available online: https://www.agricultura.it/2024/01/09/litalia-seconda-potenza-europea-per-florovivaismo-nel-2022 (accessed on 4 August 2025).
2. Available online: https://horticultureconnected.ie/news/floriculture-as-a-sector-of-italian-excellence-at-the-top-of-european-and-world-exports-urban-and-landscape-regeneration-as-future-focus/ (accessed on 4 August 2025).
3. Garibaldi, A.; Bertetti, D.; Rapetti, S.; Gullino, M.L. *Malattie delle Piante Ornamentali*; Second updated reprint of the first edition; Edagricole: Bologna, Italy, 2024; pp. 45–148.
4. Gullino, M.L.; Daughtrey, M.L.; Garibaldi, A.; Elmer, W.H. Fusarium wilts of ornamental crops and their management. *Crop Prot.* **2015**, *73*, 50–59. [CrossRef]
5. Bahadur, A. Current Status of *Fusarium* and Their Management Strategies. In *Fusarium—An Overview of the Genus*; Mahyar Mirmajlessi, S., Ed.; IntechOpen: Rijeka, Croatia, 2022. [CrossRef]
6. Mitchell, C.E.; Power, A.G. Release of invasive plants from fungal and viral pathogens. *Nature* **2003**, *421*, 625–627. [CrossRef]
7. Available online: https://gd.eppo.int/taxon/BUDDA (accessed on 4 August 2025).
8. Available online: https://www.invasiveplantatlas.org/subject.cfm?sub=3076 (accessed on 4 August 2025).
9. Stadelmann, G.; Bugmann, H.; Wermelinger, B.; Bigler, C. Spatial interactions between storm damage and subsequent infestations by the European spruce bark beetle. *For. Ecol. Manag.* **2014**, *318*, 167–174. [CrossRef]
10. Saravitz, C.H.; Joseph, C. *NCSU Phytotron Procedural Manual—For Controlled-Environment Research at the Southeastern Plant Environment Laboratory*; Technical Bulletin; North Carolina State University: Raleigh, NC, USA, 2019; p. 244. Available online: https://phytotron.ncsu.edu/procedural-manual/ (accessed on 4 August 2025).
11. Gilardi, G.; Pugliese, M.; Gullino, M.L.; Garibaldi, A. Simulated elevated atmospheric CO_2 and temperature affect the severity of bean and pelargonium rust. *Phytoparasitica* **2016**, *44*, 325–332. [CrossRef]
12. Gullino, M.L.; Pugliese, M.; Gilardi, G.; Garibaldi, A. Effect of increased CO_2 and temperature on plant diseases: A critical appraisal of results obtained in studies carried out under controlled environment facilities. *J. Plant Pathol.* **2018**, *100*, 371–389. [CrossRef]
13. Pugliese, M.; Gilardi, G.; Garibaldi, A.; Gullino, M.L. The Impact of Climate Change on Vegetable Crop Diseases and Their Management: The Value of Phytotron Studies for the Agricultural Industry and Associated Stakeholders. *Phytopathology* **2024**, *114*, 843–854. [CrossRef] [PubMed]
14. Bindi, M.; Olesen, J.E. The responses of agriculture in Europe to climate change. *Reg. Environ. Chang.* **2011**, *11*, S151–S158. [CrossRef]
15. Colombo, T.; Pelino, V.; Vergari, S.; Cristofanelli, P.; Bonasoni, P. Study of temperature and precipitation variations in Italy based on surface instrumental observations. *Glob. Planet. Chang.* **2007**, *57*, 308–318. [CrossRef]
16. Matić, S.; Tabone, G.; Garibaldi, A.; Gullino, M.L. Alternaria Leaf Spot Caused by *Alternaria* Species: An Emerging Problem on Ornamental Plants in Italy. *Plant Dis.* **2020**, *104*, 2275–2287. [CrossRef]
17. Raum, S.; Collins, C.M.; Urquhart, J.; Potter, C.; Pauleit, S.; Egerer, M. Tree insect pests and pathogens: A global systematic review of their impacts in urban areas. *Urban Ecosyst.* **2023**, *26*, 587–604. [CrossRef]
18. *Pest Risk Analysis for Thousand Cankers Disease (Geosmithia morbida and Pityophthorus juglandis)*; EPPO: Paris, France, 2015. Available online: http://www.eppo.int/QUARANTINE/Pest_Risk_Analysis/PRA_intro.htm (accessed on 4 August 2025).
19. Saracchi, M.; Sardi, P.; Kunova, A.; Cortesi, P. Potential host range of *Anthostoma decipiens* and *Endothiella* sp., agents of hornbeam blight. *J. Plant Pathol.* **2015**, *97*, 93–97.
20. Kowalski, T. Tubakia dryina, symptoms and pathogenicity to Quercus robur. *Acta Mycol.* **2006**, *41*, 299–304. [CrossRef]

21. Braun, U.; Nakashima, C.; Crous, P.W.; Groenewald, J.Z.; Moreno-Rico, O.; Rooney-Latham, S.; Blomquist, C.L.; Haas, J. Marmolejo, J. Phylogeny and taxonomy of the genus Tubakia s. lat. *Fungal Syst. Evol.* **2018**, *1*, 41–99. [CrossRef]
22. Mansfield, S.; McNeill, M.R.; Aalders, L.T.; Bell, N.L.; Kean, J.M.; Barratt, B.I.P.; Boyd-Wilson, K.; Teulon, D.A.J. The value of sentinel plants for risk assessment and surveillance to support biosecurity. *NeoBiota* **2019**, *48*, 1–24. [CrossRef]
23. Vettraino, A.M.; Li, H.-M.; Eschen, R.; Morales-Rodriguez, C.; Vannini, A. The sentinel tree nursery as an early warning system for pathway risk assessment: Fungal pathogens associated with Chinese woody plants commonly shipped to Europe. *PLoS ONE* **2017**, *12*, e0188800. [CrossRef]
24. Chang, L.; Li, Y.; Gao, Z.; Bonello, P.; Cleary, M.; Munck, I.A.; Santini, A.; Sun, H. Novel Pathogen–Plant Host Interaction *Colletotrichum jiangxiense* and *Fraxinus americana* L. (White Ash) in a Sentinel Garden in China. *Plants* **2023**, *12*, 4001. [CrossRef] [PubMed]
25. Chang, L.; Li, Y.-L.; Gao, Z.-W.; Bonello, P.; Cleary, M.; Munck, I.A.; Santini, A.; Sun, H. First report of *Epicoccum latusicollum* causing leaf spot disease on red maple (*Acer rubrum* L.) in China: Insights from a sentinel planting garden. *Crop Prot.* **2024**, *175* 106439. [CrossRef]
26. Marfleet, K.; Sharrock, S. The international plant sentinel Network: An update on phase 2. *Sibbaldia* **2020**, *2*, 105–116. [CrossRef]
27. Daughtrey, M.L.; Benson, D.M. Principles of Plant Health Management for Ornamental Plants. *Annu. Rev. Phytopathol.* **2005**, *43* 141–169. [CrossRef] [PubMed]
28. Daughtrey, M.; Buitenhuis, R. Integrated Pest and disease Management in Greenhouse ornamentals. In *Integrated Pest and Disease Management in Greenhouse Crops*, 2nd ed.; Gullino, M.L., Albajes, R., Nicot, P.C., Eds.; Springer: Cham, Switzerland, 2020.
29. van Lanteren, J.C.; Bolckmans, K.; Köhl, J.; Ravensberg, W.J.; Urbaneja, A. Biological control using invertebrates and microorganisms: Plenty of new opportunities. *BioControl* **2018**, *63*, 39–59. [CrossRef]
30. Lecomte, C.; Alabouvette, C.; Edel-Hermann, V.; Steinberg, C. Biological control of ornamental plant diseases caused by *Fusarium oxysporum*: A review. *Biol. Control* **2016**, *101*, 17–30. [CrossRef]
31. Pugliese, M.; Gullino, M.L.; Garibaldi, A. Compost suppressiveness against *Phytophthora* spp. on *Skimmia japonica* and azalea *Commun. Agric. Appl. Biol. Sci.* **2012**, *77*, 237–240. [PubMed]
32. Bardin, M.; Pugliese, M. Biocontrol agents against diseases. In *Integrated Pest and Disease Management in Greenhouse Crops*; Gullino M.L., Albajes, R., Nicot, P., Van Lenteren, J.C., Eds.; Springer: Dordrecht, The Netherlands, 2019. [CrossRef]
33. Deliopoulos, T.; Kettlewell, P.S.; Hare, M.C. Fungal disease suppression by inorganic salts: A review. *Crop Prot.* **2010**, *29* 1059–1075. [CrossRef]
34. Havlin, J.L.; Schlegel, A.J. Review of Phosphite as a Plant Nutrient and Fungicide. *Soil Syst.* **2021**, *5*, 52. [CrossRef]
35. Daughtrey, M.; Harlan, B.; Linderman, S.; Hausbeck, M.K. Coleus Cultivars and Downy Mildew. *Special Research Report #136 Disease Management*. 2014. Available online: https://endowment.org/wp-content/uploads/imported-files/136-ColeusDM-Cv-2014.pdf (accessed on 4 August 2025).
36. Suffert, F.; Suffert, M. "Phytopathological strolls" in the dual context of COVID-19 lockdown and IYPH2020: Transforming constraints into an opportunity for public education about plant pathogens. *Plant Pathol.* **2021**, *71*, 30–42. [CrossRef] [PubMed]
37. Hariharan, G.; Prasannath, K. Recent Advances in Molecular Diagnostics of Fungal Plant Pathogens: A Mini Review. *Front. Cell Infect. Microbiol.* **2021**, *10*, 600234. [CrossRef]
38. Webb, C.R.; Avramidis, E.; Castle, M.D.; Stutt, R.O.H.; Gilligan, C.A. Modelling the spread of tree pests and pathogens in urban forests. *Urban For. Urban Green.* **2023**, *86*, 128036. [CrossRef]

MDPI AG
Grosspeteranlage 5
4052 Basel
Switzerland
Tel.: +41 61 683 77 34

Horticulturae Editorial Office
E-mail: horticulturae@mdpi.com
www.mdpi.com/journal/horticulturae

www.ingramcontent.com/pod-product-compliance
Lightning Source LLC
Chambersburg PA
CBHW040833250926
51850CB00007BA/1305
* 9 7 8 3 7 2 5 8 5 7 8 9 0 *